The Grouse
of the World

The Grouse of the World

Paul A. Johnsgard

University of Nebraska Press   Lincoln and London

Library of Congress Cataloging in Publication Data
Johnsgard, Paul A.
The grouse of the world.
Bibliography: p.
Includes index.
1. Grouse I. Title.
QL696.G285J63   1983   598′.616   82–21922
ISBN 0–8032–2558–X

To the North American, European, and Soviet scientists
whose names appear in this book and whose work has
increased our understanding of these birds

Contents

Illustrations

Tables

Preface

IN THE APPROXIMATE DECADE since my *Grouse and Quails of North America* (1973) was written, there has been a continuing interest in the ecology, population dynamics, and management of gallinaceous game birds. In fact, so many new publications on the grouse and quails of North America have appeared that updating the text coverage of these two groups seemed an inordinately difficult task, and one that would add, considerable length to a book that was already relatively long. Additionally, my interests during that period had begun to focus on worldwide approaches to various bird groups, and thus the idea of doing a book on the biology of all the grouse of the world repeatedly sprang to mind. Fortuitously, in the late 1970s Dr. E. O. Höhn approached me and suggested undertaking a collaborative effort on the biology of the grouse family. Although this collaboration never fully materialized, Dr. Höhn did provide a great deal of help with the hazel grouse section, and he wrote virtually all of the initial draft of the biology section of that species account.

In 1980 I began to update the species accounts for the North American grouse that had already been published in my earlier book and to assemble materials for writing the remaining species accounts and revising the sections on comparative biology. Nearly all of these sections from the earlier book were extensively reworked to bring them up to date and to incorporate material on the Old World species. Additionally, materials dealing with the physiology of molts, digestion, and related aspects of physiology were organized to form a new chapter. I have attempted to review the literature through 1981 and have added some citations for 1982 and 1983.

The literature of the grouse family is remarkably large, primarily because of the group's great importance as game birds in those areas (North America and Europe) where game management and ornithology have traditionally been concentrated. Rather surprisingly, there has been no comprehensive summary of the entire grouse family since 1864–65, when D. G. Elliot wrote a monograph on the group, complete with imperial folio-sized hand-colored plates of all the species then known to science. Since then the literature of the grouse has increased enormously. In their classic monograph on the ruffed grouse, Bump et al. (1947) cited more than four hundred references relative to this species, and in a recent bibliography Moulton and Vanderschaegen (1974) offered more than five hundred citations. Similarly, Boyce and Tate (1979) listed nearly three hundred references on the sage grouse, Bergmann et al. (1980) more than two hundred on the hazel grouse, Severson (1978) more than two hundred on the blue grouse, Robinson (1980) nearly two hundred on the spruce grouse, and Höhn about the same number on ptarmigans. No complete bibliographies have yet been published for the sharp-tailed grouse or the pinnated grouse, but rather extensive literature surveys for all the European grouse species are found in Glutz (1973) and in Cramp and Simmons (1980).

In addition to the many people who helped me in the preparation of my earlier book, I must mention several additional sources of help with the present volume. Foremost of these is E. O. Höhn who, as I have already noted, contributed substantially to the section on the hazel grouse. New photographs were provided by Chris Knights, Cy Hampson, Hans Aschenbrenner, Yuri Pukinski, A. V. Andreev, Boris Verprintsey, Ed Bry, Robert Robel, and Robert Williams. Additionally, Dr. T. W. E. Lovel of the World Pheasant Association, Jimmy Oswald, Nick Picozzi, Robert Moss, and David Jenkins all assisted my fieldwork in Scotland. William Strunk provided me with cine footage and sound recordings of a hybrid sage × sharp-tailed grouse. Work in the American Museum of Natural History was facilitated by a travel grant from the Frank Chapman Memorial Fund. The Research Council of the University of Nebraska provided additional travel support, and several secretaries from the School of Life Sciences assisted me with manuscript typing. To all of these persons I offer my sincere appreciation.

Comparative
Biology

1

Evolution
and Taxonomy

THE EVOLUTIONARY HISTORY of grouselike birds is a fairly long one; together with other galliform (chickenlike) birds they constitute a rather generalized array of ground-dwelling forms that scratch or peck while foraging and are primarily ground-nesting birds with relatively or highly precocial young. Beddard (1898) concluded that the gallinaceous birds are an "ancient" group, having probable relationships with the tinamous on the one hand and waterfowl (Anseriformes) on the other. Some more recent ornithologists have supported the idea that the gallinaceous birds have perhaps been derived from the tinamous, but most believe they more probably are related to the Anseriformes. This latter view is primarily based on the seemingly intermediate characteristics exhibited by the screamers (Anhimidae) (Sibley and Ahlquist 1972). However, Olson and Feduccia (1980) have recently brought into question this seemingly well-accepted idea and believe that the screamers, rather than being transitional between the Galliformes and Anseriformes, are instead derived anseriform forms that do not exhibit any real galliform traits.

In any case, the fossil materials so far available for the grouse do not shed any light on this question, for such remains can be traced back only to the lower Miocene (Brodkorb 1964), by which time the major groups of Galliformes seemingly were well differentiated. All the described pre-Pliocene grouse remains are from North American localities (table 1), suggesting a North American origin for the group, but it is clear that the small sample size does not allow for any such firm conclusions. By the same time in North America the quails were well established (Johnsgard 1973), and the earliest turkey remains are from the upper Pliocene (Brodkorb 1964).

The array of extant grouse species collectively have a circumpolar distribution in the Northern Hemisphere, ranging from about 26° to 81° N latitude. Two of the species are of Holarctic distribution, seven additional species are limited to the Nearctic, and seven are confined to the Palaearctic (table 2). Evidence that North America might be regarded as the evolutionary center of the grouse includes the fact that it has more total genera and more endemic genera than does Eurasia, although the differences are slight.

TABLE 1
FOSSIL GROUSE DESCRIBED FROM NORTH AMERICA AND EURASIA

	Eurasia	North America
Lower Miocene		*Palaealectoris incertus* Wetmore
Middle Miocene		*Tympanuchus stirtoni* A. H. Miller
Upper Miocene		*Archaeophasianus roberti* (Stone)
		Archaeophasianus mioceanus (Shufeldt)
Upper Pliocene	*Tetrao macropus* Janossy	
Lower Pleistocene	*Tetrao partium* (Kretzoi)	*Tympanuchus lulli* Shufeldt
	Tetrao conjugens Janossy	
Middle Pleistocene	*Bonasa praebonasia* Janossy	*Paleotetrix gilli* Shufeldt[a]
	Tetrao praeurogallus Janossy	*Dendragapus nanus* (Shufeldt)[b]
	Lagopus cf. *lagopus* ("*atavus*")	*Dendragapus lucasi* (Schufeldt)
		Tympanuchus ceres (Shufeldt)
Total fossil genera	0	3
Total modern genera	3	2
Total fossil species	5	9
Total extant genera	4	5
Total extant species	9	9
Endemic extant species	7	7

Source: Based on Brodkorb 1964; Howard 1966; and Janossy 1976.
[a]*Dendragapus gilli* according to Jehl 1969.
[b]Not separable from *D. lucasi* according to Jehl 1969.

It is difficult to judge which of the extant genera of grouse is most like the ancestral types. Short (1967) argues that *Dendragapus* includes those species that possess a greater number of primitive features than do the species of any other extant genus. However, he also mentions two species of *Bonasa*, two of *Lagopus*, and one of *Tympanuchus* that exhibit presumably ancestral traits, leaving only *Centrocercus* as a relatively specialized genus.

I am inclined to regard *Centrocercus* and *Tympanuchus* as the most highly specialized of the extant tetraonid genera; both presumably evolved independently from forest-dwelling forms as arid habitats expanded during late Tertiary times. I similarly favor regarding the Holarctic genera *Dendragapus* and *Lagopus* as nearest the ancestral types in general morphology, with the tundra-dwelling adaptations of *Lagopus* representing a more recent development than the forest-habitat adaptations of *Dendragapus*. The Holarctic genus *Bonasa* and the Old World genus *Tetrao* can then be considered somewhat more specialized offshoots of ancestral *Dendragapus/Lagopus* stock that have remained adapted to temperate forest habitats. These ideas are summarized in figure 1, which provides a suggested evolutionary tree for the extant grouse genera and species. This diagram seems to differ considerably from that proposed by Short (1967), but actually it represents an only slightly different way of emphasizing what are essentially

TABLE 2

DISTRIBUTION OF EXTANT SPECIES OF GROUSE AND PTARMIGANS

	Entire Holarctic	North American (Nearctic)	Eurasian (Palaearctic)	Total
Total genera	1	5	4	6
Endemic genera	—	2	1	—
Total species	2	9	9	16
Endemic species	—	7	7	—

Source: Taxonomy of Short 1967.

very similar ideas. Our suggested sequences of genera are identical except for the position of *Centrocercus*, which I believe should be listed adjacent to *Dendragapus* to better emphasize its independent origin from *Tympanuchus*.

GENERAL TAXONOMIC SEQUENCE AND HIGHER CATEGORIES

Until fairly recently, the traditional American treatment of the grouse has been to designate them as a distinct family, Tetraonidae, although the 1886 American Ornithologists' Union (A.O.U.) *Check-list* also included the New World quails in this family. Familial recognition of the Tetraonidae occurred with the third edition of the A.O.U. *Check-list* in 1910 and has persisted ever since. Other major authorities who have given a corresponding ranking to the grouse include Peters (1934), Ridgway and Friedmann (1946), Wetmore (1960), and Hudson et al. (1966). But recently a number of other writers have urged a reclassification of the group as a subfamily (Tetraoninae) of the Phasianidae. Some of the authors who have supported this view include Delacour (1951), Sibley (1960), Brodkorb (1964), Holman (1964), Stresemann (1966), Short (1967), and others. Hudson et al. (1966) admit that their basis for retaining familial status for the grouse is rather weak; it apparently stems in part from the fact that the grouse genera they studied were obviously much more closely related to one another then they were to any other genera. This does not seem sufficiently strong reason to maintain the family, in my view, nor do the obviously adaptive feathered condition of the tarsus and nostrils and the pectinate toes seem to justify such separation.

The taxonomic separation of the grouse is also dependent on the degree of recognition deemed appropriate for the other major taxa of Galliformes. Thus, the New World quails, usually given subfamily rank, are generally believed to be relatively closely related to the Old World partridges, which together with the pheasants compose the central nucleus of the family Phasianidae. Further, a number of recent authorities have urged that the turkeys and guinea fowl should also probably be given no more than subfamilial recognition. However, the hoatzin (*Opisthocomus*) only very doubtfully belongs in the order Galliformes, and instead should probably be included in the Cuculiformes (Sibley and Ahlquist 1972). However, Cracraft (1981) tentatively retained the hoatzin in the Galliformes, so the question is unresolved. A proposed classification of the Galliformes follows, with the

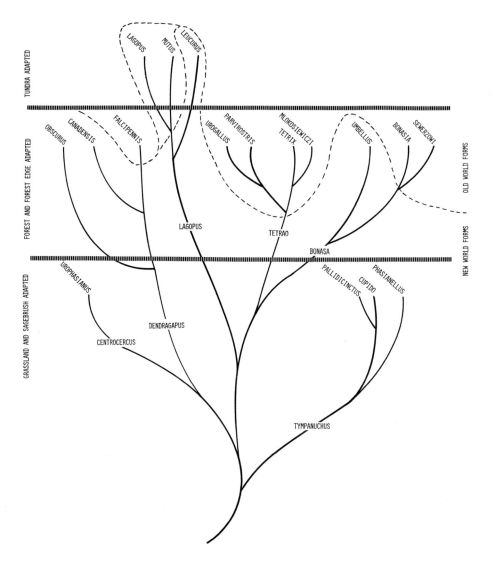

1. Suggested evolutionary relationships and ecological affinities of extant grouse and ptarmigan species. The specific status of *pallidicinctus* is doubtful.

indicated number of species in each major group based on the estimates of Bock and Farrand (1980) and my own studies.

Order Galliformes
  Superfamily Megapodea
    Family Megapodidae—megapodes or mound-builders (12 spp.)
  Superfamily Cracoidea
    Family Cracidae—chachalacas, guans, and curassows (44 spp.)
  Superfamily Phasianoidea
    Family Phasianidae—pheasantlike birds (206 spp.)
      Subfamily Meleagridinae—turkeys (2 spp.)
      Subfamily Tetraoninae—grouse and ptarmigans (16 spp.)
      Subfamily Odontophorinae—New World quails (33 spp.)
      Subfamily Phasianinae—Old World pheasants (151 spp.)
        Tribe Perdicini—Old World partridges, francolins, and quails (106 spp.)
        Tribe Phasianini—pheasants, jungle fowl, and peafowl (49 spp.)
      Subfamily Numidinae—guinea fowl (7 spp.)

GENERIC AND SPECIES LIMITS

As happens with many groups of birds that have been subject to sexual selection and selection for reproductive isolation in a polygamous or promiscuous mating system (Sibley 1957), the classification of the grouse has been confused by a plethora of generic names having little if any phylogenetic significance. Fortunately, Short (1967) has reviewed this situation thoroughly, and with respect to the North American genera has recommended eliminating both *Canachites* (= *Dendragapus*) and *Pedioecetes* (= *Tympanuchus*). At the species level, the American Ornithologists' Union (1957) has already seen fit to merge *Dendragapus franklinii* with *D. canadensis*, and *D. fuliginosus* with *D. obscurus*, as essentially allopatric populations that are best regarded as subspecies. Among the Eurasian forms, Short (1967) has advocated merging *Lyrurus* with *Tetrao*, *Falcipennis* with *Dendragapus*, and *Tetrastes* with *Bonasa*. The Scottish race of the willow ptarmigan, locally called the "red grouse," was considered by Short to be conspecific with *Lagopus lagopus*.

    The only remaining major question relative to the grouse is that posed by the "lesser" form of prairie chicken, *Tympanuchus pallidocinctus*, which is still (1982) recognized as specifically distinct by the A.O.U. Short (1967) summarized the evidence favoring the view that this population should likewise be regarded as only racially distinct from *T. cupido*, and he questioned the evidence presented by Jones (1964a) supporting species separation. More recently, Sharpe (1968) has also contributed his views, which in general agree with those of Jones. It is impossible to provide a clear-cut taxonomic resolution, and the conclusion one reaches reflects in large measure one's philosophy about the primary

TABLE 3

Ecological Distribution of Species of Extant Grouse and Ptarmigans

Habitat	Representative Old World Species	Representative New World Species
Tundra		
Alpine and subalpine	Caucasian black grouse	White-tailed ptarmigan
High arctic	Rock ptarmigan	Rock ptarmigan
Low arctic	Willow ptarmigan	Willow ptarmigan
Coniferous forest		
Dense coniferous	Sharp-winged grouse	Spruce grouse
Open coniferous and forest-edge	Black grouse	Blue grouse
Mature coniferous	Capercaillie; Black-billed capercaillie	
Hardwood, hardwood-coniferous	Hazel grouse; Black-breasted hazel grouse	Ruffed grouse
Grassland		
Prairie; steppe; brushland		Sharp-tailed grouse
Prairie; prairie-forest edge		Greater prairie chicken
Short-grass; semidesert scrub		Lesser prairie chicken
Desert scrub		
Sage; sage-grassland		Sage grouse

function of the species category. No additional evidence on the question has been gathered in this study, but *T. pallidocinctus* will not be given the space or attention accorded the better-defined species.

On the basis of these considerations, a list of the extant grouse and ptarmigan species is shown in table 3. Rather than being listed in taxonomic sequence, they have been organized according to zoogeography and the major plant communities with which they are most closely associated. A detailed identification of habitat preferences and range of ecological distributions is not possible in such a tabular comparison, but the individual species accounts in the second section of this book provide a more accurate analysis of habitat characteristics of each species. Of interest here are the substantial number of approximate ecological equivalent species in the two hemispheres (sharp-winged grouse and spruce grouse; black grouse and blue grouse; hazel grouse and ruffed grouse), which is at least in part a reflection of the taxonomic relationships of these rather closely associated forms. However, the Old World forms essentially lack the grassland and desert scrub-adapted representatives, in spite of the large areas available to such forms in Eurasia. Other galliform species, not included in this table, do of course occupy many of these habitats in Eurasia. The general geographic distributions of these vegetational communities in North America and Eurasia are shown in figures 2 and 3, which have been derived from various sources. With few exceptions these maps illustrate the distribution of potential climax vegetational types, rather than successional or disturbance community types.

	TUNDRA OR ICE AND SNOW
	CONIFEROUS FOREST
	CONIFEROUS-DECIDUOUS ECOTONE
	SOUTHEASTERN PINE FOREST
	TEMPERATE DECIDUOUS FOREST
	GRASSLAND
	GRASSLAND-SAGEBRUSH ECOTONE
	SAGEBRUSH SEMIDESERT
	CHAPARRAL (EVERGREEN HARDWOOD)
	DESERT GRASSLAND
	DESERT
	PINE-OAK UPLAND OR MONTANE FOREST
	DRY TROPICAL FOREST, SCRUB AND SAVANNAH

2. Natural vegetational communities of North America. After various sources.

3. Natural vegetational communities of Eurasia. After Eyre (1968).

TUNDRA OR ICE & SNOW
CONIFEROUS FOREST
CONIFEROUS-DECIDUOUS ECOTONE
TEMPERATE DECIDUOUS FOREST
TEMPERATE FOREST-STEPPE ECOTONE
GRASSLAND (STEPPE)
SEMIDESERT SCRUB
TROPICAL SEMI-DESERT SCRUB
DESERT

BROAD-LEAVED EVERGREEN FOREST
TROPICAL MONTANE FOREST WITH CONIFERS
MEDITERRANEAN EVERGREEN MIXED FOREST
MAQUIS, GARRIGUE & ESPARTO GRASS

TABLE 4

PROPOSED CLASSIFICATION OF EXTANT SPECIES OF GROUSE AND PTARMIGANS

Family Phasianidae: pheasantlike birds
  Subfamily Tetraoninae: grouse and ptarmigans
    Genus *Centrocercus* Swainson 1831
      1. *C. urophasianus* (Bonaparte) 1828: sage grouse
    Genus *Dendragapus* Elliot 1864
      (Subgenus *Dendragapus*)
      1. *D. obscurus* (Say) 1823: blue grouse
      (Subgenus *Canachites* Stejneger 1885)
      { 1. *D. canadensis* (Linnaeus) 1758: spruce grouse*
      { 2. *D. falcipennis* (Hartlaub) 1855: sharp-winged grouse
    Genus *Lagopus* Brisson 1760
      1. *L. lagopus* (Linnaeus) 1758: willow ptarmigan
      2. *L. mutus* (Montin) 1776: rock ptarmigan
      3. *L. leucurus* (Richardson) 1831: white-tailed ptarmigan
    Genus *Tetrao* Linnaeus 1758
      (Subgenus *Tetrao*)
      { 1. *T. urogallus* Linnaeus 1758: capercaillie
      { 2. *T. parvirostris* Bonaparte 1856: black-billed capercaillie
      (Subgenus *Lyrurus* Swainson 1832)
      { 1. *T. tetrix* Linnaeus 1758: black grouse
      { 2. *T. mlokosiewiczi* Taczanowski 1875: Caucasian black grouse
    Genus *Bonasa* Stephens 1819
      (Subgenus *Bonasa*)
      1. *B. umbellus* (Linnaeus) 1776: ruffed grouse
      (Subgenus *Tetrastes* Keyserling and Blasius 1840)
      { 1. *B. bonasia* (Linnaeus) 1758: hazel grouse
      { 2. *B. sewerzowi* (Przewalski) 1876: black-breasted hazel grouse
    Genus *Tympanuchus* Gloger 1842
      { 1. *T. cupido* (Linnaeus) 1758: pinnated grouse
      { 2. *T. phasianellus* (Linnaeus) 1758: sharp-tailed grouse

*Brackets connect superspecies groups.

An abbreviated systematic synopsis of the species included in this book follows, with subspecies excluded since they are listed under the appropriate species accounts. In the list that is presented (table 4), probable superspecies groups are connected by braces. Further, table 5 lists non-English vernacular names for the five major Eurasian grouse species to assist the reader in interpreting the European literature sources. Additional alternative vernacular names for these and non-European species are found in the individual species accounts.

TABLE 5
European Vernacular Names for Eurasian Grouse and Ptarmigans

Language	Black Grouse	Capercaillie	Hazel Grouse	Rock Ptarmigan	Willow Ptarmigan
Czech	Tetřívek obecný	Tetřev hlušec	Jeřábek lesní	Kur horský	Kur rousný
Danish	Urfugl	Tjur	Hjerpe	Fjeldrype	Dalrype
Dutch	Korhoen	Auerhoen	Hazelhoen	Sneeuwhoen	Moerassneeuwhoen
Finnish	Teeri	Metso (male) Koppelo (female)	Pyy	Kiiruna	Riekko
French	Tétras lyre	Grand tétras	Gelinotte des bois	Lagopède alpin	Lagopède des saules
German	Birkhuhn	Auerhuhn	Haselhuhn	Alpenschneehuhn	Moorschneehuhn
Hungarian	Nyirfajd	Siketfajd	Császármadár	Havasi hófajd	Sarki hófajd
Italian	Fagiano di monte	Gallo cedrone	Fracolino di monte	Pernice bianca	Pernice bianca nordica
Norse	Årfugl	Storfugl	Jerpe	Fjellrype	Lirype
Polish	Cietrzew	Gluszec	Jarząbek	Pardwa górna	Pardwa
Russian	Tetrev-kosach	Glukhar (male) Glukharka (female)	Ryabchik	Tundravaya kuropatka	Belaya kuropatka
Swedish	Orre	Tjäder	Järpe	Fjällripa	Dalripa
Spanish	Gallo lira	Urogallo	Grebul	Perdiz nival	Lagopo escandinavo

10

2

Physical
Characteristics

ALL THE GROUSE AND PTARMIGANS share a number of anatomical traits that provide the basis for their common classification within the order Galliformes. All have fowllike beaks and four toes. In all species the hind toe is elevated and quite short and thus is ill-adapted for perching. There are always 10 primaries, 15 to 21 secondaries, and 14–22 tail feathers (rectrices). Aftershafts on the contour feathers are well developed, and true down feathers are infrequent. A large crop is present and is associated with variably granivorous (seed-eating) to herbivorous (leaf-eating) diets. Egg colors range from pastel or earth tones (buff, cream, olive, etc.) to white, with darker spotting prevalent among those species having nonwhite eggs. The nest is built on the ground, and incubation is by the female alone. The young are down-covered and precocial and are usually able to fly short distances in less than 2 weeks. They are cared for by the female (most grouse) or by both parents (some ptarmigans). A number of external structural characteristics typical of grouse and ptarmigans are shown in figure 4.

Additionally, the grouse are characterized by feathered nostrils and feathering on the legs that usually extends to the base of the toes. Among ptarmigans this feathering extends to the tips of the toes in winter, but the toes become naked or nearly so in summer.

In all the grouse species the toes have marginal comblike membranes (pectinations) in winter. Males of several species of grouse have large unfeathered areas (apteria) at the sides or front of the neck, which can be exposed and enlarged by inflating the esophagus. The skin associated with these ''air sacs'' may be variously colored, or the feathers around the area may be specialized in shape or color, but the true air sac system associated with the lungs is not directly connected to these structures. An area of bare skin (eyecomb) is usually present above the eyes in mature males. Most grouse are not highly gregarious, but during fall and winter some species that migrate considerable distances may form large flocks. Grouse are usually polygamous or promiscuous, but the ptarmigans are relatively monogamous. At least as many as 21 secondaries and up to 22 rectrices are present, but in some species (ptarmigans) the central pair closely resembles the upper tail coverts, while in others (sage grouse) some upper tail coverts may easily be confused with rectrices.

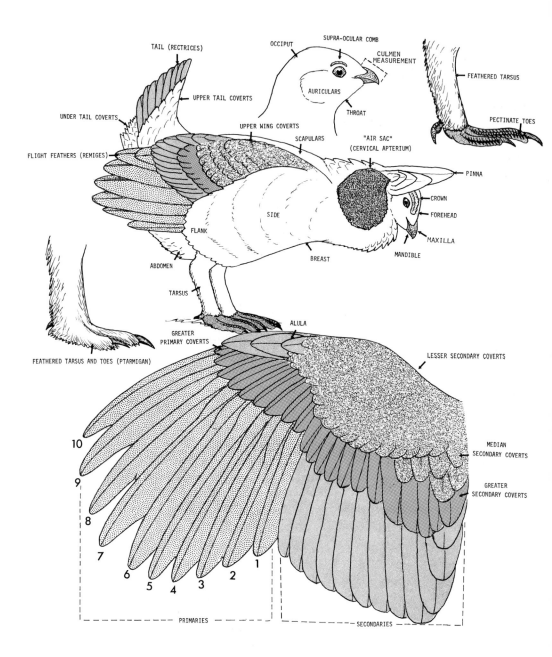

4. Body regions, wing areas, and other structural features of grouse and ptarmigans.

ADULT WEIGHTS

Weight characteristics of adults are of some interest, since they not only give the hunter an indication of the trophy value of his game but are also important anatomical adaptations to the environment. Thus, body weight in relation to the climatic conditions a species encounters, heart weights in relationship to total body weight (Hartman 1955), and body weight in relation to egg weight (Lack 1968) are all significant and provide useful indexes of ecological and physiological adaptations. A summary of adult weights is therefore provided (table 6) as they have been reported in the literature. In general the selected references represent the largest sample sizes available and do not take into account the possibility of geographic or seasonal variations in weight, which are known to occur often. Representative studies on geographic or seasonal variations in adult galliform weights include those of Gullion and Gullion (1961), Stoddard (1931), Boag (1965), Zwickel, Brigham, and Buss (1966), and Bump et al. (1947).

EGG CHARACTERISTICS

The color, markings, and other physical characteristics of bird eggs are of particular ecological interest. To some extent the physical characteristics of eggs might be expected to result from evolutionary relationships, but the requirements for concealment under existing ecological conditions are probably primary in the interpretation of egg color and patterning. Table 7 summarizes the physical characteristics of the eggs of grouse and ptarmigans. Known or estimated incubation periods are also indicated, and it may be seen that in all known cases these range from 21 to 27 days. There is no apparent relation between egg size and incubation period; the only clear example of ecological specialization in the entire group is the unusually short (21-day to 22-day) incubation period of the ptarmigans.

It is also of interest to compare egg size to the size of the adult female. This is perhaps most easily done by determining the ratio of the fresh egg's weight to that of the female (Lack 1968). Average weights of fresh eggs are not available for all the species concerned, but it is possible to calculate the volume of an egg quite accurately when its linear measurements are known. Stonehouse (1966) suggests this convenient formula:

$$\text{Volume (cc)} = .512 \times \text{length (cm)} \times \text{diameter (cm)}^2$$

Assuming that the fresh egg has an average specific gravity of 1.08 (Barth 1953), the preceding formula can be modified as follows:

$$\text{Weight (g)} = .552 \times \text{length (cm)} \times \text{diameter (cm)}^2$$

Using this formula, estimated fresh weights of eggs were calculated from the linear measurements presented in table 7 and are summarized in table 8. In addition, a calculated total estimated clutch weight, based on reported average clutch sizes (see table 11), is indicated as an index to the relative physiological drain on the female in laying an entire clutch. A female's average clutch may represent from as little as 20 to 25 percent of her own weight, as in spruce grouse and ptarmigans, to nearly 50 percent of her weight in some *Bonasa* species. Since at least several species are fairly persistent renesters, it is

TABLE 6

Adult Weights of Grouse and Ptarmigans

Species	Mean or Range of Means	Maximum Weight	References
Sage grouse			
Male	2,319–3,190 g (82–112 oz)*		Beck and Braun 1978
Female	1,364–1,745 g (48–62 oz)*		Beck and Braun 1978
Blue grouse			
Male	1,240–1,331 g (44–47 oz)*	1,425 g (50 oz)	Redfield 1973
Female	813–867 g (29–31 oz)*	1,250 g (44 oz)	Boag 1965§
Spruce grouse			
Male	492–653 g (17–23 oz)*	—	Ellison and Weeden 1979
Female	456–620 g (16–22 oz)*	—	Ellison and Weeden 1979
Sharp-winged grouse			
Adult	ca. 600 g (21 oz)	—	Dementiev and Gladkov 1967
Willow ptarmigan			
Male	535–696 g (19–25 oz)*	804 g (28 oz)	Parmelee, Stephens, and Schmidt 1967
Female	525–652 g (19–23 oz)*	749 g (26 oz)	Irving 1960
Red grouse			
Male	610–90 g (21–24 oz)*	908 g (32 oz)	Cramp and Simmons 1980
Female	550–670 g (19–24 oz)*	—	
Rock ptarmigan			
Male	466–738 g (16–26 oz)*	—	Johnsgard 1973
Female	427–701 g (15–25 oz)*	—	Semenov-Tian-Schanski 1959
White-tailed ptarmigan			
Male	323 g (11.4 oz) (24 birds)	430 g (15.2 oz)	Johnson and Löckner 1968
Female	329 g (11.5 oz) (14 birds)	490 g (17.5 oz)	G. Rogers (in litt.)
Capercaillie			
Male	3,902–4,254 g (137–50 oz)*	6,500 g (229 oz)	Semenov-Tian-Schanski 1959
Female	1,755–2,004 g (62–71 oz)*	2,210 g (77 oz)	Glutz 1973
Black-billed capercaillie			
Male	3,131 g (110.4 oz) (4 birds)	4,000 g (141 oz)	Dementiev and Gladkov 1967
Female	2,050 g (72.3 oz) (1 bird)	—	Cheng 1978
Black grouse			
Male	1,141–1,340 g (40–47 oz)*	1,900 g (67 oz)	Glutz 1973
Female	851–976 g (30–34 oz)*	1,100 g (38 oz)	Glutz 1973

Caucasian black grouse		
Male	865 g (30.5 oz) (12 birds)	1,005 g (35.5 oz)	Noska 1895
Female	766 g (27 oz) (5 birds)	820 g (29 oz)	Noska 1895
Ruffed grouse			
Male	604–54 g (21.5–23.3 oz)*	770 g (27 oz)	Nelson and Martin 1953[†]
Female	500–586 g (17.9–20.9 oz)*	679 g (24 oz)	Bump et al. 1947
Hazel grouse			
Male	362–435 g (13–15.4 oz)*	490 g (17.6 oz)	Bergmann et al. 1978
Female	370–422 g (13–15 oz)	490 g (17.3 oz)	
Black-breasted hazel grouse			
Male	278 g (9.8 oz) (3 birds)	300 g (10.6 oz)	Cheng 1978
Female	257 g (9.1 oz) (3 birds)	290 g (10.2 oz)	Cheng 1978
Greater prairie chicken			
Male	992 g (35 oz) (22 birds)	1,361 g (48 oz)	Nelson and Martin 1953[†]
Female	770 g (29 oz) (16 birds)	1,020 g (36 oz)	Nelson and Martin 1953[†]
Attwater prairie chicken			
Male	938 g (33.1 oz) (10 birds)	1,135 g (40 oz)	Lehmann 1941
Female	731 g (25.7 oz) (6 birds)	785 g (38 oz)	Lehmann 1941
Lesser prairie chicken			
Male	789 g (27.9 oz) (30 birds)	893 g (31.5 oz)	Lehmann 1941; Taylor and Guthery 1980
Female	702 g (24.8 oz) (23 birds)	779 g (27.5 oz)	Taylor and Guthery 1980
Sharp-tailed grouse			
Male	951 g (33 oz) (236 birds)	1,087 g (43 oz)	Nelson and Martin 1953[†]
Female	815 g (29 oz) (247 birds)	997 g (37 oz)	Nelson and Martin 1953[†]

*Mean weights of these species vary considerably with season and/or locality.

[†]Reported as fractions of pounds by authors.

[§]Reported in graphic form by author.

TABLE 7

Egg Characteristics and Incubation Periods of Grouse and Ptarmigans

Species	Spotting	Basic Color	Dimensions (mm)	Incubation (days)	References
Sage grouse	Moderate	Buffy green or brown	55 × 38	25–27	Patterson 1952
Blue grouse	Moderate	Buff or pale brown	48.5 × 35	24–25	Godfrey 1966
Spruce grouse	Moderate	Buff or pale rust	43 × 31	21	Pendergast and Boag 1972a
Sharp-winged grouse	Moderate	Pale brown	46.3 × 31.8	?	Dementiev and Gladkov 1967
Willow ptarmigan	Heavy	White to pale brown	43 × 31	21–22	Westerkov 1956; Jenkins, Watson, and Miller 1963
Rock ptarmigan	Heavy	White to pale brown	42 × 30	21	Godfrey 1966
White-tailed ptarmigan	Moderate	White to reddish buff	43 × 29.5	22–23	Braun 1969
Capercaillie	Slight	Pale yellowish	57 × 41	24–28	Glutz 1973
Black-billed capercaillie	Moderate	Light brown	61 × 42	24	Cheng 1978
Black grouse	Moderate	Yellowish white	50 × 36	26–27	Glutz 1973
Caucasian black grouse	Slight	Pale buff	53 × 36	20–25	Cramp and Simmons 1980
Ruffed grouse	Slight or none	Buffy white to cream	38.5 × 30	24	Bump et al. 1947
Hazel grouse	Slight	White to reddish buff	41 × 30	25	Glutz 1973
Black-breasted hazel grouse	Moderate to heavy	White to reddish buff	44 × 30.8	?	Beick 1927
Greater prairie chicken	Slight or none	White to olive buff	43 × 32.5	24–25	McEwen et al. 1969
Lesser prairie chicken	Slight or	White to buff	42 × 32.5	25–26	Coats 1955
Sharp-tailed grouse	Slight	Fawn to chocolate or olive	43 × 32	24–25	McEwen et al. 1969

apparent that a large investment of energy in a clutch is not detrimental as long as sufficient food is available.

FEATHERS AND OTHER EXTERNAL ADAPTATIONS

As in nearly all birds, the contour feathers of grouse and ptarmigans are arranged in definite tracts, or pterylae, that do not differ much among the species included. The general arrangement of these tracts is shown in figure 5. At the edges of these tracts "half-down" or semiplume feathers regularly occur, and true down feathers sometimes occur on the neck and wings. There are usually numerous long and nearly hairlike filoplumes scattered among the contour feathers; these become especially conspicuous in adult male sage grouse when they are erected during display.

The general arrangement of the feather tracts is very similar in grouse and New World quails. The major differences are that in quails the dorsal feather tract has only a small apterium and is nearly continuous with the upper cervical tract, whereas in grouse these tracts are well separated, forming a large dorsal apterium.

The number of primaries is the same (10) throughout the galliform group, but their relative lengths differ somewhat. Among the grouse the secondaries vary from 15 to 16 in the ruffed grouse, 17 in the spruce grouse, 18 in the blue, sharp-tailed, and pinnated grouse, 18 to 19 in ptarmigans, and 21 in sage grouse (Clark 1899). In most of these species the secondaries grade gradually into the scapulars and proximal coverts and thus become very difficult to count accurately. The arrangement of the wing feathers is shown in figure 4.

None of the grouse species possess elaborate crests, but several have special tracts of feathers on the neck or have unfeathered areas in this region. In the ruffed grouse, the special "ruff" feathers are borne on the lateral branches of the lower cervical tract, and there is no marked apterium between the lower and upper cervical tracts. However, the dozen or so feathers making up the pinnae of the pinnated grouse are similarly borne on each side of the upper cervical tract, below which is a large apterium (Clark 1899). In the greater prairie chicken this apterium is yellowish, presumably because of subcutaneous fat, whereas in the lesser prairie chicken it is more reddish. The sharp-tailed grouse has a similar apterium that appears reddish to violet when expanded by esophageal inflation, but this species lacks specialization of the feathers above and below. The sage grouse lacks lateral neck spaces, but there is a large and somewhat oval apterium on each side of the neck, situated quite low and somewhat frontally. These areas are about 45 by 25 mm in older males, and about 25 by 13 mm in females (Brooks 1930). The bare skin of males is olive gray but appears yellowish when expanded during display. The lower and laterally adjacent breast feathers of male sage grouse are curiously bristly, which was once thought to be a result of wear, until Brooks (1930) discovered that newly grown feathers have the same appearance. They evidently produce the rasping or squeaking sound made when the foreparts of the wings are brushed over the lower breast during display (Lumsden 1968).

Although the blue grouse lacks such specialized feathers on the neck, males do expose

TABLE 8

RELATION OF ADULT FEMALE WEIGHTS TO ESTIMATED EGG AND CLUTCH WEIGHTS

Species	Estimated Egg Weight (g)	Percentage of Female Weight	Average Clutch Size	Percentage of Female Weight
Sage grouse	44	3.4	7.4	25.2
Blue grouse	33	3.6	6.2	22.4
Spruce grouse	23	4.2	5.8	24.4
Sharp-winged grouse	26	ca. 4.3	8	ca. 34.6
Willow ptarmigan	23	3.3	7.1	23.1
Rock ptarmigan	21	4.1	7.0	28.7
White-tailed ptarmigan	21	6.4	5.2	33.3
Capercaillie	48	2.7	8	21.6
Black-billed capercaillie	54	2.6	8?	20.8
Black grouse	32	3.5	8	28
Caucasian black grouse	31	4.0	6.0	24.0
Ruffed grouse	19	3.8	11.5	43.7
Hazel grouse	19	4.7	9.0	42.7
Greater prairie chicken	24	3.1	12.0	37.2
Lesser prairie chicken	24	3.3	10.7	35.3
Sharp-tailed grouse	24	2.9	12.1	35.1

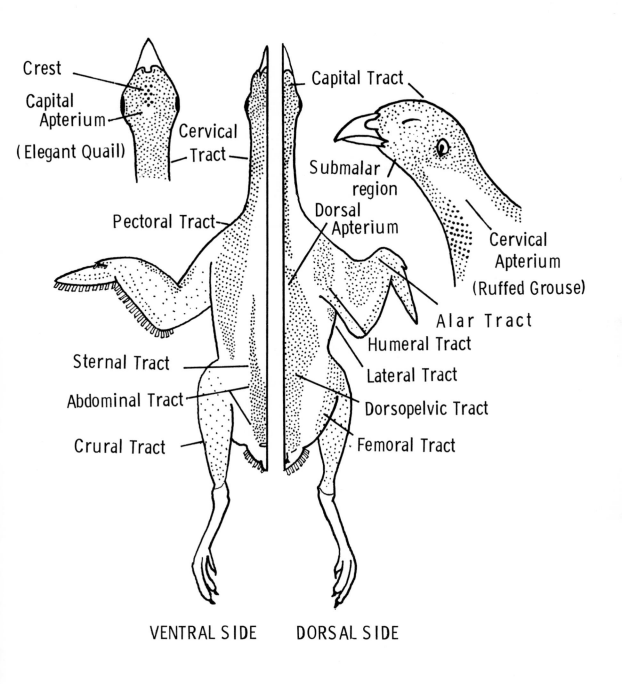

Crest

Capital Apterium

(Elegant Quail)

Cervical Tract

Capital Tract

Pectoral Tract

Submalar region

Dorsal Apterium

Cervical Apterium

(Ruffed Grouse)

Alar Tract

Humeral Tract

Sternal Tract

Lateral Tract

Abdominal Tract

Dorsopelvic Tract

Crural Tract

Femoral Tract

VENTRAL SIDE   DORSAL SIDE

5. Feather tracts of grouse and ptarmigans (primarily after Clark 1899).

rounded areas of the neck during "hooting," which are emphasized by the whitish bases of the surrounding neck feathers. The exposed skin in these areas varies from (in the interior races) thin and flesh-colored, changing to purplish red when expanded, to (in the coastal races) highly thickened, gelatinous, corrugated, and deep yellow. These conditions presumably result from subcutaneous fat deposits, which are less evident during the nonbreeding periods (Brooks 1926).

As I have mentioned, the sharp-tailed grouse lacks specialized neck feathers associated with display, but Lumsden (1965) has found that the tail feathers are unusually well developed in this species and are related to the tail-rattling noises made during display. The rectrices in males are very stout basally but taper rapidly. Ventrally the shaft projects in two keels, but dorsally it is rounded and projects only slightly. The outer webs of the vanes are stiff and curve sharply downward, and the inner webs are also thickened. Each clicking sound is produced by lateral feather movements, during which the inner web catches on the ventrally projecting shaft of the inwardly adjacent feather web, and after some resistance the two disengage, producing a click. Simultaneously the curved outer webs brush over the dorsal surface of the next outwardly adjacent feather, producing a scraping sound. Additional nonvocal noise in males of these species may be produced by foot stamping, and to a lesser degree the same can be said for the pinnated grouse. In the greater, Attwater, and lesser prairie chickens tail-spreading or tail-clicking noises that are taxon-typical occur during display (Sharpe 1968).

The number of tail feathers (rectrices) varies considerably among the species and genera of grouse. Short (1967) notes that in ptarmigans, spruce grouse, sharp-winged grouse, and hazel grouse the usual number of rectrices is 16. There are normally 18 in the other grouse except for the blue grouse and the sage grouse, which regularly vary from 18 to 20. However, the total observed variation in 168 specimens of blue grouse was from 16 to 22, while in the ruffed grouse there was a range of from 14 to 20 among 396 specimens. As Short noted, such variation obviously makes the application of rectrix number for taxonomic purposes extremely limited.

3

Molts and Plumages

UNDERSTANDING THE MOLTS and plumages of the grouse and ptarmigans is of great importance to the applied biologist, for they provide valuable clues to the age and sex of individual birds without internal examination. They thus offer a means of analyzing the sex and age composition of wild populations, which provides basic indexes to past and potential reproductive performance and probable mortality rates. Additionally, molts and plumages are generally species-specific traits that have resulted from pressures of natural selection over a long period in a particular habitat and climate. The ecology of the species is of major importance in this regard; species occurring in northerly regions may undergo their molts more rapidly than those in southerly ones, or, as in the case of the willow ptarmigan, certain races may even lack particular plumages that occur in areas with different climates.

From the time they hatch, all grouse and ptarmigans exhibit a series of specific plumages, separated by equally definite molts, that are comparable in nearly all species. The only known exception to this occurs in the genus *Lagopus*, which is unique in having an extra molt, and thus a supplementary plumage, intercalated between the summer and winter plumages. This special case will be dealt with as required; the following summary covers the basic sequences and terminology of molts and plumages found in the world's grouse and ptarmigans.

NATAL PLUMAGE

All galliform birds hatch covered with a dense coat of down that insulates and camouflages the precocial young, which typically leave the nest shortly after hatching. This natal plumage generally is extremely similar among related species and, because of the lack of known selective pressures for rapid divergence in downy patterns during speciation, often

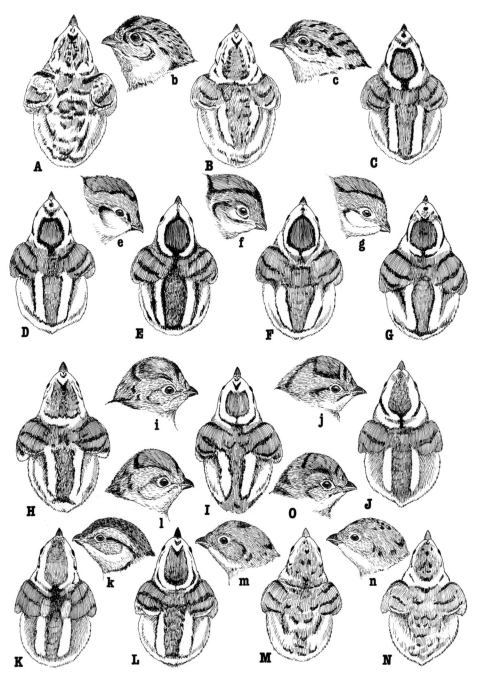

6. Natal down patterns of grouse and ptarmigans. A = sage grouse, B = blue grouse, C = sharp-winged grouse, D = spruce grouse, E = rock ptarmigan, F = willow ptarmigan, G = white-tailed ptarmigan, H = capercaillie, I = black grouse, J = hazel grouse, K = ruffed grouse, L = black-billed capercaillie, M = greater prairie chicken, N = sharp-tailed grouse, O = Caucasian black grouse.

provides more valuable clues to evolutionary relationships than do adult plumage patterns (fig. 6).

POSTNATAL MOLT AND JUVENAL PLUMAGE

Virtually at the time of hatching, or at least within the first week of life, the first indications of the juvenal plumage become apparent through the emergence of the secondary and inner primary feathers and the rectrices. The two outermost juvenal primaries and the innermost of the juvenal secondaries appear later than those near the middle of the wing. All native galliform species have 10 primaries, counted outward from the most proximal one; the number of secondary feathers is somewhat greater and varies among species, with the innermost secondaries sometimes designated "tertiaries" (although like typical secondaries they insert on the ulna rather than on the humerus). The secondaries are counted inward, from the feather nearest the first of the primaries (which insert on the bones of the hand). The third secondary is typically the first to emerge, followed by progressively more proximal ones, and the two outermost ones often emerge at about the same time as those near the proximal end. While the primaries and secondaries are growing, all the upper greater coverts begin to grow. The upper coverts for the two outermost primaries start growing before their associated primaries and possibly serve as functional substitutes for these flight feathers, which typically are delayed in development. Correlated with this, the ninth upper primary covert of the juvenal plumage is often notably more pointed and larger than are the adjoining coverts.

The juvenal remiges (primaries and secondaries) and rectrices are scarcely fully grown before they begin to be pushed out by the remiges and rectrices of the next plumage, but during the short time they are present the rest of the body is being transformed from down-covered to covered with contour feathers. This transformation, called the *postnatal molt*, is a complete molt. The feathers that replace the natal down are called *juvenal* feathers, and the associated age category is called the *juvenile* stage.

As the juvenal remiges and rectrices are appearing in the manner described above, other juvenal feathers begin to emerge on both sides of the breast and backward toward the flanks. Shortly, juvenal feathers also appear on the crown, base of the neck, scapular region, and upper legs, spreading toward the back. The greater and lesser upper wing coverts are fully grown before their associated remiges and are followed by the median coverts. These upper coverts appear in advance of the lower coverts (Dwight 1900). Before all the juvenal feathers have appeared throughout the head region, the first signs of the next (postjuvenal) molt will be evident in the loss of the inner juvenal primaries and the emergence of new (first-winter) primaries in their places. This occurs as early as 18 days after hatching in the willow ptarmigan and blue grouse, and it occurs within the first month of life in most or all species. The juvenal primaries are molted outward at 5- to 10-day intervals for the inner ones, and at increasing intervals for the outer ones (table 9). The two outer juvenal primaries (numbers 9 and 10) will have just completed their growth shortly

TABLE 9

AVERAGE AGE (IN DAYS) AT START AND COMPLETION OF GROWTH OF FIRST-WINTER
PRIMARY FEATHERS IN GROUSE AND PTARMIGANS

	Primary Number[a]								
	1	2	3	4	5	6	7	8	
	A/B	A/B	A/B	A/B	A/B	A/B	A/B	A/B	Reference
Sage grouse									
Males	24/35	29/42+	34/49+	40/56+	47/77+	59/84+	74/105+	102/140+	Pyrah 1963
Females	24/42	27/42	33/49+	38/56+	45/63+	54/77+	66/91+	91/126+	
Blue grouse*	17/38	23/44	26/50	35/59	41/68	47/80	59/95	71/113	Zwickel and Lance 1966
Spruce grouse*	19/-	21/-	27/-	34/-	38/-	49/-	53/-	80/-	Robinson 1980
Red grouse	18/-	22-23/-	26/-	29-30/-	34-35/-	40-42/-	49-51/-	65-69/-	Kolb 1971
Willow ptarmigan*	18/-	25/-	30/-	35/-	40/-	46/-	53/-	65/91	Westerskov 1956
White-tailed ptarmigan	17/-	22/-	26/-	31/-	36/-	43/-	53/-	68/-	Giesen and Braun 1979
Capercaillie*	17/44	25/53	32/66	37/75	42/84	52/98	60/110	81/133	E. Sutter, in litt.
Black grouse*	15/40	20/45	26/53	32/65	37/74	44/85	53/98	70/116	E. Sutter, in litt.
Ruffed grouse									
New York	14/45	20/49	27/63	35/68	42/77	49/83	61/98	74/119	Bump et al. 1947
Ohio	23/46	27/54	34/54	40/68	47/73	56/88	68/102	85/124	Davis 1968
Greater prairie chicken	28/56	35/56	41/64	49/77	55/84	60/84+	70/-	82/126+	Baker 1953

[a]Counting from the inside; A = starts growth; B = grown.

*Based on average age of molting juvenal primaries; data for black grouse and capercaillie are from small samples and are weighted averages.

before the eighth juvenal primary is dropped. Except in occasional instances these two outer juvenal primaries normally are not molted in the species under consideration here. In the ring-necked pheasant this does occur, but it is not typical of the introduced Old World partridges (*Perdix* and *Alectoris*). An important difference between the New World quails and the grouse occurs in association with the postjuvenal molt of the eight inner primaries. In the grouse (as well as in *Perdix* and *Alectoris*) the associated juvenal greater upper primary coverts are also molted in the postjuvenal molt, whereas in all the New World quails so far studied the juvenal greater upper wing coverts of these primaries are not molted but rather are held through the winter and spring until they are molted in the annual (postnuptial) molt (van Rossem 1925; Petrides 1942).

The juvenal secondaries, as well as all the juvenal rectrices, are also rapidly lost at about the time the juvenal primaries are being shed. The juvenal rectrices may be dropped almost simultaneously, as in the rock ptarmigan (Salomonsen 1939); molted from the lateral follicles toward the middle ones (centripetally), as in most or perhaps all grouse species; or molted from the central follicles outward (centrifugally), as in various New World quails (Raitt 1961; Ohmart 1967). The gray partridge also has an imperfectly centrifugal postjuvenal molt of the rectrices (McCabe and Hawkins 1946). The juvenal secondaries are lost in about the same sequence as they emerged; starting from the third, the molt proceeds inward, with the two outermost secondaries dropping as the more proximal secondaries are being lost.

Body feathers of the juvenal plumage are surprisingly similar throughout the typical grouse. Typically white or pale buffy shaftstreaks are conspicuous, especially on the upperparts; these often expand near the tip of the feather to form distinctive hammer-shaped markings. Apparently only the ptarmigans lack these distinctive juvenile markings. Usually the sexes are nearly identical in this plumage.

POSTJUVENAL MOLT AND FIRST-WINTER PLUMAGE

The postjuvenal molt (or "prebasic," according to Humphrey and Parkes, 1959) gradually replaces the juvenal body feathers with the more distinctly species-specific feathers of the first-winter (or "basic") plumage. The postjuvenal molt is virtually complete in all the species considered here, involving all the body feathers and all the flight feathers except the two outermost (primaries 9 and 10) and their coverts.

Because the outer two juvenal primaries are retained during the postjuvenal molt, they will normally be carried until the bird's next complete molt, which occurs after the next breeding season. This is evidently the case in all grouse and ptarmigans. Because of their relatively long persistence, the outer two primaries are usually subject to considerable fading and wear; they thus provide a basic method of estimating age. Their value for this is limited by possible difficulties in estimating their wear relative to that of the more proximal primaries and by occasional aberrations in wing molt. This latter problem may result from a precocious molting of one or both of the outer juvenal primaries during the first fall (as regularly occurs in pheasants) or from an abnormally arrested molt in which the juvenal remiges are retained longer than normal.

The gradual loss of the eight inner juvenal primaries, and their replacement by primaries of the first-winter plumage, provides an excellent method of estimating the ages of young grouse and ptarmigans between about 3 and 15 or 20 weeks of age, by which time the last of these primaries will have completed their growth. Growth rates of representative species of nearly all the genera of grouse and ptarmigans have been studied, and for these age can be estimated by determining the extent of primary replacement and growth during this period (table 9). Undoubtedly there may be some population variations in growth rates of these species, and hand-reared birds may develop at somewhat different rates than wild ones, but the availability of such aging criteria is extremely valuable for backdating probable hatching periods by examining young birds.

First-winter secondaries replace juvenal secondaries at the same time the primaries are replaced or slightly later, and by the time the last of the juvenal secondaries have been shed the young bird will be well into the first-winter plumage. By the time the bird is 4 or 5 months old it should have grown all its first-winter flight feathers and lost all its juvenal feathers other than those few wing feathers that are carried through the winter. With the loss of its juvenal body feathers the bird can be classified as an immature rather than a juvenile. Except in the ptarmigans, no further molt will occur until at least the following spring among the species considered here. However, the three ptarmigan species present a special case, for which an additional plumage stage and molt cycle must be mentioned.

SUPPLEMENTARY POSTJUVENAL MOLT AND SUPPLEMENTARY PLUMAGE

In at least most forms of ptarmigans, a special plumage situation exists. One unique fact is that when the two outer juvenal primaries emerge (at 2 or 3 weeks of age) they do not resemble the other brownish primaries but rather have the white vanes typical of first-winter primaries. Indeed, Salomonsen (1939) considers them first-winter rather than juvenal primaries, but this is to accept the view that a major evolutionary difference exists between the primary molt of ptarmigans and all other grouse, and it seems more reasonable to believe that the coloration of the two outer juvenal primaries has only been adaptively modified in the genus *Lagopus* in relation to ecological requirements for concealment. The postjuvenal molt of the body feathers of young ptarmigans likewise begins unusually early, at about 4 or 5 weeks of age, and the first feathers of preliminary winter plumage begin to appear. These initially consist of vermiculated or mottled feathers rather than pure white ones. Some juvenal feathers are retained for a time, including ones on the throat, breast, and hindneck. As this postjuvenal molt is being completed, a second stage of molt ("supplementary postjuvenal molt") begins, which replaces the last of the juvenal body feathers and also replaces the grayish or brownish feathers grown during the earlier stages of the postjuvenal molt with new white feathers. The body plumage held during the first winter thus includes both some of those feathers acquired during the preliminary postjuvenal molt, such as those on the abdomen, under tail coverts, under wing coverts, legs, and toes, as well as others acquired during the later or supplementary postjuvenal molt, all of which are white (Salomonsen 1939).

PRENUPTIAL MOLT AND FIRST NUPTIAL PLUMAGE

In most grouse relatively little and possibly no additional molting occurs after the assumption of the first-winter plumage and the first breeding season. Dwight (1900) reported a correspondingly restricted chin and head molt in species of the genera *Tympanuchus* and *Bonasa*, and possibly but not definitely in species of *Dendragapus*.

In the ptarmigan species there can be no question about the occurrence of a prenuptial molt ("pre-alternate" according to the classification of Humphrey and Parkes 1959) and distinctive nuptial (or "alternate") plumage. The extent of this molt may vary with age, sex, and latitude, but at this time the males first become markedly different from females. The male willow ptarmigan thus assumes its characteristic rusty brown upperparts, while the male rock ptarmigan acquires vermiculated grayish feathering, and females of both become decidedly barred. In the female molt may proceed somewhat more rapidly and be more extensive than in the male. However, at least in the rock ptarmigan, both sexes retain through the summer at least some portions of the preceding winter plumage, including feathers of their legs, toes, under wing coverts, and some upper wing coverts (Salomonsen 1939). The white-tailed ptarmigan is similar to the rock ptarmigan in this regard (Clait Braun, pers. comm.).

POSTNUPTIAL MOLT AND SECOND-WINTER PLUMAGE

Except in those few species such as sage grouse in which sexual maturity may not invariably be attained the first year, the bird will normally have attempted to breed while still in its first nuptial plumage. The timing of the following postnuptial molt is generally associated with endocrine changes related to changes in gonadal activity. In any case it is typical for all the species considered here to begin a complete body molt in late summer, with the males generally somewhat in advance of the females. At this time the primaries will begin to be molted in outward sequence from the first through the tenth, the secondaries will begin to be molted gradually from one or two molt centers, and the rectrices will begin a gradual or rapid molt. The adult tail molt of grouse, like that of most pheasants (Beebe 1926), is generally centripetal, as reported by Bendell (1955*b*) for blue grouse and by Bergerud, Peters, and McGrath (1963) for willow ptarmigan, but it may be virtually simultaneous, as indicated by Salomonsen (1939) for the rock ptarmigan and Stoneberg (1967) for the spruce grouse. However, Baker (1953) mentions a greater prairie chicken specimen that was undergoing an apparent centrifugal tail molt.

At the same time as the wing and tail feathers are being molted, the body feathers are being renewed, approximately in the same order that they originally grew in during the postjuvenal molt. Except in the ptarmigans, all the feathers that grow in during the postnuptial molt will be carried through the following winter as second-winter plumage. It is of interest that only at this time will the last traces of the juvenal plumage be lost—namely, the two outermost primaries and their coverts.

SUPPLEMENTARY POSTNUPTIAL MOLT AND SUPPLEMENTARY WINTER PLUMAGE

As noted, the ptarmigans differ from the other grouse in the postnuptial molt sequence, and they exhibit an early or preliminary postnuptial fall molt in adults that corresponds to the early postjuvenal molt of young birds. In this mixed white and grayish plumage adult male rock ptarmigans closely resemble females, and both can hardly be differentiated from immature birds (although the old birds will be replacing their two outer primaries at this time). This stage is referred to as the preliminary second-winter plumage. A few body feathers will still be retained at this time from the summer plumage, including (in males) some greater wing coverts or tertiaries and some mantle or hindneck feathers. Females retain many lower breast or flank feathers, some inner median and greater coverts, some tertiaries, and some scattered upper breast, throat, and mantle feathers. These summer feathers, plus the grayish fall feathers just acquired and including some feathers of the upperparts, some flank feathers, the tertiaries, and some upper wing coverts, are now quickly replaced with white feathers by a special supplementary postnuptial molt (Salomonsen 1939). Observations by Høst (1942) of captive willow ptarmigans clearly indicate the importance of photoperiod not only in regulating the timing of molt in willow ptarmigans but also in influencing the pigmentation of the new feathers. Høst found that by exposing birds in winter plumage to artificially long photoperiods starting in November, he could induce the precocious assumption of the spring nuptial plumage, and he even stimulated a female to lay a clutch of eggs in December and January. One of the males that had acquired a nuptial plumage at the beginning of February was then exposed to a 7-hour photoperiod, upon which it molted directly back into a white winter plumage without passing through an intervening fall plumage. However, five birds that had their daylight reduced in August passed through a short fall plumage before assuming their winter plumage.

SECOND NUPTIAL PLUMAGE

The second nuptial plumage is acquired in the same manner as the first, and later plumages and their intervening molts are repetitions of the earlier ones. Once the juvenal outer primaries have been lost in late summer, it is generally almost impossible to distinguish birds in their second fall of life from older age categories.

Figure 7 summarized the foregoing information with respect to the rock ptarmigan (based mostly on data provided by Salomonsen 1939). The relatively great complexity of the ptarmigan plumages and the compression of the natal and juvenal plumages into the minimum possible time spans are apparent in this diagram.

ADULT MOLTS

After the birds attain adulthood, molting patterns in grouse and ptarmigans tend to repeat their earlier cycles. The primaries are molted in simple outward sequence, starting with the innermost at about the end of the mating season. At least in the capercaillie, the primary molt is prolonged over at least 5 months, with no more than two primaries growing at the same time (Castroviejo 1975). Secondaries in adult grouse typically begin molting from the third and progress inward, with a second center of molt usually occurring almost simultaneously that progresses inward from the innermost long secondaries to the shorter "tertiaries" (Bergmann et al., 1978; Castroviejo 1975). The two outermost secondaries are considerably delayed in their molt and are molted inward. The wing coverts are molted in the same sequence as their associated remiges but usually slightly in advance of them, thus protecting the developing flight feathers. However, at least in two species of ptarmigans the two outermost primary coverts are molted in reverse sequence, with the outermost covert dropping before the ninth primary covert. This may be a method of protecting the leading edge of the wing while the rest of the feathers are still growing (Castroviejo 1975). The tail feathers are typically molted in centripetal sequence, usually beginning when the primary molt is about half-completed. The horny covering of the bill is shed gradually, starting at the edges of both mandibles and progressing toward the culmen. Likewise, during the latter part of the spring the pectinated edges of the toes are shed, and their comblike profile is lost until the structures grow back in the fall. At least in

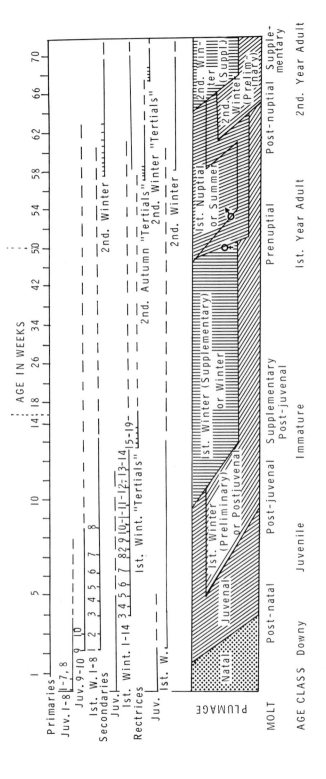

7. Sequence of molts and plumages of the rock ptarmigan. From Johnsgard (1973).

the capercaillie this molt begins at the base of the toes and progresses outward. At about the same time the feathers on the lower tarsi are lost, from back to front and downward (Castroviejo 1975). In ptarmigans these pectinated scales are supplemented by long, branched downy feathers that are also molted during spring, and likewise the replaced claws are approximately 30 percent shorter in summer than those grown during the winter (Höhn 1977).

4

Physiological Traits

BECAUSE OF THE GREAT PROBLEMS involved in maintaining large numbers of grouse in captivity, the birds are extremely difficult subjects for physiological experiments. However, the group poses a number of interesting and unusual physiological adaptations, and a few of these, involving vocal physiology, digestive physiology, and physiological aspects of molting, are worthy of summary here.

VOCAL APPARATUS AND SYRINGEAL SOUND PRODUCTION

In addition to various nonvocal sounds, grouse can also produce sound by internal means, including use of the syrinx in conjunction with the inflation and deflation of the esophageal "air sacs" in various grouse, and use of the syrinx alone in all species of grouse and ptarmigans.

The anatomy of the syrinx in grouse is relatively simple and is very similar to that in the domestic fowl, as described by Myers (1917) and Gross (1964). Figure 8 shows the syringeal anatomy of the domestic fowl and a representative species of grouse.

The syrinx of gallinaceous birds is tracheobronchial; that is, it occurs at the junction of the trachea and the paired bronchi. The syrinx consists of a variable number of partially fused tracheal rings, collectively called the tympanum. The bony structure at the junction of the trachea and the two bronchi is the pessulus; it provides important support for the two parts of tympaniform membranes. One such pair consists of the external tympaniform membranes between the fused tympanum-pessulus complex and the first pair of bronchial rings. A second pair of internal tympaniform membranes is situated medially between the pessulus and the second pair of bronchial rings. The tension on these membranes can probably be increased either by stretching the neck or by pulling the trachea forward through the action of the tracheolateralis muscles. The tension can also be reduced by contracting the sternotrachealis muscles, which insert anterior to the syrinx on the sides of the trachea. When these latter muscles are in normal state of tension the internal and

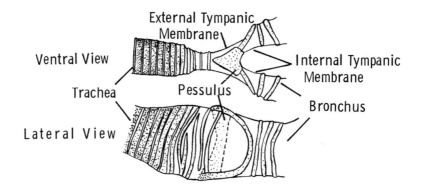

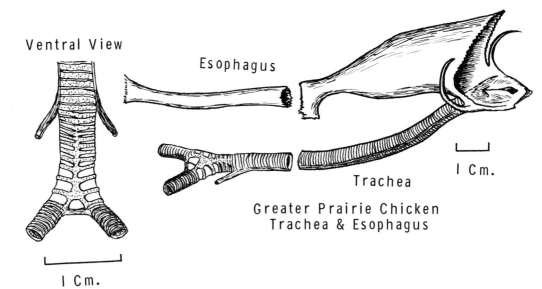

8. Syringeal anatomy of the domestic fowl (*above*, after Gross 1964) and a male greater prairie chicken (*below*, original).

external tympaniform membranes are held well apart, and air can pass unimpeded between them. When the muscles are contracted, however, the membranes are brought closer together, and air resistance builds up pressure in the bronchi, lungs, and air sacs. As air then passes outward between the membranes they are drawn more closely together and set into vibration (the Bernoulli effect), thus producing sound (Gross 1964).

In domestic fowl at least, the frequency (pitch) of the induced vibrations and associated sound production can be increased by stretching the tympaniform membranes and thus increasing membrane tension (Harris, Gross, and Robeson 1968). Simply changing the length of the tracheal tube within the limits imposed by anatomy evidently has little effect on the fundamental frequency established by the surface dimensions and tension of the tympaniform membranes. Since smaller species have narrower tympaniform membranes, it is not surprising that the fundamental frequencies of their calls average somewhat higher than the corresponding calls of large relatives (Sutherland and McChesney 1965). The fundamental frequency of vocalizations in similar-sized species having essentially identical syrinxes is thus regulated by the tension of the tympaniform membranes, which vibrate at a rate proportionate to the square root of their tension (Harris, Gross, and Robeson 1968).

In species such as the turkey that have an interbronchial ligament posterior to the tympaniform membranes, posterior movement of the trachea and syrinx caused by contraction of the sternotrachealis muscles cannot force the bronchi back very far and instead alters the shape of the syrinx. Specifically, the internal and external tympaniform membranes are pushed closer together and the latter are stretched, thus increasing the fundamental frequency (Gross 1968).

It is also clear that two fundamental frequencies are sometimes simultaneously produced by individual birds. This "internal duetting" can be explained theoretically by assuming that each side of the syrinx can operate independently of the other (Greenwalt 1968), or perhaps there is a simultaneous activation of the internal and external tympaniform membranes under differential tension.

Few if any of the vocal sounds produced by grouse are pure tones; rather, in addition to a basic or fundamental frequency generated by the vibration of the tympaniform membrane, there are usually also a considerable number of higher overtones or harmonics that are progressive multiples of the fundamental frequency. These harmonics are of varying loudness, or amplitude, since they are differentially amplified or dampened by the resonating characteristics of the tracheal tube and pharynx. The acoustical effect of the trachea, oral cavity, and beak is thus to tune the bird's vocalizations to a resonant frequency that serves to sharpen the pitch and perhaps to reduce the number of harmonics (Harris, Gross, and Robeson 1968).

There is no direct relation between the fundamental frequency of a vocalization (which is regulated by the vibrations of the tympaniform membranes) and the resonant frequency, which is determined by physical characteristics such as the length of the tracheal tube and its associated resonating structures. The resulting sound is therefore a composite of these two independently determined acoustic characteristics. Although as an individual animal matures the growth of its syrinx and trachea results in a concomitant lowering of both the fundamental frequency and the resonant frequency, those two variables can also have contrasting effects. For example, during neck stretching the increased tension on the tympaniform membranes increases the fundamental frequency, whereas stretching of the tracheal tube lowers the resonant frequency. Gaunt and Wells (1973) have emphasized that the fundamental frequencies of bird vocalizations are evidently not directly coupled to resonating characteristics of the trachea, though they did not exlcude the possibility that air sacs or structures such as tracheal bullae act as resonant filters. Not mentioned by them but presumably included in this category would be inflatable esophageal "air sacs."

By means of a simple formula, the expected resonant frequency and its associated harmonics can readily be calculated for a tracheal tube of any length. Harris, Gross, and Robeson (1968), for example, compared such calculated frequencies with the observed frequencies they generated by using differing lengths of an excised trachea and syrinx from a domestic fowl. They concluded that the trachea and bronchi combine acoustically to form a single resonant tube, and the formula they used indicates that they assumed that the combined structures represent a close-tube acoustical system. However, Sutherland and McChesney (1965) made somewhat similar calculations for calls recorded from live individuals of two species of geese and concluded that the vocal apparatus had resonance characteristics more closely related to those of an open tube than a closed tube. Thus, in an open-tube sound system only the odd-numbered harmonics above the resonant frequency should be expressed, whereas in a closed-tube system both the even-numbered and odd-numbered harmonics will be amplified. A diagram showing calculated resonant frequency curves for acoustical tubes of various lengths has been presented in an earlier publication (Johnsgard 1973). Comparing these curves with the harmonic patterns produced by quail and grouse species (Johnsgard 1973) will illustrate the point that an open-tube acoustic system appears to be present in grouse and quail vocalizations. We may conclude, therefore, that the fundamental frequencies of these birds' vocalizations result from the vibration rates of the tympaniform membranes but that the relative amplitudes of the fundamental frequencies as well as their associated harmonics are differentially amplified or dampened according to the resonance characteristics of the tracheal tube and pharynx.

In the rock ptarmigan the resonance aspects of the vocalizations of males may be influenced not only by the inflatable esophagus, but also by an inflatable membrane about the size of a small walnut that expands to fill the area between the trachea and the esophagus (MacDonald 1970). In the male capercaillie there is a tracheal loop in the crop region, so that the trachea is about a third longer than the neck (Grzimek and Müller-Using, in Grzimek 1972).

Male grouse of many species can inflate their esophageal "air sacs" during sound production. This vocal process has been studied by Gross (1928) for the pinnated grouse, and presumably the same principle applies to the other species. When this species "booms," the beak is closed, the tongue is raised upward against the roof of the mouth, and the internal nares become blocked. The glottis thus open directly in front of the esophagus, and the latter fills with air passing out of the trachea. The expanded anterior end of the esophagus then becomes part of the resonating structure, and the total length and volume of the sound chamber are considerably increased. This combination of the trachea and esophagus is acoustically similar to that of a cylindrical tube and an associated expansible chamber. The resonant frequency of such a combination of tube and cavity is inversely proportional to the volume of the cavity (Harris, Gross, and Robeson 1968). This clearly accounts for the low fundamental frequency of such calls (under 200 Hz). Besides having the obvious visual signal value associated with the inflation of the unfeathered neck region, these low-frequency sounds have considerably greater carrying

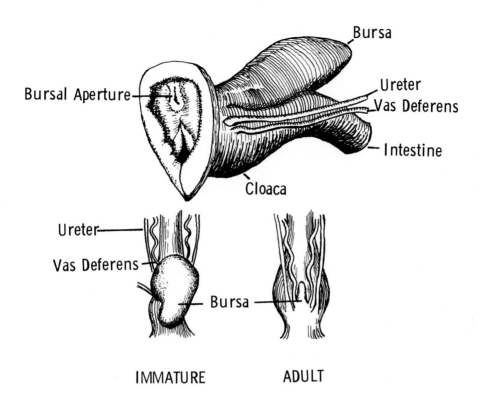

IMMATURE   ADULT

9. The bursa of Fabricius in lateral view (*above*) and dorsal view (*below*).

power than do high-frequency sounds of the same amplitude. Alfred Gross (in Bent 1932) has mentioned that the booming sounds of the heath hen sounded softer than did the bird's more typical calls, yet carried considerably farther. The ecological value of booming is thus clearly apparent.

DIGESTIVE PHYSIOLOGY

In common with many other gallinaceous birds, grouse and quails possess a blind sac on the dorsal wall of the cloaca, called the bursa of Fabricius (fig. 9). In younger birds this typically opens directly into the cloaca; in sexually mature birds it regresses in size and may completely disappear. The bursa does not always open into the cloaca and instead may be occluded by a thin membrane, so its presence cannot in all cases be detected by probing. The function of the bursa is now known to be antibody production (Warner and Szenberg 1964), and its removal or inactivation interferes with immunological processes. Since the relative size and activity of the bursa decrease with age, this structure has been used as a supplementary means of estimating age in gallinaceous birds. Gower (1939) indicates its usefulness through the first November of a bird's life in determining the ages of ruffed, sharp-tailed, pinnated, and spruce grouse, as well as the gray partridge. It also has some limited value in estimating ages of California quail (Lewin 1963), but the age-related differences in upper primary covert coloration in this group are obviously much more convenient.

In most or all of the Galliformes, another outpocketing of the digestive tract occurs at the junction of the small and large intestines. Here a pair of ceca occur that vary greatly in length among different species but are particularly long in the grouse. Apparently these ceca primarily provide a place for the bacterial breakdown of cellulose and similar fibrous materials that cannot be handled by the bird's digestive enzymes. Leopold (1953) surveyed the relative development of these ceca in grouse and quail species and found that most North American grouse with adult weights of about 500 g have ceca averaging about 44 cm in length, with the sage grouse having the longest ceca (68–78 cm) of all species studied. By comparison, although gray and chukar partridges also weigh nearly 500 g, their cecum lengths averaged about 17 cm. Adult quail, ranging in weight from about 170 to 250 g, had cecum lengths of from 10 to 17 cm. Additionally, grouse, which are generally herbivorous, exhibited somewhat longer total intestine lengths than did quails, which are largely granivorous. In a comparison of the ceca of Japanese quails (*Coturnix coturnix*) and spruce grouse, Fenna and Boag (1974) noted that in combined length they respectively equaled 35 and 76 percent of the small intestine and internally differed in lumen size, length of villi, and degree of internal folding. They proposed that the major function of the ceca is to sort intestinal contents into two groups, the cellulose-rich material that is rather quickly discarded from the intestines, and the nutrient-rich residue that is retained in the ceca for further digestion and absorbtion. The length of the cecum in sprice grouse increases in fall, as their diet is progressively dominated by conifer needles (Pendergast and Boag 1973).

Robinson (1980) reported that the needles of jack pines, the major winter food of spruce grouse in eastern North America, contain from 8 to 9 percent protein, or close to the maximum reported for the needles of various conifer species. However, this is considerably below the protein levels of most summer foods of the species. Mineral contents of conifer needles are likewise relatively low, but spruce grouse and other conifer-dependent grouse probably survive the winter simply by consuming a relatively large amount of food. In fact, captive spruce grouse have been found to lose weight and eventually die while eating 90 g of dry-weight spruce needles (or more than 180 g fresh weight) daily, a remarkable total daily intake of food for a bird weighing in the neighborhood of 600 g (Ellison 1972). Probably only about 30 percent of the potential food energy of the food is actually extracted, and thus in 40 g (dry weight) of needles the potential energy equivalent of 213 kcal is diminished to less than 100 (Pendergast and Boag 1971*b*).

Pauli (1978) reported that the rate of digestibility of Norwegian spruce (*Picea abies*) needles by the black grouse is 39.8 percent of intake, or considerably higher than for either bilberry stems or larch sprouts. This grouse evidently prefers winter foods that are rich in total sugars and thus provide easily available energy, whereas in summer it selects foods rich in protein in association with demands for egg production and molting. The approximate 40 percent digestion of winter foods by wild black grouse is approximately equal to that of herbivorous mammals such as deer and hares, but is of higher efficiency than that of domestic fowl (*Gallus gallus*) or coots (*Fulica atra*), according to Zbinden (1980). Like the black grouse, the hazel grouse in the Swiss Zura also prefers winter foods with high sugar content (Zbinden 1979), while the rock ptarmigan population in the Swiss Alps eats a variety of winter plants, nearly all relatively high in sugar. However, plants with a high protein content, such as *Vaccinium myrtillus*, seem to be preferred by that species in winter. The stems of *Vaccinium* are especially heavily utilized and provide a well-balanced supply of food. In late winter and early spring its buds are also important and provide a rich supply of crude protein.

ENDOCRINE AND ENVIRONMENTAL CONTROLS OF MOLTING

Timing of the postnatal and postjuvenal molts can probably be regarded as age dependent, progressing as rapidly as food supplies and general bodily development allow. Additionally, in the ptarmigans, ecological requirements for color changes related to the seasons may place special demands on molt timing. Hewson (1973) and Watson (1973) studied molt cycles in captive and wild groups of Scottish rock ptarmigans respectively. Both observed that males molted earlier than females during the fall molt, that females molted earlier and more completely into the white winter plumage, and that, although males began their spring molt ahead of females, the females soon caught up and molted just as fully. Hewson reported that birds kept indoors and exposed to higher winter temperatures grew more darkly pigmented winter feathers, and Watson observed that wild birds not only were whiter in colder winters but also turned dark earlier in milder winters. The darker winter plumage in milder winters apparently resulted not only from the birds' retaining more of the autumn feathers, but also from their growing relatively darker winter feathers. It is probable that both the timing and the duration of molting in ptarmigans are ultimately adaptively adjusted to climate, with local races being appropriately adjusted to the local light cycle by proximate controls. There may also be individual genetic variation in birds that affects individual variations in molting patterns (Watson 1973).

Although the thyroid hormone is most commonly implicated in molt initiation, recent evidence (summarized by Sturkie 1965 and by Lofts and Murton 1968) indicates that molting may be relatively independent of thyroid activity, or at least that the increased metabolic activity associated with molting may not indicate direct thyroid control. Juhn and Harris (1955, 1968) found that injected progesterone can initiate molt in adult female domestic fowl, and that prolactin stimulates molt in capons when given alone or in conjunction with progesterone. Shaffner (1955) and Adams (1956) also reported on the molt-stimulating effects of progesterone. Jones (1969) found that progesterone injection alone did not stimulate defeathering associated with incubation-patch development in the California quails, but that this hormone in conjunction with prolactin had such effects.

Although molt might be initiated by progesterone alone in various year-round breeders, this effect evidently does not occur among seasonally breeding birds (Kobayashi 1958). In seasonally breeding forms there may instead by a synergistic relationship between progesterone and thyroxin relative to molt control, since Kobayashi found that thyroidectomy inhibited molt induction through progesterone treatment in such species.

Zwickel and Dake (1977) noted that in the blue grouse the primary molt timing is related to reproductive status, and that it is apparently inhibited by reproductive activities. Possibly it is initiated only when there is a reduction in the levels of some hormone that acts as an inhibitor. However, late-molting birds accelerate their molt and "catch up" with earlier molters. Perhaps the birds are "primed" to molt annually by some external clue such as photoperiod. The timing of the completion of the molt is evidently related to physiological preparation for winter rather than for migration (Zwickel and Dake 1977).

In addition to direct endocrine controls, external factors such as photoperiod changes may thus be additional regulators of molt, as suggested by Høst's (1942) early experiments with willow ptarmigan. Lofts and Murton (1968) have reviewed the evidence on this point and have confirmed that at least some north-temperate photoperiodic species of birds require a postnuptial exposure to reduced photoperiod, not only to regain their photosensitivity relative to reproduction but also for the normal temporal completion of their molt. Stokkan's (1979) studies on the willow ptarmigan indicated that the molt from winter to breeding plumage in males is daylight dependent, but that the assumption of breeding plumage requires the presence of testosterone. In castrated males the breeding plumage was omitted, producing a slow molt from the winter to the summer (postnuptial) plumage. During short days neither testosterone treatment nor castration has any effect on molting. However, in intact birds molting of the winter plumage was observed to occur as soon as 8–9 days after exposure to increased day lengths.

In contrast to the willow ptarmigan, testicular hormones evidently suppress the onset of molt in the rock ptarmigan from the white winter to the pigmented summer plumage. Thus a male that had been implanted with testosterone in late June remained in its winter plumage throughout most of the summer and continued to exhibit sexual behavior (MacDonald 1970). By late July the bird had only partially molted into the plumage typical of late summer, supporting the view that testosterone may have a neutralizing effect on the hormone that normally initiates feather growth in this species.

Outside the breeding season, it is apparent that hormones other than testosterone or estrogen must influence the timing of molt and the control of pigment deposition in species having seasonally variable plumages, if indeed such processes are regulated hormonally. Höhn and Braun (1977) observed that willow ptarmigans that had been plucked grew pigmented replacement feathers if they had been injected with any of a variety of pituitary hormones or thyroxine.

HEART/BODY WEIGHT RATIOS AND COLD OR DROUGHT TOLERANCES

Quantitative data are still lacking, but it appears that grouse may have considerably larger hearts relative to body size than do other gallinaceous birds. In general, smaller birds have relatively larger hearts than do larger ones (Hartman 1955). Yet Johnson and Lockner (1968) report that the three species of ptarmigans have heart sizes ranging from 0.87 to 1.85 percent of body weight. It is quite possible that the relatively large hearts of ptarmigans are related to their migratory movements (Irving 1960); among the three ptarmigan species the relative heart size is not correlated with altitudinal distribution (Johnson and Lockner 1968). Weeden (1979) reported that among the Alaskan species of grouse and ptarmigans the rock ptarmigan has the largest relative heart size (2 percent of body weight), with the willow ptarmigan and white-tailed ptarmigans smaller (1.4 and 1.1 percent respectively). On the other hand, the blue and ruffed grouse have heart/body ratios that are among the lowest known in birds (0.4–0.5 percent). He concluded that within the grouse different species have evolved very different cardiovascular strategies to meet similar energy demands, and that heart/body ratios are but a small part of the total problem. Pulliainen (1980) also reported that the rock ptarmigan has a significantly larger heart/body ratio than does the willow ptarmigan in each age-class of both sexes, and that the hazel grouse has the lowest heart/body ratio (0.434) of any of the five European species studied. He attributed this to their different living conditions during winter, the hazel grouse spending much time inside snow burrows, whereas the rock ptarmigan typically spends the winter outside the snow. He did not, however, try to measure the presumably very different insulating values of the plumages of these two species. In the hazel grouse the weight of the feathers represents about 22 percent of the adult weight, while in the rock ptarmigan the proportion is only about 13 percent, a rather surprising statistic (Semenov-Tian-Schanski 1959).

The insulating values of ptarmigan plumage appear to be among the highest of any bird species yet studied (Sturkie 1965). Apparently arctic species have this high capacity for insulation because the barbules at the tips of the contour feather are unusually soft and have extended processes that cling to adjacent feathers when erected, thus trapping air (Irving 1960). The feathered legs and toes of ptarmigans probably both serve as snowshoes (Höhn 1977) and provide insulation.

Although it might be expected that the wide-ranging ptarmigans should conform to Bergman's rule (that warm-blooded animals should increase in volume and weight at higher latitudes), this appears to be true only for the rock ptarmigan, which is largest and heaviest in the Spitsbergen race, averaging over 700 g in both sexes and reportedly even attaining a weight of 1 kg at times, about 35 percent heavier than more southerly populations. On the other hand, the willow ptarmigan does not exhibit this trend (Semenov-Tian-Schanski 1959). Apparently the major difference in the winter ecology of these species is that the rock ptarmigan winters in areas where there are strong winds that clear the snow. The willow ptarmigan, with its heavier beak, is adapted to browse from trees and shrubs that reach above deep snow cover, and it perhaps is better able to use snow burrows to avoid winter temperature extremes. The white-tailed ptarmigan exhibits a trend counter to Bergman's rule, being smallest in Alaska and largest in the southern end of its range (Clait Braun, pers. comm.).

A final interesting aspect of grouse physiology is their water requirements. Regrettably little is known of this situation in most species, but probably only the sage grouse might be expected to exhibit physiological adaptations allowing minimal water intake. Edminster (1954) states (without supporting evidence) that young and old sage grouse can go many days without water, but that birds will travel considerable distances to water and that good populations occur only where a supply of water is available. With regard to the ruffed grouse, Bump et al. (1947) report that three of six adult birds died in less than 4 days when deprived of both food and water, but twelve birds all survived 9 days when given water but no food.

5

Hybridization

As is apparent from the review of Gray (1958), natural hybridization among the grouse has been documented on many occasions. Gray reported some thirty-six types of alleged or apparent hybridization involving at least one species of grouse. Although of course most of these were intrafamily hybrids with other grouse species, several alleged cases of hybridization with domestic fowl (*Gallus gallus*), partridge (*Perdix perdix*), ring-necked pheasant (*Phasianus colchicus*), silver pheasant (*Lophura nycthemera*), and even the turkey (*Meleagris gallopavo*) are among the examples summarized by Gray. The authenticity of some of these latter hybrid records is doubtful at best, and they have little if any relevance to the general problem of isolating mechanisms and ecological interactions among wild grouse. Thus this review will deal only with natural interspecific hybridization within the grouse subfamily Tetraoninae.

As may be seem from table 10, at least sixteen types of natural interspecific hybridization have thus far been reported in the Tetraoninae, involving twelve of the sixteen species Short and I accept. In theory, these sixteen species should be capable of producing some 120 different hybrid combinations, assuming that every species had an opportunity to hybridize with every other one. This, however, is impossible on the basis of distribution, and obviously only those species that are geographically and ecologically sympatric somewhere in their ranges have any opportunity for natural hybridization (Johnsgard 1982). This situation greatly restricts hybridization and reduces the theoretical opportunities from 120 to probably somewhere in the neighborhood of twenty-five to thirty combinations, depending on the accuracy of available range maps. Thus, the sixteen reported probable hybrid combinations actually represent perhaps at least half of the total natural hybrid combinations that could have occurred under natural conditions. Seven of the sixteen combinations involve intrageneric hybrids on the basis of the taxonomy used here; thus most combinations are actually intergeneric crosses. However, if frequency of individual cases is considered, then by far the majority are intrageneric, as might be expected. In fact, given the known range patterns, nearly all potential intrageneric hybridizations have actually been reported, with the single obvious exception of the black grouse × black-billed capercaillie. The white-tailed ptarmigan has also not yet been implicated in hybridization, but it is questionable whether it comes into local contact with any other ptarmigan species except perhaps the rock ptarmigan.

TABLE 10

S<small>UMMARY OF</small> R<small>EPORTED</small> C<small>ASES OF</small> H<small>YBRIDIZATION</small> (×)
<small>AMONG</small> S<small>PECIES OF</small> G<small>ROUSE AND</small> P<small>TARMIGANS</small>

	Blue grouse	Spruce grouse	Willow ptarmigan	Rock ptarmigan	Black grouse	Capercaillie	Black-billed capercaillie	Hazel grouse	Ruffed grouse	Sharp-tailed grouse
Sage grouse	×	—	—	—	—	—	—	—	—	×
Blue grouse		×	—	—	—	—	—	—	—	×
Spruce grouse			×	—	—	—	—	—	×	—
Willow ptarmigan				×	×	×	—	×	—	—
Rock ptarmigan					×	—	—	×	—	—
Black grouse						×	—	×	—	—
Capercaillie							×	—	—	—
Greater prairie chicken								—	—	×

INTRAGENERIC HYBRIDS

Dendragapus

The only opportunities for intrageneric hybridization within this genus are between the blue grouse and the spruce grouse, which are fairly extensively sympatric in western North America, from Yukon Territory south through British Columbia, western Alberta, northern Idaho, and western Montana. So far, only a single specimen of this combination has been reported (Jollie 1955). This individual was shot in Benewah County, Idaho, where both of the parent species are relatively rare, and where the habitat consists of heavy forests of pines, firs, cedars, and other conifers. The hybrid was a young male, and no information is available on its behavior or fertility.

These two species are not isolated ecologically and must come into fairly frequent contact during the breeding season. Males exhibit a considerable number of plumage and display differences, but females are distinctly similar, and it seems probable that males would not discriminate between females of the two forms.

Lagopus

As noted earlier, the white-tailed ptarmigan has so far not been implicated in hybridization, but both of the other two ptarmigans have hybridized on several occasions. The willow ptarmigan (including the red grouse) and rock ptarmigan have extensive geographic overlap both in North America and in Eurasia. Apparent hybrids between them have been reported from Norway and Sweden (Kihlen 1914; Schaanning 1920; Gray 1958), and also from Great Britain (Collette 1886). However, there is no convincing example from North America. Todd (1963) mentioned one specimen from Labrador that he thought might be an abnormally colored willow ptarmigan or possibly a hybrid. Harper (1953) also described a subadult male ptarmigan collected in Keewatin that had intermediate bill depth measurements but unusually low weight and wing measurements. He concluded that it must be a hybrid or a highly aberrant willow ptarmigan. These two species certainly exhibit extensive local contacts over wide areas, although their habitat preferences do tend to maintain ecological segregation between them. Further, they establish relatively monogamous pair bonds, which also probably promotes reproductive isolation.

Tetrao

Three Eurasian species of *Tetrao* exhibit widespread overlap, namely the black grouse, capercaillie, and black-billed capercaillie. As noted earlier, the black-billed capercaillie has not yet been reported to hybridize with the black grouse, although such hybridization seems quite likely, but the other two hybrid combinations are well verified. The combination of the black grouse and capercaillie has been known to occur in Norway and Sweden from as early as 1744 and is common enough there to be given a specific vernacular name (rakkelfugl). Farther south in Europe it is called the rackelhahn and rackelhane. Hybrids have also been produced in captivity, and both sexes appear to be reproductively active. Backcrosses between the male F_1 hybrid and the capercaillie have been bred, but F_1 female hybrids are evidently infertile even though they sometimes lay eggs (Gray 1958). Hybridization is believed to sometimes result from conditions of range expansion when the capercaillie moves into new areas. Females are apparently more mobile than males and thus may move into an area first. On finding no male capercaillies, they mate with male black grouse. In other areas where male capercaillies have been heavily hunted the females may also be prone to hybridize. More than two hundred individual examples of this hybrid combination are known, and probably only the reduced fertility of the F_1 generation and the ecological differences of the parent species prevent more extensive gene exchange (Mayr 1942). Apparently the males of the F_1 hybrids often join leks of black grouse; because of their large size, they easily dominate the resident male black grouse and often send them fleeing. The birds display more actively than do the black grouse, and their loud, raucous call (the basis for their vernacular name) is faintly reminiscent of that of the male capercaillie (Müller-Using, in Grzimek 1972).

Similarly, natural hybridization between the capercaillie and the black-billed species is known to occur in the rather limited areas of geographic sympatry between these two very closely related forms. In one area of overlap in western Siberia 12 percent of the courting male birds were identified as hybrids (Kirpichev 1958, cited by Short 1967). As recent observations by Andreev (1979) indicate, the male behavior patterns of these two species are actually very similar, especially posturing and general lek behavior. However, acoustic characteristics are quite different in the two species, and there are a few plumage and soft-part coloration differences that may have some significance as isolation mechanisms.

Tympanuchus

In North America, hybridization between the sharp-tailed grouse and the greater prairie chicken has been previously surveyed (Johnsgard and Wood 1968) and has been documented for every state and province where there is natural contact between these species. These include four Canadian provinces from Ontario to Alberta and eight states (the Dakotas, Colorado, Nebraska, Iowa, Minnesota, Wisconsin, and Michigan). The highest known incidence of hybridization was reported from Manitoulin Island, Ontario, where the two species rather recently came into contact and where from 5 to 25 percent of the total population may be of hybrid origin. On the Great Plains the incidence of hybridization is considerably lower, but in Nebraska the probable minimum rate is between 0.3 and 1.2 percent of the combined population (Johnsgard and Wood 1968). In a study area in western Minnesota the rate of hybridization increased from 1.0 to 3.7 percent as the ratio between the two species increased. In that area, studies of hybrids indicated that their displays were intermediate in form between those of the parent species and may have repulsed females (Sparling 1979, 1980). However, apparently both the F_1 hybrids and backcrosses are fertile, and thus reproductive isolation is largely dependent upon behavioral mechanisms in areas of local sympatry.

INTERGENERIC HYBRIDS

Centrocercus × Tympanuchus

The sage grouse has a distribution in North America that closely approximates that of various species of sagebrush (*Artemisia* spp.), and it exhibits little ecological or geographic overlap with any other grouse species. The only species with which it is fairly widely sympatric is the sharp-tailed grouse, and over much of eastern Montana and parts of adjacent Wyoming the two species are in contact. Two hybrids were obtained from central Montana in 1969 (Eng 1971) in an area of transitional habitat between these two species. More recently, a hybrid male was found on a sharp-tailed grouse display ground in Sheridan County, Wyoming, during March 1979 (Williams 1979). This bird performed several sharptail-like displays on this ground, which was about a quarter mile from a sage grouse strutting ground. Later the bird was seen in company with two sage grouse hens, to which it also displayed.

Although sage grouse and sharp-tails are lek-forming types, their plumages and sexual display patterns are distinctly different, and the considerable difference in their adult body weights also militates against successful hybridization. Further, the two species are probably not very closely related, since it is likely that the sage grouse is more closely allied to the blue grouse and other ''forest'' grouse than to the ''prairie'' grouse of the genus *Tympanuchus* (Johnsgard 1973). It seems likely that reproductive isolation between *Centrocercus* and other grouse is largely maintained by ecological and geographic isolation, complemented by the distinctive structural features and behavioral aspects of male display.

Dr. Clait Braun (pers. comm.) has informed me that specimens of hybrid sage × blue grouse have been obtained in Utah, but this hybrid combination is still undescribed in the literature.

Rather surprisingly, the hybrid sage × sharp-tailed grouse observed by Williams (1979) exhibited such typical features of sharptail display as cooing, running parallel, and flutter-jumping. However, its most frequent display approached the typical strutting behavior of sage grouse, judging from film sequences. In this display the tail is partly fanned, the wings are suddenly lowered, and a short, gutteral croak is uttered (fig. 10). The bird remains essentially in place, sometimes rotating slightly, but unlike either parent species no large ''air sac'' is exposed during display.

Lagopus × *Tetrao*

In addition to its contacts with the rock ptarmigan, the willow ptarmigan is also sympatric with the black grouse, the capercaillie, and the hazel grouse. Rather surprisingly, wild hybrids involving all three of these combinations have been described. The most frequently reported involve the black grouse; Gray (1958) lists twenty-one references that relate to such hybrids. According to her, wild hybrids have ''often'' been reported from Norway, and additionally a brood of seven hybrids was hatched in the Stockholm Zoo, most of which survived at least 2 weeks. There is some indication of reduced fertility in this cross, since males sometimes exhibit only a rudimentary right testis and in females the ovary is poorly developed. The cross has apparently occurred in both directions (reciprocal hybridization), and in Norway and Sweden is suffciently common to have a special vernacular name, ''rype-orre.'' Collette (1886) reported locating at least twelve speci-

10. Display sequence of a male sage × sharp-tailed grouse, based on 16 mm cine sequences by William Strunk. Elapsed time (in tenths of seconds) is indicated.

1. Sage grouse, male. Photo by Ed Bry.

2. Sage grouse, male. Photo by Ed Bry.

3. Sage grouse lek. Photo by author.

4. Sage grouse lek. Photo by author.

5. Sage grouse male. Photo by Ed Bry.

6. Dusky blue grouse, male. Photo by author.

7. Dusky blue grouse, female. Photo by author.

8. Dusky blue grouse, male. Photo by author.

9. Dusky blue grouse, male. Photo by author.

10. Dusky blue grouse, male. Photo by author.

11. Canada spruce grouse, male. Photo by author.

12. Franklin spruce grouse, male copulating with dummy. Photo by C. G. Hampson.

13. Canada spruce grouse, female. Photo by author.

14. Sharp-winged grouse, female on nest. Photo by Yuri Pukinski.

15. Sharp-winged grouse, male. Photo by Yuri Pukinski.

16. Sharp-winged grouse, displaying male. Photo by Yuri Pukinski.

17. Sharp-winged grouse, male. Photo by Yuri Pukinski.

18. Sharp-winged grouse, chick. Photo by Yuri Pukinski.

19. Newfoundland willow ptarmigan, male in spring. Photo by author.

20. Newfoundland willow ptarmigan, male in spring. Photo by author.

21. Newfoundland willow ptarmigan, female in spring. Photo by author.

22. Newfoundland rock ptarmigan, female in spring. Photo by author.

23. Newfoundland rock ptarmigan, male in spring. Photo by author.

24. White-tailed ptarmigan, female in spring. Photo by author.

25. White-tailed ptarmigan, male in spring. Photo by author.

26. Capercaillie, male and harem of receptive females. Photo by Chris Smith.

27. Capercaillie, male and harem of receptive females. Photo by Chris Smith.

28. Capercaillie, female. Photo by Hans Aschenbrenner.

29. Black-billed capercaillie, female and chicks. Photo by A. V. Kretchmar.

30. Black-billed capercaillie, displaying male. Photo by A. Andreev.

31. Capercaillie, head of displaying male. Photo by author.

mens from Sweden and at least twenty-two from Norway. He also noted that they are "not unusual" in Russia and mentioned that one possible example involving *L. l. scoticus* is known from Scotland. According to Dresser (1876), male willow ptarmigans sometimes attend the leks of black grouse, which probably accounts for the frequency of the hybrid combination.

A seemingly less likely hybrid combination is between the willow ptarmigan and the capercaillie, but these two species have also apparently hybridized repeatedly in Norway (Grieg 1889). At least three specimens of the combination have been preserved (*Ibis*, 1894, p. 447). As with the black grouse, ecological separation between the willow ptarmigan and capercaillie should normally provide complete reproductive isolation, and the patterns of mating (monogamous versus complete promiscuity) also make this an extremely unlikely match.

The rock ptarmigan, in addition to hybridizing with the willow ptarmigan, is sympatric with the black grouse and the hazel grouse and has hybridized with both of these species. It is also geographically sympatric with the capercaillie and black-billed capercaillie but nevertheless has not yet been reported to hybridize with either of these species. Of the two hybrid combinations that have been described, the one involving the black grouse has been more frequently reported. In Norway this cross is known as the "fjeldrype-orre" (Collette 1898; Schaanning 1920), and several apparent examples have been reported from there.

Lagopus × *Bonasa*

Equally surprising is the hybrid cross between the willow ptarmigan and the hazel grouse, since isolation is promoted by their ecological preferences, although they are geographically sympatric over an extremely broad range. Collette (1886) described a male specimen with "well defined" testes from Sweden, and this cross is reportedly "not common" in Scandinavia (Gray 1958). Both species have essentially monogamous pair bonds, which would seem to allow for "correction" of incipient pairing mistakes between the time of pair formation in late fall or winter and breeding the following spring.

Geographic overlap between the rock ptarmigan and hazel grouse is extremely slight; besides occurring in the Alps, it may also locally exist in central Norway, northern Finland, and especially in northern and northeastern Siberia. However, the more widespread overlaps are mostly areas where exact distributions are uncertain, and it seems unlikely that actual ecological contacts between these two species would be frequent. Gray (1958) mentions three apparent hybrids of this combination, but I have not been able to see either original citation she provided.

Dendragapus × *Tympanuchus*

The blue grouse is extensively sympatric with the ruffed grouse and to a lesser extent with the sharp-tailed grouse. There are several known hybrids of the former combination (Ouellet 1974; Tufts 1975), and a single example of the latter cross has been reported (Brooks 1907). The blue grouse and sharp-tailed grouse exhibit considerable geographic sympatry in the Yukon Territory, northeastern British Columbia, south-central British

Columbia, north-central Washington, and few areas of local contact in Idaho, Colorado, and possibly elsewhere, but ecologic differences tend to keep the species fairly well separated. Nevertheless, male blue grouse do often move into fairly exposed areas when displaying, and perhaps in such circumstances come into local contact with female sharp-tailed grouse. The only known example of this cross was obtained near Osoyoos, British Columbia, near the Washington border, and Brooks believed it was the result of a male blue grouse mating with a female sharptail. The hybrid was sexually active, since it had been observed the previous spring displaying in the company of sharp-tailed grouse. Although no hybrids have been reported between the spruce grouse and the sharp-tailed grouse, these two species are locally sympatric, and a mixed group has been observed in which sharp-tailed males were displaying (during fall) to female spruce grouse (Krahn 1980).

Dendragapus × Lagopus

The spruce grouse is sympatric with several North American grouse in addition to the blue grouse. These include the sharp-tailed grouse, the ruffed grouse, and the willow ptarmigan. The areas of geographic sympatry are extensive in all three cases, but the best case of resulting hybridization is with the willow ptarmigan (Lumsden 1969). At least three specimens of natural hybrids of this combination have been reported, and two of them came from the Hudson Bay area of Ontario, where spruce stands near rivers that provide spruce grouse habitat are in close proximity to heath and lichen communities that support willow ptarmigans. The last of the known hybrids came from York Factory, Manitoba, which is also near Hudson Bay and probably represents similar habitat interdigitation. No information is available on the possible fertility or the sexual activity of this cross. The mating systems of the two species are rather different, with the spruce grouse essentially promiscuous while the willow ptarmigan establishes monogamous pair bonds, so this would tend to reduce further the probability of frequent hybridization.

Dendragapus × Bonasa

In North America the ruffed grouse is widely sympatric with the spruce grouse and the blue grouse, but so far hybrids have been described only for the spruce grouse. Ouellet (1974) reported on a hybrid specimen that had been shot in Champlain County, Quebec, in an area of intensive logging, where spruce grouse habitat and probably their numbers were declining and ruffed grouse were thriving. Tufts (1975) later pointed out that this cross had been documented several times in the late 1800s from Nova Scotia, and that one specimen from there was known to be still extant.

Tetrao × Bonasa

Both the black grouse and the capercaillie are extensively geographically sympatric with the hazel grouse, but only the former is believed to have hybridized with this species. One such male hybrid was described by Dresser (1876), and Pleske (1887) described and illustrated a male and female of this combination. The combination has also been illustrated by Schaanning (1920–23). The combination is probably relatively rare, but it indicates a surprising potential for breakdown of reproductive isolation, considering that one of the parent species is a promiscuous, lek-forming type and the other exhibits a monogamous pair-bonding breeding system. Dresser (1876) suggested that it most probably results when a wandering unmated hazel grouse male encounters a female black grouse rather than the reverse.

ECOLOGICAL AND BEHAVIORAL CORRELATES OF HYBRIDIZATION

Of the fifteen kinds of interspecific hybrids summarized here, six are among strictly North American species, eight involve strictly Old World species, and the remaining case consists of the two Holarctic ptarmigans. In only one situation of probable extensive intrageneric sympatry (black grouse and black-billed capercaillie) is there still no definite evidence of natural hybridization, while in five other cases of intrageneric sympatry the rate of hybridization ranges from relatively rare (willow and rock ptarmigans) to extremely frequent (sharp-tailed grouse and greater prairie chicken). There are two situations of extensive intergeneric sympatry in North America where hybridization is unreported but might be expected (ruffed grouse with blue grouse, and spruce grouse with sharp-tailed grouse), while in Eurasia extensive intergeneric sympatry without known hybridization occurs between the hazel grouse and the two species of capercaillies, and also with the Siberian sharpwing. This last species in turn is likewise sympatric with the black-billed capercaillie, but hybridization is unknown. Any hybridization involving these East Asian grouse could well go undetected, as, for example, the apparently fairly common hybridization between the capercaillie and black-billed capercaillie was unreported until fairly recently.

If the cases of known grouse hybridization are arranged in descending order of apparent frequency, some interesting trends emerge, as indicated in the following summary:

Hybrid Combination	Occurrence of Hybrids	Mating Types
Sharp-tailed grouse × prairie chicken	From under 1.0–3 percent of combined population, rarely to 25 percent	Lek × lek
Capercaillie × black-billed capercaillie	Up to 12 percent of males in one local area	Lek × lek
Black grouse × capercaillie	More than 200 specimens known	Lek × lek
Willow ptarmigan × black grouse	At least 34 specimens known	Monogamous × lek
Willow ptarmigan × hazel grouse	Many specimens known	Monogamous × monogamous
Rock ptarmigan × black grouse	Several specimens known	Monogamous × lek
Rock ptarmigan × willow ptarmigan	Several specimens known	Monogamous × monogamous
Willow ptarmigan × capercaillie	Several specimens known	Monogamous × lek
Black grouse × hazel grouse	Several specimens known	Lek × monogamous

Ruffed grouse × spruce grouse	Several specimens known	Promiscuous solitary × promiscuous solitary
Willow ptarmigan × spruce grouse	Three specimens known	Monogamous × promiscuous solitary
Rock ptarmigan × hazel grouse		
Sage grouse × sharp-tailed grouse	Three specimens known	Monogamous × monogamous
Blue grouse × spruce grouse	Two specimens known	Lek × lek
Blue grouse × sharp-tailed grouse	One specimen known	Promiscuous solitary × promiscuous solitary
Blue grouse × sage grouse	One specimen known	Promiscuous solitary × lek
	Hybridization alleged	Promiscuous solitary × lek

Thus, lek species are involved in fourteen of the combinations, promiscuous but solitary species are associated with six combinations, and monogamous species are involved in eleven of the combinations. This is rather surprising and suggests that monogamous species are little if at all more immune to possible hybridization than are lek-forming species, particularly inasmuch as there are six species of lek-forming grouse and probably only four monogamous species. However, all three of the most frequently occurring hybrid combinations involve pairings by two lek-forming species, as might be expected.

6

Reproductive Biology

THE REPRODUCTIVE POTENTIAL of animal species is a compound result of numerous behavioral and physiological characteristics, most of which can be considered species typical. These include such things as the time required to attain reproductive maturity, the number of nesting or renesting attempts per year once maturity is attained, the number of eggs laid per breeding attempt, and the number of years adults may remain reproductively active. These traits place an upper limit on the reproductive potential of a species, which is never actually attained. Rather, the actual rate of increase will only approach the reproductive potential, being limited by such things as the incidence of nonbreeding, mortality rates of adults, decreased hatching success resulting from infertility, predation, or nest abandonment, relative rearing success, incidence of renesting and clutch sizes of renests, and similar factors that affect reproductive efficiency. The relative involvement of the male in protecting the nest or the young may also influence hatching or rearing success. Among those species in which the male does not participate in nesting, the relative degree of monogamy, polygamy, or promiscuity may strongly influence reproductive ecology and population genetics. Although many of these considerations will be treated under the accounts of the individual species, a general comparison of the grouse and ptarmigans as a whole is worth considering here, to see if any general trends can be detected.

AGE OF SEXUAL MATURITY AND INCIDENCE OF NONBREEDING

Bump et al. (1947) reported that nonbreeding by wild female ruffed grouse varied from none in most years to over 25 percent in some years. Weeden (1965b) found no indications of female nonbreeding in wild rock ptarmigans, although Maher (1959) found some evidence of nonbreeding in wild willow ptarmigans. Stanton (1958) reported that 25 percent of yearling female sage grouse failed to produce eggs, and Bendell and Elliott (1967) found that 25 percent of 38 yearling female blue grouse were nonbreeders, compared with 4 percent of 69 adult females. Yearling male blue grouse are nonterritorial according to these authors. Yet in this species, as in several other grouse, the highly promiscuous mating system allows for effective fertilization of all females by a relatively small proportion of fully mature males.

NUMBER OF NESTING OR RENESTING ATTEMPTS PER YEAR

No instances of double brooding have been reported for any species of grouse, and, indeed, known examples of renesting when nests are lost after incubation has begun are hard to find. Giesen and Braun (1979c) found two renests of the white-tailed ptarmigan and believed that renesting was fairly common. Weeden (1965b) reported only one known case of renesting in rock ptarmigans but noted that 3 percent of 228 nests and broods were late-hatching. Jenkins, Watson, and Miller (1963) mention that among Scottish red grouse definite renesting occurs in some years, and the clutch sizes in second nesting attempts are sometimes smaller than in first ones. They noted that 5 of 7 marked birds laid again after their eggs were taken. Patterson (1949) estimated that a small incidence of renesting probably occurs in sage grouse, and Peterson (1980) reported renesting to be fairly common. Stoneberg (1967) found no indication of renesting in the spruce grouse, and so far only 2 definite cases of renesting in the blue grouse have been reported (Zwickel and Lance 1965). Renesting by ruffed grouse is apparently infrequent (Bump et al. 1947), with probably fewer than 25 percent of the unsuccessful females attempting to renest (Edminster 1947). Ammann (1957) reported that no more than 10 percent of young sharp-tailed grouse hatched in Michigan could have resulted from renesting. Nests of the greater and lesser prairie chickens show a decline in clutch size toward the end of the nesting season (Hamerstrom 1939; Baker 1953; Copelin 1963), suggesting a certain incidence of renesting, but until recently only in the Attwater prairie chicken had any verified cases been reported (Lehmann 1941). However, Robel et al. (1970) found that 3 of 14 radio-tracked greater prairie chicken females renested, one of them making two renesting attempts.

PARTICIPATION OF THE MALE IN INCUBATION
AND DEFENDING THE BROOD

Since the availability of the male influences the likelihood of successful renesting and allows for possible double brooding, a summary of male participation in breeding is of some interest. Among the grouse, no cases of male incubation have been reported. However, the male willow ptarmigan actively defends the nest and brood (Dixon 1927; Conover 1926). In the rock ptarmigan the male rarely stays with the brooding female and does not defend the brood (Weeden 1965b), or if present may desert them when they can fly or even earlier (Bannerman 1963). However, some instances of active brood defense have been seen by MacDonald (1970). In the white-tailed ptarmigan the male plays no part in the incubation or care of young (Choate 1960).

52

TABLE 11
REPORTED CLUTCH SIZES UNDER NATURAL CONDITIONS

Species	Normal Range	Mean Clutch Size	References
Sage grouse	7–13	7.39 (154 nests)	Patterson 1952
Blue grouse	3–9	6.37 (118 nests)	Zwickel 1975
Spruce grouse	77–10	5.8 (Nova Scotia, 39 nests)	Tufts 1961
	4–9	7.54 (Alaska, 26 nests)	Ellison 1974
Sharp-winged grouse	—	8 (1 nest)	Dementiev and Gladkov 1967
Willow ptarmigan	2–15	7.1 (Scotland, 395 nests)	Jenkins, Watson, and Miller 1963
		10.2 (Newfoundland, 106 nests)	Bergerud 1970b
		9.7 (USSR, 61 nests)	Semenov-Tian-Schanski 1959
Rock ptarmigan	3–11	7.0 (Alaska, 101 nests)	Weeden 1965b
		6.6 (Scotland, 148 nests)	Watson 1965
White-tailed ptarmigan	3–9	5.2 (11 nests)	Choate 1963
	2–8	5.9 (48 nests)	Giesen, Braun, and May 1980
Capercaillie	5–12	8.3 (Belorussia, 75 nests)	Dolbik 1968
		6.7 (USSR, 110 nests)	Semenov-Tian-Schanski 1959
		7.11 (Norway, 102 nests)	Wegge 1979
		7.07 (Finland, 267 nests)	Rajala 1974
Black-billed capercaillie	5–8	—	Dementiev and Gladkov 1967
Black grouse	7–10	7.7–8.3 (Finland, 1,313 nests)	Helminen 1963
		8.8 (Belorussia, 113 nests)	Dolbik 1968
		7.7 (USSR, 71 nests)	Semenov-Tian-Schanski 1959
Caucasian black grouse	2–10	6.0 (10 nests)	Dementiev and Gladkov 1967
		6.1 (13 nests)	Vitovitch (in Cramp and Simmons 1980)
Ruffed grouse	6–15	11.5 (1,473 nests)	Bump et al. 1947
Hazel grouse	7–11	9.4 (Finland, 40 nests)	Glutz 1973
		8.3 (Belorussia, 137 nests)	Dolbik 1968
Sharp-tailed grouse	5–17	12.1 (36 nests)	Hamerstrom 1939
Greater prairie chicken	5–17	12.0 (66 nests)	Hamerstrom 1939
Lesser prairie chicken	6–13	10.7 (7 nests)	Copelin 1963

CLUTCH SIZES AND EGG-LAYING RATES

The rate at which egg laying in birds occurs presumably depends on how rapidly follicles can be ovulated and associated albumen secreted by the female, and for the species under consideration here this generally averages slightly more than one day per egg. Some estimates for various grouse species are 1.1 days per egg for the rock ptarmigan

53

(Westerskov 1956), 1.3 days per egg for the sage grouse (Patterson 1952), and 1.5 days per egg for the ruffed grouse (Edminster 1947). Among the European grouse species, such as the capercaillie, the egg-laying interval ranges from 1.2 to 2.2 days per egg, while in the black grouse it is from 1.5 to 2.0 days per egg (Cramp and Simmons 1980).

It is difficult to be confident about clutch-size data, for not only do these figures tend to be influenced by the generally smaller clutches laid late in the season by renesting females, but also there may be considerable geographic variation in the average sizes of first clutches in various parts of the range. Thus, clutch-size figures for the white-tailed ptarmigan from Montana are quite different from observations made in Alaska. Nonetheless, since information on average clutch sizes is of such basic importance in calculating reproductive potentials of these species, a summary of published information on clutch sizes is provided (table 11). Among the grouse the smallest average clutch sizes occur among the ptarmigans and the coniferous-forest-dwelling species, while the ruffed grouse and the prairie- and grassland-dwelling species of *Tympanuchus* have clutches of about a dozen eggs. Interestingly, the sage grouse falls closer to the species of *Dendragapus* in its average clutch size (and also in the appearance of its eggs) than to the prairie grouse.

EGG HATCHABILITY AND HATCHING SUCCESS

All available evidence from field studies indicates that the incidence of infertility and embryonic death is probably so low among wild populations as to be almost insignificant. The most extensive observations available for any grouse species are those of Bump et al. (1947), which include data from more than 5,000 ruffed grouse eggs. This and other studies indicate that in general more than 90 percent of the eggs laid under these conditions are fertile and capable of hatching (table 12). The actual percentage of eggs that hatch, however, is invariably less, ranging from about 90 percent to as little as 15 or 20 percent, depending on the rate of nest desertion and predation (table 12). Substantial brood mortality usually occurs during the first month or so, further reproductive success (table 13).

THE EVOLUTIONARY SIGNIFICANCE OF CLUTCH-SIZE VARIATIONS

The question of the adaptive significance of the considerable variations in average clutch sizes for the species under consideration here (from about 5 to 16 eggs) has recently been discussed by Lack (1968). He concluded that average clutch size in these species is generally inversely related to egg size; that is, species that have relatively small clutches typically lay relatively large eggs. The apparent advantage, for species with precocial young, of producing large eggs is that the young can be hatched at a relatively advanced and less vulnerable stage so they can begin feeding for themselves and soon become independent of the parent. In this group, therefore, natural selection has seemingly compromised by allowing the largest clutch size that can be produced by the energy

TABLE 12

EGG HATCHABILITY AND HATCHING SUCCESS UNDER NATURAL CONDITIONS

Species	Hatchability of Eggs	Percentage of Nests Hatching	References
Sage grouse	—	44.2% of 533 nests	Hickey 1955
Blue grouse	ca. 98% of eggs in 36 nests*	75% of 36 nests 57% of 164 nests	Bendell 1955a Zwickel and Carveth 1978
Spruce grouse	91% of eggs in 21 nests	81% of 26 nests	Ellison 1974
Willow ptarmigan	79.6% of 2,603 eggs*	82.5% of 395 nests*	Jenkins, Watson, and Miller 1963
Rock ptarmigan	90% of 147 eggs (Scotland) 94% of 393 eggs (Alaska)	— 65% of 86 nests	Watson 1965 Weeden 1965a
White-tailed ptarmigan	— 88.1% of 177 eggs	70% of 11 nests 56.7% of 60 nests	Choate 1963 Giesen, Braun, and May 1980
Capercaillie	—	83% of 323 nests	Höglund (cited in Glutz 1973)
Black grouse	—	31% of 16 nests	Ellison 1979
Ruffed grouse	95.6% of 5,392 eggs (1st nests)* 92% of 480 eggs (2 nests)*	51.4% of 1,431 nests	Bump et al. 1947
Sharp-tailed grouse	88.2% of 136 eggs* 92% of 324 eggs	40% of 176 nests 55% of 56 nests	Ammann 1957 Sisson 1976
Greater prairie chicken	90.9% of 343 eggs* 83% of 177 eggs	46% of 165 nests 50% of 29 nests	Ammann 1957 Sisson 1976

*Calculated from data presented by authors.

reserves of the female while retaining an egg size adequate for the young to be hatched at a stage sufficiently advanced to favor their survival.

Assuming that natural selection fixes a relatively inflexible optimum egg size for each species (which can conveniently be estimated as the weight of the egg in proportion to the adult female's weight), the physiological drain on a laying female may thus be regarded as this constant multiplied by the average clutch size. It should also be noted that, among all birds, smaller species tend to lay relatively larger eggs than do larger ones, apparently reflecting the minimum investment of energy needed to produce a viable egg. Lack (1968) believed that average clutch size in the gallinaceous birds must therefore be limited either by the number of eggs that the incubating bird can effectively cover, which he rejects, or by the average food reserves of the female as modified by the relative egg size. He suggests that the latter explanation best accounts for the variations in clutch sizes to be found in this group.

Lack makes a number of additional observations about clutch sizes in the pheasantlike birds. First, he notes that among related species clutch sizes tend to be smaller in southern latitudes than in more northerly latitudes. Thus tropical forms are likely to have smaller average clutches than related species of the same size breeding in temperate or arctic

TABLE 13

ESTIMATES OF EARLY BROOD MORTALITY UNDER NATURAL CONDITIONS

Species	Mortality Estimates	References
Sage grouse	From 32% to 54% loss reported in three studies	Hickey 1955
	Average brood size reduced from 5.56 in June to 2.33 by August (48% brood loss)	Keller (in Rogers 1964)
Blue grouse	Estimated 67% brood mortality by August	Bendell 1955a
Spruce grouse	Survival varied from 16% to 40% over 4 years of study (mortality 60–84%)	Robinson 1980
Willow ptarmigan (red grouse)	Average 52% of young from successful nests reared by August (48% brood mortality)	Jenkins, Watson, and Miller 1963
Rock ptarmigan	Average 20.2% brood loss among 208 broods by late July	Weeden 1965a
	Average brood size reduced to 3.6 young at 10–12 weeks	Watson 1965
White-tailed ptarmigan	Approximate 33.1% brood loss among 41 broods in first 8 weeks	Chote 1963
Capercaillie	Average brood size reduced from 8.5 to 3.4 in first 100 days (58% brood mortality)	Glutz 1973
Black grouse	Average loss of 34% of eggs and/or young of successful hens between laying and August	Ellison 1979
	Survival rate from eggs laid to juveniles alive in August 8.5–40.0% (total mortality 60–91.5%)	Angelstam 1979
Ruffed grouse	Average brood mortality averaged from 60.9% (11-year average) to 63.2% (13-year average) in two areas	Bump et al. 1947
Hazel grouse	Brood size in July 5.38–6.74, compared with clutch size of 8.41–8.46 (17–37% mortality of eggs and/or young of successful hens)	Siivonen 1952
Sharp-tailed grouse	Average brood size reduced from 8.7 to 4.6 young (47% loss)	Hart, Lee, and Low 1952
Greater prairie chicken	Average brood size reduced from 8.0 to 6.6 young (17.5% loss)	Baker 1953
	Brood mortality of 46%	Yeatter 1943
Attwater prairie chicken	Approximate 50% mortality in 1st month; 12% later	Lehmann 1941

regions. Second, Lack detected no clear correlation between clutch size and the habitat or pair-bond characteristics of the species. He noted that only a weak positive correlation exists between egg size and incubation period, but he did not consider other possible influences on incubation periods such as the length of the breeding season.

As may be noted from table 8, there is only a weak inverse relationship between the average weight of the egg in proportion to that of the female and the average clutch size in the species under consideration here. It may be seen that such species as the spruce grouse, rock ptarmigan, and white-tailed ptarmigan tend to have small average clutches and fairly large relative egg sizes, whereas the ruffed grouse, sharp-tailed grouse, and two prairie chicken forms have large clutches and smaller relative egg sizes. It is of interest, however, that the three ptarmigan species lay eggs of nearly the same size and that their average clutch sizes are nearly the same, though they have markedly different adult weights. One might expect the willow ptarmigan to have a considerably larger average clutch size than the white-tailed ptarmigan.

If no strong case can be made for food reserves of the female as a major factor possibly limiting clutch size, alternate or supplementary factors must be considered. One possibility, that clutch size is limited by the number of eggs the adult can effectively incubate, is unpromising inasmuch as the large-bodied grouse typically produce smaller clutches than do most of the much smaller quail. It might be noted, however, that the grouse must cover their eggs more effectively, since they are mostly cool-temperate to subarctic breeders, whereas the breeding distributions of quails are more southerly and their eggs are less likely to be chilled during incubation. It seems unlikely that a ptarmigan could effectively incubate a dozen or more eggs, and each day that is invested in producing another egg not only reduces the time available for incubation and rearing of the young but also exposes the untended nest to possible predation that much longer.

If indeed the length of the breeding season is significant, and if the danger of chilling the eggs increases when the clutch size exceeds a number related to the size of the adult in proportion to the egg, then average clutch sizes should increase as breeding distributions are arranged from arctic or alpine areas to warmer ones, rather than the opposite, as Lack has suggested. It is difficult to pick representative figures on frost-free periods for the habitats of the species in question, but it might be argued that among the North American grouse the species might be arranged in a northerly, or alpine, to southerly, or warm-temperate, series as follows: white-tailed ptarmigan, rock ptarmigan, willow ptarmigan, spruce grouse, blue grouse, sage grouse, ruffed grouse, sharp-tailed grouse, pinnated grouse. Except for the sage grouse, which commonly breeds in parts of Utah, Nevada, and Wyoming that have frost-free seasons of 100 days or less, this series closely agrees with a progressively increasing average clutch size. Among the Eurasian species, a similar latitudinal sequence might be rock ptarmigan, willow ptarmigan, capercaillie, black-billed capercaillie, black grouse, hazel grouse, Caucasian black grouse. Except for the Caucasian black grouse, which is low-latitude but alpine, the average clutch sizes also seem to increase with decreasing latitude.

Since nearly all the species of quail breed sufficiently far south that the length of the breeding season is probably not a significant factor affecting their clutch sizes, it seems that some other factor, such as food reserves or predation, might play a role. Provided that adequate food is available, it is quite evident, from studies of captive quail, that females can continue to lay eggs at approximate 1.5 - day intervals almost indefinitely. Instead, the factors limiting clutch sizes in these species might perhaps be the maximum number of eggs that the adult can effectively incubate or the increasing dangers of losing the entire clutch to predators during every day the nest is left untended during the egg-laying period. Thus an average clutch of from 10 to 15 eggs may require about 20 days to complete, and each passing day increases possibility of their discovery by predators. Lack has dismissed the possibility that predation can effectively limit clutch sizes in birds, pointing out that for it to be fully effective the predation rate must exceed the rate of laying, or approximate 1 egg per day. Yet, since predators usually destroy entire clutches or at least often cause desertion of the nest, they may become equally effective whenever the daily likelihood of predation exceeds the inverse of the then existing clutch size. As clutch size increases, fixed daily predation levels therefore become increasingly effective as a potential limiting factor, especially for species that are relatively defenseless or do not attempt to guard the nest before incubation starts.

Figure 11 presents the calculated effects of various daily predation levels on species that lay one egg per day, assuming a constant daily predation rate during the egg-laying period causing destruction or desertion of the entire clutch. For species that average a 2-day interval between eggs, the indicated effects would be doubled (thus a 5 percent daily predation rate would have the effect of the 10 percent rate shown in the figure). The diagram demonstrates that species suffering a 20 percent daily predation level (20 percent of all initiated nests being destroyed each day) cannot effectively increase their clutch size after the third day of laying, and selection would thus favor the evolution of a clutch size of only 3 or 4 eggs. Similarly, those species exposed to a 10 percent daily predation loss cannot increase their effective clutch size beyond the eighth day. Species having predation levels of 5 percent per day can increase their effective clutch size only through the fourteenth to eighteenth day of laying, after which it levels off at 8 eggs. Predation levels of less than 2 percent per day during egg laying are probably ineffective in keeping clutch sizes below the physiological limits of the female of the maximum number that can effectively be incubated, at least among species that lay an average of one egg per day. Bump et al. (1947) found that 38.6 percent of 1,431 ruffed grouse nests were broken up, 89 percent of the disruption being attributable to predators. Six studies summarized by Gill (1966) provide nest destruction estimates on 503 sage grouse nests, which averaged 47.7 percent losses (with a range of 26 to 76 percent). Recently, Ricklefs (1969) has calculated daily natural nest mortality rates for a number of North American game birds from data summarized by Hickey (1955) (see chap. 7). These calculated nest mortality rates for fifteen studies averaged 2.96 percent per day (with a range of 1.55 to 4.66 percent). This admittedly represents a minimum estimate, since the figures are based on the entire nesting period (egg laying plus incubation), whereas most nests are not found until the

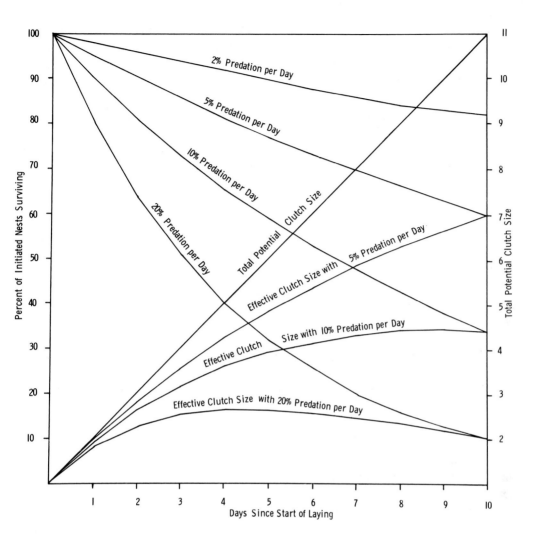

11. Theoretical effects of various predation levels during the egg-laying period on effective clutch size, assuming an egg-laying rate of one per day and predation of the entire available clutch. From Johnsgard (1973).

nesting period is partly over. If, in addition, it is true that in galliforms the mortality rates from predation are higher before incubation begins than afterward, it is clear that such preincubation predation rates might have a significant role in influencing clutch size. Additionally, the need to complete a clutch rapidly and to lay no more eggs than can effectively be warmed by the female might represent a significant factor in limiting clutch sizes of arctic- or alpine-breeding species but is progressively less important for the more temperate-breeding forms. Limiting factors affecting clutch sizes of temperate-breeding species of grouse might be related to the number of eggs an adult can effectively incubate and to predation levels during the relatively long egg-laying period, both of which would tend to allow fairly large rather than relatively small clutch sizes.

7

Population Ecology and Dynamics

LIKE OTHER ANIMALS, grouse and ptarmigans exist as natural populations dependent upon particular habitats, and they vary in density from the absolute minimum populations that have permitted past survival to fairly dense populations that may approach or even temporarily exceed the carrying capacity of the habitat. Each species may also have an upper limit on the density of the population—a saturation point—that is independent of the carrying capacity of the habitat but is determined by social adaptations. Within the population as a whole, individual birds or flocks may have home ranges, geographical areas to which their movements are limited and within which they spend their entire lives. Part of the home range may be defended by individuals so that conspecifics of the same sex are excluded for part or all of the year; such areas of localized social dominance and conspecific exclusion are called territories. Among species lacking discrete territories and in which the social unit is the covey or flock rather than the pair or family, dominance hierarchies, or peck orders, may serve to integrate activities in the flock. These behavioral adaptations and habitat relationships play important roles in population ecology and will be considered in detail in the individual species accounts. However, a preliminary survey may help provide generalizations that will be useful when considering individual species.

Natural populations, whatever their densities, have definable structures in terms of the individuals that make up the population unit. Thus their sex composition, as defined by sex ratios, and their age composition, as similarly defined by age ratios, provide important information on the proportion of the total population that are potential breeders. The fall age ratio, readily determined by the number of immature birds appearing in hunters' kills, also provides the best information available to the field biologist about the success of the breeding season immediately past.

A final important characteristic of natural populations is the rate at which population recycling occurs, which in turn depends upon the mortality and survival rates characteristic of it. Mortality and survival are opposite sides of the same coin; as mortality rates increase, average survival probabilities decrease and life expectancy (or mean longevity) consequently decreases. Mortality rates can thus be used to determine a statistical measure of life expectancy among individuals of a population, and these data are of basic significance to the field biologist. Regardless of the actual mortality rate, all animals in a population must eventually die; the length of time required for a virtual 100 percent turnover of a population age-class is called the turnover rate. This figure corresponds to the maximum possible longevity that may be attained by 1 percent or less of the individuals in that population.

POPULATION DENSITIES

Since virtually all the species of concern here are game birds, information on estimated population densities may be found scattered widely through the technical literature. However, these figures are often not completely comparable; different census techniques may give different results for the same species, to say nothing of their effects on different species, and year-to-year fluctuations in the same population must be taken into account. In addition, census data for some species (such as strongly territorial or lek-forming grouse) are most readily obtained during spring, whereas fall or winter data may be more readily obtained for species that form flocks and are most conspicuous at that time. Further, some census figures are based on territorial males per unit area, while others consider both sexes. Since the sex ratios of adult populations often differ considerably from 50:50, it may be impossible to make the data exactly comparable.

Surprisingly little information is available on minimum tolerable population sizes in the grouse, as Hickey (1955) pointed out. These may vary considerably among various species; solitary species such as ruffed grouse and spruce grouse can perhaps tolerate quite low population densities, whereas highly social species such as the socially displaying grouse may have definite minimum thresholds of survival imposed by the physiological stress or inadequate behavioral stimulation of sparse populations. In general, however, the reproductive potential of most grouse species is so great that populations that are drastically reduced by some means have the biological potential for rapid recovery as long as habitat conditions are favorable. Rather marked population fluctuations are in fact common among certain grouse, particularly the arctic populations of ptarmigans and the more temperate populations of ruffed grouse, sharp-tailed grouse, and greater prairie chicken. Estimates of average population density for these species, at least in areas where major fluctuations are prevalent, must necessarily take these variations into account. The existence and possible causes of these periodic population fluctuations are much too complex and controversial to be considered here, and several review papers (such as Hickey 1955 and Watson and Moss 1979) have dealt with the problem.

It seems evident that, whereas populations may exist over a wide range of densities at the lower limits, upper population densities of a species may have a definite limit. To some degree this is ultimately habitat imposed, the limiting factors being available food, nesting sites, winter cover, predation, and other density-dependent variables. In addition, territory size may establish a maximum density where the habitat might otherwise be capable of supporting a larger number of birds. Even in the absence of actual territorial boundaries the level of intraspecific fighting among reproductively active individuals may force mutual avoidance, causing the population to spread out over the widest available habitat. To the extent that maximum population densities are the result of such species-typical behavior traits rather than habitat variations, they should be fairly constant for a species in different parts of its range. If, on the other hand, maximum densities are primarily a reflection of the differential carrying capacities of the various habitats a species occupies, they are likely to vary considerably between areas and from year to year in the same area.

In spite of difficulties, for the reasons mentioned earlier, in finding comparable data, it is of interest to compare estimated population densities of the species concerned here. These are in general late winter, spring, or adult breeding population figures (table 14).

FLOCKING AND COVEY BEHAVIOR

Among the grouse, perhaps the best-known examples of flocking and covey formation are to be found among sharp-tailed grouse and pinnated grouse during late fall and winter. These migratory movements, often involving large flocks, were once conspicuous in such midwestern states as Minnesota, Iowa, and Missouri (Bent 1932). Hamerstrom and Hamerstrom (1951) describe late fall "packs" of sharp-tailed and pinnated grouse that often numbered in the hundreds, sometimes as many as 400 birds. Similar fall packs of spruce grouse once occurred, and migratory flocks of willow ptarmigans numbering in the thousands have been noted (Bent 1932). Likewise, rock ptarmigans congregate in relatively large flocks during their seasonal movements to and from their breeding grounds.

HOME RANGES AND TERRITORIES

Most grouse and ptarmigans are fairly mobile, but relatively few undertake long migrations. Vertical migrations are known to occur in such mountain-dwelling species as white-tailed ptarmigan and blue grouse, and in the blue grouse the winter range is at a higher altitude than is the summer range. The arctic-breeding rock and willow ptarmigans perform definite seasonal migrations in some areas (Bent 1932), and Hamerstrom and Hamerstrom (1949, 1951) have summarized data on seasonal movements of the sharp-tailed grouse and pinnated grouse. The home ranges of these fairly mobile species must be the largest of any of the grouse, but detailed data are still lacking. Hamerstrom and Hamerstrom (1951) reported that band returns indicated sharp-tailed grouse movements of up to 21 miles, but most returns were obtained within 3 miles of the point of banding. A few transplanted sharptails were also found to have moved more than 20 miles before

TABLE 14

Some Reported Population Densities in Favorable Habitats (Acres per Bird)

Species	Density	References
Sage grouse	51 acres per male on strutting grounds in spring, Wyoming	Patterson 1952
	13–21 acres per bird during fall in best habitats, Colorado	Rogers 1964
Blue grouse	9 acres per adult male, summer averages, British Columbia	Fowle 1960
	2.3–7.7 acres per male on summer range, British Columbia	Bendell and Elliott 1967
	2.5 acres per female; 1.3 acres per male, British Columbia	Bendell 1955*a*
Spruce grouse	128 acres per territorial male, Montana	Stoneberg 1967
	64–90 acres per male (30% of males territorial), Alaska	Ellison 1968*b*
Willow ptarmigan	3.2–12.3 acres per male in spring, Alaska	Weeden 1965*b*
	4.5–9.0 acres per pair in spring, Scotland	Jenkins, Watson, and Miller 1963
Rock ptarmigan	56–109 acres per male, spring, Alaska	Weeden 1965*b*
	4.9–24.7 acres per territorial pair (peak year), Scotland	Watson 1965
	19.8–74 acres per territorial pair (low year), Scotland	Watson 1965
White-tailed ptarmigan	12.8–42 acres per adult in summer, Montana	Choate 1963
Capercaillie	10.6–32 acres per bird in winter, Scotland	Moss, Weir, and Jones 1979
Black grouse	62–137 acres per male in spring, France	Ellison 1979
	49 acres per male in spring, Switerland	Pauli 1974
	55–89 acres per bird in August, Sweden	Angelstam 1979
Ruffed grouse	8–38 acres per adult during breeding season, New York	Edminster 1954
	13.5–30 acres per adult in spring, New York	Bump et al. 1947
	3.4 acres per adult in spring (based on nests), Michigan	Palmer 1954
Hazel grouse	8–13 acres per bird in spring, Poland	Wiesner et al. 1977
Sharp-tailed grouse	45 acres per bird in spring, Michigan	Ammann (in Edminster 1954)
	16–25.6 acres per bird in late summer, Saskatchewan	Symington and Harper 1957
Greater prairie chicken	10–42.7 acres per bird (summary of four studies)	Trippensee 1948
Lesser prairie chicken	17–38 acres per adult male in spring, Oklahoma	Davison 1940

TABLE 15

REPORTED MOVEMENTS OF BANDED PRAIRIE CHICKENS IN WISCONSIN

	Number of Movements	Miles								
		0–1	1–2	2–3	3–4	4–5	5–6	6–7	7–8	Over 8
All female movements	400									
% of total		6.8	21.0	20.7	16.0	10.0	9.0	4.0	6.0	6.5
Cumulative %		6.8	27.8	48.5	64.5	74.5	83.5	87.5	93.5	100
All male movements	1,055									
% of total		25.8	30.3	23.7	10.2	4.5	2.7	1.6	0.3	0.8
Cumulative %		25.8	56.1	79.8	90.0	94.5	97.2	98.8	99.9	100
Male movements from "home" booming ground										
Immatures	318									
% of total		18.9	25.5	24.2	14.5	7.2	5.7	3.5	0.3	0.3
Cumulative %		18.9	44.4	68.6	83.1	93.0	96.0	99.5	99.9	100
Adults	223									
% of total		27.8	29.6	28.3	9.4	2.2	2.2	0.4	0.0	0.0
Cumulative %		27.8	57.4	85.7	95.1	97.3	99.5	100	100	100

Source: Data of Hamerstrom and Hamerstrom 1973; male movements from "home" booming ground based on maximum distance reported at any time and at indicated age.

being shot. Fewer recoveries were obtained for the pinnated grouse, which is apparently the more mobile of the two species. Two banded greater prairie chickens moved as far as 29 miles, and one moved approximately 100 miles (Hamerston and Hamerstrom 1949). In a more recent summary, Hamerstrom and Hamerstrom (1973) found that female greater prairie chickens tend to be considerably more mobile than males, and that among males the immatures are more mobile than adults (table 15). Robel et al. (1970) used radio tracking to determine that greater prairie chicken ranges varied from less than 200 acres in late summer to more than 500 acres during fall and spring, with adult males having maximum monthly ranges of more than 1,200 acres during March. Home-range data for the other species of grouse are equally difficult to obtain, partly because of problems in distinguishing home ranges (occupied but not defended areas) from areas of territorial defense in these species. Males of the forest-dwelling grouse may occupy a fairly large range and establish territorial limits only where they encounter other males, so that possibly no firm distinction between home ranges and territories may be made (MacDonald 1968). In the spruce grouse, males may occupy home ranges of 10 to 15 acres, or occasionally as little as 3 acres (Stoneberg 1967), but both Stoneberg and MacDonald (1968) found that males spend most of their time within a small portion of their home ranges. Ellison (1968b) reported that territorial adult males remained on areas of 5 to 9 acres in early May, within which display occurred and territorial behavior was seen. All adult males but only some yearlings held territories, and yearlings' territories ranged in size up to 21 acres. Other nonterritorial immatures occupied "activity centers" of 6 to 16 acres, but they sometimes moved more than a mile away from these centers. Nondisplaying or nonterritorial immature males have also been noted in ruffed grouse, blue grouse, and sage grouse. In late May and June the territorial males developed larger home ranges of up to 61 acres, and the nonterritorial birds wandered over areas of from 270 to 556 acres (Ellison 1968*b*).

In the ecologically similar blue grouse, territory sizes appear to average somewhat smaller. Boag (1966) and Mussehl (1960) estimated territory size in this species to be from 1 to 2 acres, and Blackford (1963) provides diagrams indicating that 8 territories averaged about 5 acres in size. Bendell and Elliott (1967) reported that territories were about 1.5 acres when blue grouse populations were high and from 5 to 11 acres when populations were low. About 30 percent of the males on the breeding range consisted of nonterritorial and wandering immature males. With regard to the forest-dwelling ruffed grouse, Marshall (1965) stated that one male remained within a 10-acre area during April and May, while Eng (1959) pointed out that males usually stayed within 100 feet of their drumming logs during this period.

In the case of the open-country ptarmigans, several studies on breeding distribution patterns have been done. Weeden (1959) estimated that the territories of willow ptarmigans may range from 3.5 to 7 acres, and the data of Jenkins, Watson, and Miller (1963) suggest that breeding densities of red grouse in Scotland may allow territories of approximately this size, since from 16 to 40 males occupied territories on a 138-acre study area over a 4-year period. Similarly, Watson (1965) reported that populations of rock ptarmigans in favored habitats might have territories of 1.2 to 3.5 hectares (3 to 8.1 acres). Schmidt (1969) indicated that the average territory of white-tailed ptarmigans in Colorado is from 16 to 47 acres (with smaller "areas of maximum use"), and Choate (1963) indicated that in Montana this species' territories average about 40 by 100 yards, or slightly less than an acre.

Lek-forming grouse have the smallest territories of any of the species concerned here. Dalke et al. (1960) indicated that in the sage grouse the master cocks had territories 40 feet or less in diameter (or 0.03 acre). Lumsden (1965) indicated that the central territories of sharp-tailed grouse were approximately 15 by 25 feet (or 0.01 acre), while peripheral ones were larger. Robel (1965) indicated that territories of male greater prairie chickens varied from 23.6 to 106.5 square meters (or 0.006 to 0.026 acres), and Copelin (1963) stated that territories of the lesser prairie chicken were only about 12 to 15 feet in diameter (or 0.002 to 0.004 acres).

VARIATIONS IN POPULATION DENSITIES

Population densities of grouse are highly variable among species and habitats but probably exceed one pair per hectare only under the best late-summer conditions. The three ptarmigans often have spring and summer densities of from 10 to 50 hectares per pair or per territorial male, and about the same is true of the three grassland and sage-adapted species of grouse. Densities of the forest-dwelling grouse seem to be highly variable, with breeding populations in North America ranging from about 2 to 90 males per square kilometer. There are regional and temporal variations in population densities that are often quite large and still of controversial origin; the apparent cyclical fluctuations of such species as ruffed grouse have received considerable attention but have yet to be explained adequately, as is indicated in the individual species accounts. Such factors as variations in weather, food, disease, and predation levels have all been advanced with varying success to account for these density changes. There also seem to be behavioral differences in aggressiveness, breeding behavior, and perhaps such things as dispersal tendencies that are associated with population densities, although whether these are the causes or effects of differences in population densities is still to be established (Watson and Moss 1979).

SEX RATIOS AND AGE RATIOS

The importance of data about the sex and age composition of game bird populations can scarcely be exaggerated. Such data are generally easy to obtain for the species under consideration here, since reliable techniques for determining sex and age are available for most species. Sex-ratio data may provide useful indications of a species' relative reproductive efficiency. For example, adult (or "tertiary") sex ratios in strictly monogamous species such as most quails should clearly be as near 1:1 as possible in order to achieve efficient reproduction, whereas in highly promiscuous or polygamous species a sex ratio strongly favoring females probably represents the most efficient reproductive structure for the population. Nearly all the data available for grouse (except sage grouse and blue grouse) indicate that sex ratios diverge from nearly equal numbers of the sexes at hatching to ratios favoring males in the adult population (table 16). A slight excess of males in renesting species may not be undesirable, inasmuch as it may ensure that sexually active males will be available to fertilize renesting females whose mates have already reached a postreproductive condition. On the other hand, males of polygamous or promiscuous species may be selectively harvested without significantly reducing the reproductive potential of the population. Among such species in which only a single sex is hunted, changes in prehunting and posthunting sex ratios provide a valuable means of calculating population sizes (Davis, in Mosby 1963).

The acquisition of age ratio data is at least as important to biologists as the knowledge of sex ratios in wild populations. Hickey (1955) reviewed the history of age ratio studies and their application for wildlife biologists. He also summarized the data then available for age ratios of gallinaceous birds. In table 19 additional age ratio data are summarized, for the most part chosen to supplement rather than duplicate the figures provided by Hickey.

Age ratio data have two immediate applications. One such application is as a means of estimating survival rates for relatively short-lived species, without the necessity of

TABLE 16

SOME REPORTED SEX RATIOS IN GROUSE AND PTARMIGANS
(PERCENTAGE OF MALES IN POPULATION)

Species	Age-Class	Percentage Male	Sample Size	References
Sage grouse	Immatures	45.3	2,693	Patterson 1952*
	Adults	29.6	1,964	Patterson 1952*
	Mixed ages	40.0	7,355	Rogers 1964
Blue grouse	Immatures	50.0	—	Boag 1966
	Adults and subadults	40.0	—	Boag 1966
Spruce grouse	Immatures	48.3	766	Lumsden and Weeden 1963*
	Adults	55.3	423	Lumsden and Weeden 1963*
Willow ptarmigan	Adults	55.9	2,211	Jenkins, Watson, and Miller 1963*
Rock ptarmigan	Adults	58.5	1,545	Watson 1965*
Capercaillie	Immatures	43.8	—	Rajala (cited in Glutz 1973)
	Adults	40.0	12,609	Rajala 1974
Black grouse	Immatures	43.5	69	Ellison 1978
	Adults	47	16,097	Rajala 1974
Ruffed grouse	Immatures	51.2	17,577	Dorney 1963*
	Adults	54.6	5,365	Dorney 1963*
Sharp-tailed grouse	Immatures	56.0	2,108	Ammann 1957
	Adults	60.0	889	Ammann 1957
Greater prairie chicken	Immatures	54.9	306	Baker 1953
	Adults	60.7	637	Hamerstrom and Hamerstrom 1973
Lesser prairie chicken	Immatures	50.3	1,351	Campbell 1972
	Adults	67.0	1,096	Campbell 1972

*Calculated from data presented by the authors.

marking birds individually and gathering recapture or recovery data. Marsden and Baskett (1958) used the technique of assuming that the percentage of immature birds in the fall hunting sample represented an estimate of the annual mortality rate of adults, and indeed these estimates are generally in close agreement with mortality estimates based on data from banded birds as summarized by Hickey (1955).

The second and more generally applicable use of age ratios is to supplement the evidence obtained from nesting and brood counts about the relative success of the past breeding season. By comparing the number of immature birds in the fall population with the number of adults (or adult females, as is done by some investigators), one can estimate breeding productivity. Thus, a ratio of 50 percent inmmatures to 50 percent adults in the fall kill sample would suggest a breeding season productivity of 100 percent, while a ratio of 75 percent immatures to 25 percent adults would provide a productivity factor of 300

TABLE 17

Some Reported Fall and Winter Age Ratios in Grouse and Ptarmigans
(Percentage of Immatures in Population)

Species	Percentage Immature	Sample Size	References
Sage grouse	57.89	4,657	Patterson 1952
	51.4	7,355	Rogers 1964
Blue grouse	61–69	3,829	Zwickel, Brigham, and Buss 1975
	57–65	—	Hoffman et al. (cited in Bendell (1955b)
Spruce grouse	64.4	1,189	Lumsden and Weeden 1963*
Willow ptarmigan	72	5,266	Bergerud 1970b
	59	52,360	Myrberget 1974
Rock ptarmigan	57	785	Alaska Game Bird Reports
White-tailed ptarmigan	33–47	—	Choate 1963
	35.5	765	Hoffman and Braun 1977
Capercaillie	59.7	33,373	Rajala 1974
	34.3	—	Wegge 1979
Black grouse	56	—	Helminen 1963
	60	44,139	Rajala 1974
Ruffed grouse	77	22,942	Dorney 1963*
Hazel grouse	60.7–75	—	Gajdar (cited in Bergmann et al. 1978)
Sharp-tailed grouse	69.3	25,517	Hillman and Jackson 1973
	63.5	16,283	Johnson 1964*
Greater prairie chicken	50.2	604	Baker 1953
	63.0	1,709	Hamerstrom and Hamerstrom 1973
Lesser prairie chicken	53.2	932	Lee 1950
	55.2	2,347	Campbell 1972

*Calculated from data presented by the authors.

percent. The ultimate limit on such productivity factors is determined by the average clutch size of the species, and the difference between the actual productivity ratio and the potential one (assuming an equal sex ratio in adults) might provide an estimate of the reproductive efficiency of the population.

Reported age ratio data for as many species of grouse and ptarmigans as possible are summarized in table 17. It should be apparent that such data are likely to vary considerably in different years or under different ecological conditions. Nevertheless, such data provide sample figures for interspecies and intraspecies comparisons and for illustrating the theoretical relation just mentioned between clutch size and potential productivity. When tertiary sex ratio data are available, the possibility of inserting a correction factor based on the percentage of adult females in the breeding population is of course desirable.

It has been emphasized that populations of animals can vary in density, in spatial distribution patterns (territoriality favors dispersion, sociality favors clumping), and in sex and age composition. Not only can the population be analyzed for immature and adult components, but the adults themselves have age composition characteristics, with the relative frequency of the various age-classes depending on the rate at which animals die. It is possible to gather such mortality information only by marking individuals (preferably while they are still young enough so one can determine their exact age at the time of marking), releasing them, and resampling the population at later times to determine how long the marked individuals survive. A review by Farner (1955) provides the theoretical concepts and practical methods required in performing such investigations with birds, and it is beyond the scope of this short review to discuss them here. A few ideas, however, are so basic to understanding this aspect of population dynamics that they must be considered individually.

The relative rate at which individuals in a population die is usually expressed as an annual mortality rate (M), which is the ratio of those individuals dying during a year to the number that were alive at the beginning of the twelve-month period, whatever its starting point. The annual survival rate (S) is the opposite ratio: the proportion of the animals still surviving at the end of a twelve-month period to those that were alive at its start. Thus, $S + M = 1.0$, or $S = 1.0 - M$. Some examples of estimated survival rates appear in table 18. The total population may be subdivided into different age-classes according to the year in which each individual was hatched. The population thus consists of varying numbers of one-year-olds, two-year-olds, etc. For the species under consideration here, all the individuals in a single age-class will probably have actual ages within two or three months of one another, depending on the length of the breeding season. Each breeding season thus generates a new cohort of birds that have hatched during the same year and constitute a single age-class. The time required for an entire cohort of hatched young to be essentially eliminated from the population is referred to as the turnover period or turnover rate. This is perhaps properly estimated on the basis of time required for 100 percent of the age-class to be reduced to 1 percent of the original cohort, but practice varies in this regard (Hickey 1955). The means proposed by Petrides (1949) for calculating an expected turnover rate is based on the assumption that the mortality rate is constant for all ages. It is therefore convenient to define the initial cohort as, for example, the birds alive at the start of the first October following hatching to avoid the problems of the higher mortality rates usually associated with the first few months of life. Obviously, turnover periods whose starting points consist of 100 percent of the immatures surviving to fall will be longer than those based on a cohort of newly hatched young. Even shorter would be the turnover rates based on 100 percent of the potential young, in the form of total eggs laid. Although this last basis for defining a cohort is rarely if ever used in practice, it has one theoretical advantage. That is, by starting with the eggs laid rather than with some later stage, it is possible to introduce differential rates of prehatching, juvenile, and adult mortality rates in

TABLE 18

SOME REPORTED ANNUAL SURVIVAL RATES IN GROUSE AND PTARMIGANS

Species	Survival Rate (%)	References
Sage grouse		
Yearling females	35	Wallestad 1975
Adult females	40	Wallestad 1975
Juveniles	15	Wallestad 1975
Adult males	48	Wiley 1973b
Blue grouse		
Adult males	69–75	Zwickel and Bendell 1972
Adult females	69–72	Zwickel and Bendell 1972
Adults (both sexes)	72	Bendell and Elliott 1967
Juveniles (both sexes)	39–60	Zwickel and Bendell 1972
All age-classes	64.4	Brown and Smith 1980
Spruce grouse		
Adults (both sexes)	35–75	Bendell and Zwickel 1979
Juveniles (both sexes)	54–79	Keppie 1975b
Adult males	72	Boag et al. 1979
Adult females	63	Boag et al. 1979
Willow ptarmigan		
Both sexes (Norway)	22–23	Hagen (cited in Hickey 1955)
Both sexes (Scotland)	ca. 35	Jenkins, Watson, and Miller 1967
Both sexes (Newfoundland)	28	Bergerud 1970b
Rock ptarmigan		
Both sexes, all ages[a]	38.3	Weeden and Theberg 1972
White-tailed ptarmigan		
Males	75	Hoffman and Braun 1977
Females	72.7	Hoffman and Braun 1977
Capercaillie		
First-year males (Scotland)	41	Moss, Weir, and Jones 1979
Adult males	61	Moss, Weir, and Jones 1979
First-year females (USSR)	33	Semenov-Tian-Schanski 1959
First-year males	46	Semenov-Tian-Schanski 1959
Older males	59	Semenov-Tian-Schanski 1959
Black grouse		
Both sexes (Finland)	40	Rajala 1974
Both sexes (Finland)	40–60	Helminen 1963
Adult males (Netherlands)	66	Vos 1983
Ruffed grouse		
Adult males	47	Gullion and Marshall 1968
Greater prairie chicken		
Males (all ages)	47	Hamerstrom and Hamerstrom 1973
Females (all ages)	44	Hamerstrom and Hamerstrom 1973
Lesser prairie chicken		
Males	32	Campbell 1972
Sharp-tailed grouse		
Both sexes	40	Ammann 1957
Both sexes	20.5–29.6	Robel et al. 1972

[a]August to spring only, 10-year average.

the construction of a survivorship curve, which not only provides a more realistic view of population diminution but also introduces the possibility of calculating the rate of egg replacement potential in the adult age-class of the resulting survivorship series. This must be based on average clutch size estimates, knowledge of possible nonbreeding rates in younger age classes, and tertiary sex ratio information, but it provides a useful means of estimating the population regeneration potential of species having varying mortality rates of eggs, juveniles, and adults. Some examples of such calculations are presented in figures 12 and 13.

One of the most useful statistics that can be derived on the basis of known and constant mortality rates is an estimate of further life expectancy as of a prescribed initial date or age. Thus, a life expectancy figure may be defined as of the date of hatching, the date of fledging, or some later chosen time. In general, it is perhaps best designated for birds as the earliest age at which juvenile mortality rates have decreased to the point where they become virtually identical with adult mortality rates. This may be as early as the first September or October after hatching or possibly even a year later. In any case, the further life expectancy for any age-class is in effect the length of time required to reduce the number of surviving individuals of that age-class by 50 percent. The expectation of further life is thus an estimated mean after lifetime, or a mean longevity as of a selected initial date. Farner (1955) has suggested that an estimate of a mean after lifetime can conveniently be calculated, by using the following formula, if the mean annual mortality rate is known and if the mortality rate of the included age-classes do not differ significantly from the overall mean mortality rate:

$$\text{Mean after lifetime} = \frac{0.4343}{\log_{10} S}$$

If the selected initial date from which a mean after lifetime is calculated is chosen as some point following hatching itself, then of course the estimated mean after lifetime is not the same as the average life-span. Rather, the average life-span (or mean total longevity) will be somewhat less than the sum of the mean after lifetime estimate and the interval between hatching and the initially selected date, with the difference dependent on the higher mortality rates between hatching and the initially selected date. Lack (1966) has provided a convenient formula for computing further life expectancies in years by the following method, in which M equals the annual mortality rate:

$$\frac{2-M}{2M}.$$

Recently, a valuable contribution by Ricklefs (1969) has concentrated on the significance of mortality rates of eggs and young, and he has provided a ready method of estimating short-term (weekly, daily, etc.) mortality rates for these important stages in the life cycle. He found that such mortality rates can be calculated by the equation:

$$m = \frac{-(\log_e P)}{t},$$

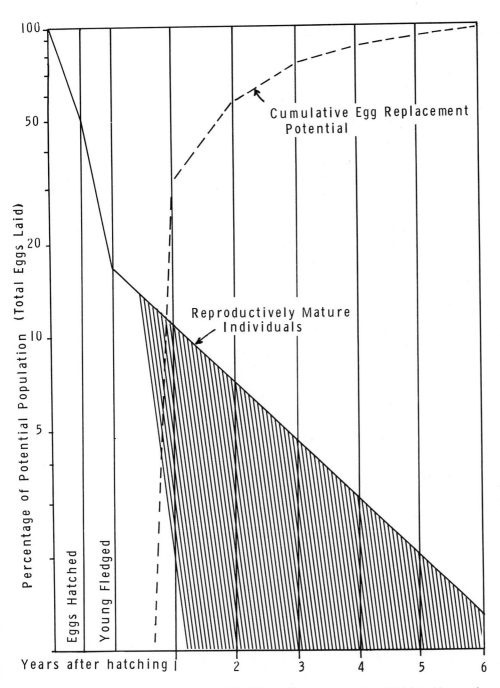

12. Survival curve and egg-replacement potential of female blue grouse based on field data. (Assumptions are of a 50 percent hatching and 33 percent rearing success, 62 percent annual survival after first fall, and an average clutch of 6.2 eggs, with 25 percent of the first-year females nonbreeders.) After Johnsgard (1973).

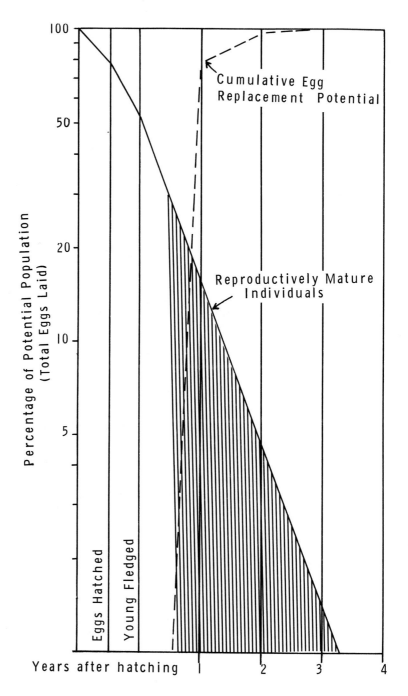

13. Calculated survival curve and egg-replacement potential of the willow ptarmigan. (Assumptions are of a 77 percent hatching and 33 percent rearing success, a 44 percent annual survival rate of both sexes after the first fall, and an average clutch of 7.1 eggs.) After Johnsgard (1973).

where m equals the mortality rate per unit of time (t) and P equals the proportion of nests or individuals surviving the total period considered, again assuming that mortality rates are constant throughout the entire period. As noted in the previous chapter, daily nest mortality rates are generally between 2 and 4 percent, whereas chick mortality rates are considerably lower (Ricklefs 1969).

An equally useful formula is that proposed by Petrides (1949) for estimating the turnover period, this term being defined as the time required to reduce an original age-class cohort of 100 percent to its virtual elimination from the population. Such an effective end-point might be 5 percent, 1 percent, or 0.1 percent, depending on one's views. Petrides reported that the turnover period can readily be calculated by the following formula, again assuming that the mortality rates of different age-classes do not vary significantly from the overall annual mortality rate:

$$\text{Turnover period (years)} = \frac{\log_{10} \text{ of surviving fraction of cohort}}{\log_{10} S} + 1 \ .$$

If 1 percent is chosen as the surviving fraction of the cohort that represents the virtual elimination of an age-class from the population, then the formula can be restated simply as:

$$\frac{-2.0}{\log_{10} S} + 1 \ .$$

Table 19 presents some calculated mean after lifetimes (usually after the first fall of life) and estimated turnover periods among various species for which annual mortality estimates have been reported. In some cases these estimates of mean after lifetimes differ slightly from those reported by the original authors, the variations being the result of different techniques or assumptions, but in general the estimates are very close to those published earlier for these species.

Such calculated turnover periods should provide at least a general estimate of potential natural longevity, as represented by the oldest age-class that might be encountered in natural populations. Potential natural longevity is likely to be less than potential longevity under ideal conditions, such as optimum conditions of captivity. Table 20 presents some reported estimates of mean after lifetimes and records of unusual longevity for wild or captive individuals. It might seem that 4 or 5 years represents close to the potential natural longevity of most grouse and ptarmigan species, but available mortality rates of a few species (especially blue grouse and white-tailed ptarmigan) indicate that it might be considerably longer than this.

TABLE 19
Some Longevity Estimates Based on Reported Survival Rates

Species	Survival Rates (%)	Mean Longevity after First Fall[a]	Maximum Longevity and Turnover Period[b]	References
Blue grouse				
Males	72.5	3.1 years	15.3 years	Zwickel 1966*b*
Females	62.0	2.09 years	10.6 years	
Willow ptarmigan				
Both sexes	30.0	10 months	4.8 years	Jenkins, Watson, and Miller 1963
White-tailed ptarmigan				
First year	37.0	0.99 year	Overall (57.9% *S*) 9.4 years	Choate 1963
After first year	71.0	2.92 years	After first year 14.4 years	
Ruffed grouse				
Males (after first winter)	47.0	1.25 years	6.76 years	Gullion and Marshall 1968
Sharp-tailed grouse	40.0	1.10 years	6.0 years	Ammann 1957
Greater prairie chicken	46.0	1.2 years	6.8 years	Hamerstrom and Hamerstrom 1973

[a]Method of Farner (1955, p. 409).
[b]Method of Petrides (1949), using 1% of original cohort as end-point.

TABLE 20
Some Longevity Estimates and Maximum Longevity Records

Species	Longevity
Sage grouse	One banded female survived 8 years (Wallestad 1975). Returns of marked birds returning to strutting grounds one year later varied from 5 to 21 percent over 3 years (Dalke et al. 1963).
Spruce grouse	One banded female survived 5½ years; one male lived at least 7½ years (Robinson 1980).
Willow ptarmigan (red grouse)	Seven birds at least 4 years old were recovered from 12,050 banded (Jenkins, Watson, and Miller 1963). One banded bird survived at least 7 years and 8 months (Cramp and Simmons 1980).
White-tailed ptarmigan	Twelve of 36 females and 16 of 31 males lived at least 5 years. Estimated mean longevity after first summer, 3.02 years; estimated maximum longevity of 13–15 years (Choate 1963).
Capercaillie	Three individually recognizable males survived 9, 11, and 12 years (Glutz 1973). One male survived to 13.5 years (Müller 1979).
Black grouse	One banded female survived for at least 5.5 years (Glutz 1973).
Ruffed grouse	Maximum known survival of 94 months by 1 of 978 marked birds. Mean life-span of 8.56 months for immature females; 8.63 months for immature males. Mean life-span of birds banded as adults was 25.3 months for males, 23.7 months for females (Gullion and Marshall 1968). One bird survived to 11 years (Grange 1948).
Hazel grouse	One banded bird survived 7 years and 3 months (Rydzewski, cited in Cramp and Simmons 1980).
Sharp-tailed grouse	One bird at least 7.5 years old from 93 banded birds. Mean longevity after full growth estimated from 1.51 years (females) to 1.61 years (males) (Amman 1957).
Greater prairie chicken	One bird, out of 942 banded, observed in eighth year after banding (Hamerstrom and Hamerstrom 1973).
Heath hen	The last surviving male on Martha's Vineyard was probably 8 years old when it died in 1932.

8

Social Behavior
and Vocalizations

ONE OF THE MOST COMPLEX and fascinating aspects of grouse and ptarmigan biology is the birds' social behavior, particularly that related to reproduction. Natural selection in some galliform groups such as quails and partridges seemingly has favored the retention of a strong monogamous mating system, with the associated advantages of maintaining the pair bond through the breeding season. This system allows the male to help protect the nest, possibly participate in incubation, and later help care for the brood. Only in the ptarmigans do the grouse seem to approach this system, and even there it is weakly developed. Rather, among the grouse selection has apparently more strongly favored the retention of a strong male premating advertisement behavior, at the expense of the male's participation in subsequent nesting and brood care, and there is instead a direct relation between a male's capacity to establish and advertise a favorable territory and his ability to reproduce successfully. This territoriality perhaps results mainly from the wide variation among males in aggressiveness and reproductive vigor, but it is also possible that in these species the control of resources (mating opportunities or actual defense of breeding areas) in relation to the population density may be more significant for the species' survival than are any advantages of brood protection. Thus premating territorial behavior among males is conspicuous in all the grouse species.

EVOLUTIONARY TRENDS IN GROUSE SOCIAL BEHAVIOR

The size of the male's territory and the length of time he defends it vary considerably among grouse. At one possible extreme, throughout the breeding season the male defends fairly large territory within which a single female nests and where he defends her and her brood. From this one may trace the progressive development of smaller territories that are defended only until after females are fertilized, in a system where females do not nest within the territorial boundaries and the males do not defend them or their broods. This trend toward the evolution of a polygamous or promiscuous mating system is associated with many parallel evolutionary trends. There is an increased pressure on males to enhance their attraction value to females; thus a tendency exists for more elaborate or more conspicuous sexual signal systems among males. Since they no longer must remain near the female and the nest, pressures for protective coloration are countered by those of sexual selection, and increased behavioral and plumage dimorphism is to be expected.

Not only can conspicuousness in male sexual displays be enhanced by increase in body size and by elaborate visual and acoustical signals in an individual male, but such effects can be multiplied through the aggregation of several males. These counterpressures— those favoring the maintenance of definite and maximum territorial areas as a factor of reproductive success and those favoring the aggregation of several displaying males in a limited area to increase the likelihood of female attraction and reduce the danger of predators to individual males—have led directly to the evolution of arena behavior in several grouse species. This unlikely form of communal display, in which individual male territories are closely adjacent, are relatively small, and serve only as mating stations, can evolve only under certain conditions. First, the males must be totally freed from defending areas large enough for the females to nest within and also from defending the female during incubation and brooding. Next, the reproductive efficiency of a group of males must be greater than that of single males, either because of their greater attraction for females or because the assembled males are safer from predators than are males displaying solitarily. Further, to assure assortative mating there must be enough individual variation in aggressiveness among males that territorial size or location is directly related to breeding success; these variations are perhaps most likely among species that require two years or more to attain full reproductive development. In addition, if male display aggregations are to develop it must be advantageous for the less successful males to associate with the more successful ones. It may be argued that such early experience increases the male's chances of holding a larger or more centrally located territory that will be more reproductively efficient later in his lifetime. Peripheral males participating in arena displays may be regarded as apprentices that reproductively benefit more from such experience than they would from establishing independent and solitary territories.

Since arena displays among grouse might logically be expected to evolve more readily in open-country habitats than in heavily forested ones, open-country and polygamous species are preadapted for the evolution of arena behavior. It seems quite probable that the arena behavior of sage grouse evolved independently from that of the prairie grouse (*Tympanuchus*), and the corresponding behavior of the Caucasian black grouse and Eurasian black grouse may also have evolved independently. This last species is actually a woodland edge form, but its arena displays occur in open heaths. The communal leks of the black grouse were the earliest of the arena displays of the grouse studied, and the term lek is now generally applied to arena behavior of all grouse. Koivisto (1965) suggested that *display ground* be used to describe the general topographic location in which social display is performed, *arena* be used to indicate the specific area (the collective territories), and *lek* more broadly applied to both the birds and their arena. Similarly, the term *lekking* can be used to indicate the general process of communal male display in grouse.

To illustrate how arena behavior may have gradually evolved from more typical territorial behavior, I will mention a series of representative grouse species that provide reference points along this behavioral spectrum.

Of all the grouse, the willow ptarmigan's (red grouse population) actions come closest to the presumed ancestral (or most generalized) type of reproductive social behavior. In this species males establish fairly large territories in fall (at least in nonmigratory populations). These individual territories are largest for the most aggressive males, and many young or inexperienced males may be unable to establish territories, especially in dense populations. The female is attracted to a displaying male, and a firm pair bond is formed. Sometimes males form pair bonds with two females and may breed with both. Territorial displays and defense continue after the pair bond is established, but such activities diminish during the nesting season. At that time the male defends the female and nest, and after hatching he remains with the female and her brood. After the brood is reared the territorial boundaries are again established.

In the rock ptarmigan and also in the white-tailed ptarmigan, the pair bond is typically established in the spring. At least in the rock ptarmigan, two or three females may sometimes be associated with a single territorial male, but there is little or no evidence of polygamy or promiscuity in the white-tailed ptarmigan. The male continues to defend the territory while the female is incubating, although with reduced intensity, and the territory is abandoned about the time of hatching. The female and young may perhaps remain in the male's territory but he only infrequently accompanies them, and he usually takes no part in defending the young. In the rock ptarmigan the male reestablishes his territory in the fall, whereas in the white-tailed ptarmigan this evidently does not occur until spring (Watson 1965; Choate 1963). However, some males do return to their territories in the fall (Clait Braun, pers. comm.).

In the apparently monogamous European hazel grouse the male reportedly establishes his territory in the fall, with those in optimum habitats being the most successful in attracting females. A male usually remains on his territory, defending both it and the female during incubation and brooding periods, but only atypically performs distraction displays or utters warning calls to the female (Pynnönen 1954). Some observers have nonetheless reported seeing males attending broods with females.

In the blue grouse there exists a stage clearly intermediate between the extreme of a monogamous or nearly monogamous pair bond associated with a territory large enough for rearing a brood and the other extreme of complete promiscuity and territorial defense limited to an area that will attract females and provide a mating station. Other North American species that fall into this general category are the ruffed grouse and the spruce grouse, but the blue grouse will serve as an example.

Because of their winter migration, blue grouse males probably first establish territories in spring. Although these areas may cover several acres, hooting is limited to particular places within the territorial boundaries. The home ranges occupied by females associated with territorial males may overlap the boundaries of several male territories. The typical mating system of blue grouse may thus be considered polygamous or promiscuous (Bendell 1955*c*; Bendell and Elliott 1967), but in local populations at least some birds may form strong pair bonds that persist until after the young hatch (Blackford 1958, 1963). The location of the female's nest is not associated with the male's hooting sites, and the male does not defend the nest or the brood. In general, males' hooting sites are well separated and their territories are not contiguous, but in a few cases apparently communal male displays involving four or more males have been observed (Blackford 1958, 1963). Males remain on their territories until their late-summer migration, well after active territorial defense ceases.

The forest-dwelling capercaillie provides a slightly more advanced stage in the evolution of communal displays, judging from such reports as those of Lumsden (1961*b*). He studies an arena with three territories (varying from 300 to 1,000 square yards in area) that did not have contiguous boundaries but were separated by 20 to 40 yards. Four nonterritorial males visited the arena, all apparently yearlings; they performed partial sexual displays and sometimes threatened one another but were ignored by the territorial cocks, between whose territories they moved at will. Up to 9 females visited the display ground at one time, and of 13 copulations seen 12 were performed by a single male. Dementiev and Gladkov (1967) found that 66 display grounds contained 630 males, collectively averaging 9.5 males per display ground (individual averages ranging from 2 to 12 males). However, Hjorth (1970) does not consider the capercaillie a lek-forming species, and Vos (1979) regards it as transitional between the forest-displaying solitary forms and the arena-forming species.

In the related black grouse, the seasonal maximum number of males occupying a display ground averages about 9 and ranges from 3 to 26, the strongest one or two of which ("first-class" males) occupy relatively central territories (Koivisto 1965). The territories of black grouse males are nearly contiguous and range in size from 100 to 400 square meters (Kruijt and Hogan 1967). Koivisto (1965) estimated that territories in this species may range from 2 to 200 hundred square meters, with no significant differences in the sizes of territories of first-class and second-class males. Immature males, which make up about one-third of the population, either are nonterritorial and are not tolerated by territorial males or occupy small and peripheral territories ("third-class" males). Koivisto believed that the primary survival value of these immature birds for the group is their tendency to warn the actively displaying males of danger. He found that there is a direct relationship between age and hierarchical position in the arena, the first-class males being mature birds that are the most fit for reproduction and also are the most successful in attracting females. Of 47 copulations he observed 56 percent were performed by first-class males. The value to the species of such assortative mating and the relative protection first-class males gained from the presence of the other categories of males appeared to Koivisto to be the primary evolutionary advantages of communal male display.

Among the North American grouse, corresponding arena behavior occurs in the pinnated grouse, sharp-tailed grouse, and sage grouse. In both the pinnated grouse and the sharp-tailed grouse, the average number of male birds occupying display grounds in general equals or exceeds the number reported for the black grouse. Copelin (1963) indicates that in the display grounds he studied the number of male lesser prairie chickens ranged from 1 to 43, and active grounds averaged 13.7 males over an eleven-year period. Robel's greater prairie chicken study area (1967) had from 17 to 25 resident males present in a three-year period. He found (1966) that 10 marked territorial males defended areas of from 164 to 1,069 square meters (averaging 518 square meters), and that the 2 males defending the largest territories in two years of study accounted for 72.5 percent of 54 observed copulations.

Numbers of male sharp-tailed grouse on display grounds vary considerably with population density in Nebraska; leks of both this species and the pinnated grouse average approximately 10 males, but they sometimes exceed 20 and occasionally reach 40 or more. Hart, Lee, and Low (1952) reported that up to 100 male sharp-tailed grouse were observed on display grounds in Utah, but the average on 29 grounds was 12.2 males. Evans (1961) confirmed that females select the most dominant males for matings, and Lumsden (1965) reported that on a display ground he studied one male accounted for 76 percent of the 17 attempted or completed copulations seen. Scott (1950) concluded that the social organization of sharp-tailed grouse is more highly developed than that of the pinnated grouse but not as complex as that of the sage grouse.

The sage grouse provides the final stage in this evolutionary sequence; it exhibits a higher degree of size dimorphism than any other species of North American grouse (adult weight ratio of females to males being 1:1.6–1.9), the display areas have a larger average number of participating males, and the central territories are among the smallest of any grouse species. Scott (1942) was the first to recognize the hierarchical nature of the territorial distribution pattern and to describe first-rank or master cocks, which were responsible for 74 percent of the 174 copulations he observed. Dalke et al. (1960) reported that the territories held by master cocks were often 40 feet or less in diameter, and Lumsden (1965) showed the territorial distribution of 19 males that exhibited an average distance from the nearest neighbor of about 40 feet. In Colorado, 407 counts of strutting grounds indicated an average maximum number of 27.1 males present (Rogers 1964). Patterson (1952) provided figures indicating that 8,479 males were counted over a three-year period on Wyoming display grounds, averaging about 70 males per display ground. Patterson reported one ground containing 400 males, and Scott's observations (1942) were made on a ground of similar size. Lumsden (1968) found that individual birds may have strutting areas that overlap those of other males, and that, though entire groups of males may move about somewhat, the relative positions of the males remain the same. Futhermore, large sage grouse leks may have several centers of social dominance, and Lumsden suggests that these should be called conjunct leks. He believes that yearling males are not tolerated by old males in the center of the lek but can move about fairly freely near the edges of the arena. They probably do not normally establish territories until their second year, when they may become "attendant" males with territorial status. The remarkably large size and complex social hierarchy of sage grouse leks, as well as their extraordinarily complicated strutting performances, seem to qualify this species as representing the ultimate stage in evolutionary trends discernible through the entire group. Since sage grouse are ecologically isolated from all other grouse species and are known to have hybridized only rarely, it seems that these complex behavioral adaptations are the result of intraspecific selective pressures rather than the need for reproductive isolation from related forms.

A possible index of the intensity of sexual selection in promoting sexual differences in behavior and morphology of the sage grouse was indicated earlier as weight differences between adult males and females that approach ratios of 1:2. Corresponding ratios can readily be calculated for the other grouse species from table 6 in chapter 2. For the essentially monogamous ptarmigan species these female-to-male weight ratios range from about 1:1 to 1:1.09. For the blue grouse, spruce grouse, and ruffed grouse they range from 1:1.1 to 1:1.33, and in the prairie grouse they range from 1:1.14 to 1:1.31. These data suggest that the intensity of sexual selection insofar as it might affect weight differences in the sexes is about the same in the lek-forming prairie grouse as in the non-lek-forming but polygamous or promiscuous forest-dwelling species. Data presented in table 6 and reported by Dementiev and Gladkov (1967) indicate weight ratios for the black grouse of from 1:1.27 to 1:1.38, and for capercaillie the estimated ratio is 1:2.28, even higher than in sage grouse. Berndt and Meise (1962) report the adult weight ratio of females to males in the capercaillie to be from 1:2.08 to 1:2.25. This species and the black-billed capercaillie are by considerable measure the largest of the grouse, and the ecological implications of both total body size and sexual differences in body size of these two species are still obscure. However, Moss (1980) has recently suggested that the unusually large size of male capercaillies is perhaps related to the observation that hens seem to prefer the best fighters when selecting males for copulation, and thus sexual selection is likely to have favored unusually large and aggressive males.

ACOUSTIC SIGNALS IN GROUSE

The feather specializations found in the sharp-tailed grouse that are related to tail rattling have been mentioned in chapter 2. Similar tail rattling occurs in male sage grouse, tail clicking noises are made by pinnated grouse, and a tail swishing display occurs in Franklin spruce grouse, involving both alternate and simultaneous spreading of the rectrices (MacDonald 1968). Likewise foot stamping sounds are made by males of many species; these are perhaps most apparent in the sharp-tailed grouse, but they also occur in pinnated grouse, willow ptarmigan ("rapid stamping" of Watson and Jenkins 1964), and probably other species.

More interesting nonvocal sexual signals used by male grouse are the drumming and clapping sounds made by various species, which apparently represent variably specialized or ritualized territorial flights. A rapid survey of the grouse with respect to such variations is instructive.

The territorial display flights of male ptarmigans serve as a starting point from which the increasingly specialized variations of the other species may be derived. In the red grouse (willow ptarmigan), Watson and Jenkins (1964) reported that the bird (either sex) ''flies steeply upwards for about ten meters, sails for less than a second, and then gradually descends with rapidly beating wings, fanned tail, and extended head and neck. On landing, its primaries often touch the ground, and it then stands high with drooping wings, bobbing its body and fanning its tail in and out.'' Calling occurs during the ascent and descent and after landing, with the loudness of the call and length of the flight varying with the bird's relative dominance.

Schmidt (1969) described the ''scream flight'' display of the white-tailed ptarmigan, and Choate (1960) reported once seeing a male white-tailed ptarmigan fly upward in a nearly vertical flight, hovering, screaming, and gliding down in a single spiral, then landing with another scream about 35 feet from the starting point. This kind of flight was reported by Bent (1932) for the rock ptarmigan, in which the male flies upward 30 or 40 feet, than floats downward on stiff wings until he is near the ground, when he checks his descent and may sail up again, calling loudly. MacDonald (1970) has described this display of the rock ptarmigan in considerable detail and related its seasonal appearance and strength to territorial advertisement and attachment.

In the eastern Canadian and Alaskan forms of spruce grouse an apparently corresponding aerial display occurs as the male flies steeply downward out of a tree he is using as a display perch, stops his descent about 4 to 8 feet above the ground, then descends rapidly with strongly beating wings (Lumsden 1961a; Ellison 1968b). In the Franklin spruce grouse males fly vertically up to a perch, slowly and with whirring wings. They may then rush forward along the branch and spread the wings and tail, make three of four drumlike wingbeats while standing upright, or perform an aerial wingclap display (MacDonald 1968). In this display the bird takes flight and at some point pauses in midair with a deep wingstroke, following which he sharply strikes his wings together above his back and drops to the ground, with a second wingclap after landing.

Short (1967) noted that males of Franklin spruce grouse have outer primaries that are more indented and more closely approach those of the Siberian sharp-winged grouse than they do those of the eastern race *canadensis*; thus it is probable that similar whirring sounds are made during aerial displays of the Siberian species (Hjorth, 1970).

Corresponding drumming flight behavior is found in the blue grouse (Wing 1946). Bendell and Elliott (1967) report that a ''flutter flight'' occurs in both sexes of the sooty blue grouse (*fuliginosus*) but that the noise produced is a ripping sound and apparently is not so elaborate as in the interior populations such as *richardsonii* and *pallidus*. Blackford (1958, 1963) reports that individuals (both sexes) of the former race perform a wing flutter (or flutter-jump) display some 8 or 10 inches off the ground. Males make more extensive drumming flights; they may also exhibit a fairly sharp whipping of the wings on alighting in a tree, and they sometimes produce a wing clap, consisting of a single loud wing note, presumably made in the same manner as by Franklin spruce grouse. In typical drumming flights the male jumps from his display perch, flies strongly upward with whirring wings, and returns after a horseshoe-shaped flight course to a point near where he started (Blackford 1963). Aerial rotations during display flights may also occur (Wing 1946; Blackford 1958).

The well-known drumming display of ruffed grouse appears to be an exaggerated version of the drumming movements of the Franklin spruce grouse or a ritualized drumming flight in which the male has substituted wing beating for flight. No actual flight displays are known to occur in this species, but the hazel grouse exhibits both wing flapping displays and display flights with associated wing noise (Bergmann et al. 1978). Male vocalizations in the hazel grouse include a territorial whistling "song," alarm calls, calls associated with aggression and fright, and several other call types (Bergmann et al. 1978).

The typical flutter-jump display, in which males make short, nearly vertical flights with strongly beating wings and sometimes associated vocalizations, appears to be an alternate evolutionary modification of the territorial song flights of ptarmigans. Typical flutter-jump displays occur in the prairie grouse and black grouse (Hamerstrom and Hamerstrom 1960), as well as in the capercaillie (Lumsden 1961b). Flutter-jumps of capercaillie, which have loud wing noises, are performed without associated vocalizations. Male sharp-tailed grouse only rarely utter calls at the start of these flights, which nonetheless are conspicuous in their open-country habitat. The pinnated grouse may utter calls before, during, or after the display, and the black grouse utters hissing sounds during flutter-jumping. The sage grouse completely lacks a flutter-jump display, judging from all recent observations.

In summary, it appears that the visually and acoustically conspicuous territorial flights of ptarmigans have, in the forest-dwelling grouse, been replaced by drumming, fluttering, or whirring flights, wing clapping noises, and sedentary wing drumming displays. In most of the lekking grouse they have been restricted to short and often quiet flutter-jumps, which are visually conspicuous in these open-country birds but are limited in length to the typically small territories.

As a final point, these aerial displays occur in both sexes of ptarmigans, are more common and better developed in males than in females of *Dendragapus* species, and are performed only by males in the lek-forming species of grouse. Ultimately, in the heavy-bodied sage grouse with its closely packed leks, the flutter-jump display has been lost altogether. Lumsden (1968) has suggested that the rotary wing movements made during strutting may represent the vestigial remnants of the sage grouse's flutter-jump display.

The summary of major male social signals of grouse (table 21) may be compared with figures 14 and 15, depicting representative display postures of twelve grouse species,

TABLE 21

SUMMARY OF MAJOR MALE SOCIAL SIGNALS IN REPRESENTATIVE GROUSE SPECIES

Species	Major Male Display Features		Major Male Acoustical Signals	
	"Air Sacs"	*Eye Comb*	*Vocal*	*Nonvocal*
Sage grouse[b]	Yellowish	Yellow	*Wa-um-poo* Grunting	Wing rustling Tail rattling Air sac "plop"
Blue grouse[c]	Yellow to reddish	Yellow to reddish orange	Hooting *Oop call*	Wing clapping Wing drumming
Spruce grouse[d]	None	Red	Hooting Snoring	Wing clapping Wing drumming
Willow ptarmigan (red grouse)[e]	None	Red	Hissing *Kohwayo/Kohway/ korow/kok/ka* etc.	Rapid stamping (audible?)
Ruffed grouse[f]	None	Orange (small)	Hissing	Wing drumming
Pinnated grouse[g]	Yellow to red	Yellow	Booming or Gobbling, Cackling *Pwoik,* etc.	Tail snapping Foot stamping
Sharp-tailed grouse[h]	Purplish to red	Yellow	Cooing Cackling *Lock-a-lock*	Tail rattling

[a]"Strutting" refers to high intensity ground display, tail cocking and wing drooping present in all species.
[b]Based on Lumsden 1968.
[c]Based on Brooks 1926 and others.
[d]Based on Lumsden 1961a and MacDonald 1968.

although it should be emphasized that these postures are not homologous in all cases. Rather, the drawings illustrate species-specific plumage characteristics that probably provide significant visual signals during display. Reported hybrid combinations of these species are also indicated.

One of the most complete surveys of grouse vocalizations is that of Jenkins and Watson (1964) for the red grouse (table 22). In this monogamous species, as in the rock ptarmigan (Watson 1972), there are about twelve adult calls that are common to both sexes. Two lek-forming species of prairie grouse are also included in the table. Data on the sharp-tailed grouse are based on the observations of Lumsden (1965), whose study did not

TABLE 21 *continued*

SUMMARY OF MAJOR MALE SOCIAL SIGNALS IN REPRESENTATIVE GROUSE SPECIES

Major Male Display Postures and Movements		
Aerial	Strutting[a]	*Other Displays*
None	Tail fanned equally	Shoulder spot
Drumming flight Flutter-jump	Tail fanned, tilted strongly	Short run with head low
Drumming flight	Tail fanned, "swished" laterally	Head jerk with squatting Foot tramping
Flight song	Tail fanned, tilted strongly	Waltzing (circling) Rapid foot stamping Bowing Walking in line Crouching with head wagging
None	Tail fanned, tilted slightly	Short run Rotary head shake
Flutter-jump	Tail spread, snapped shut	Shoulder spot Circling Nuptial bow Running parallel
Flutter-jump	Tail slightly spread, shaken rapidly Wings spread	Shoulder spot Circling Nuptial bow and posing Running parallel Foot stamping (dancing)

[e]Based on Watson and Jenkins 1964
[f]Based on Bump et al. 1947 and others.
[g]Based on Sharpe 1968 and Hamerstrom and Hamerstrom 1960.
[h]Based on Lumsden 1965.

include possible female parental calls but is otherwise apparently comprehensive. Vocalizations of the pinnated grouse are generally so similar to those of the sharp-tailed grouse that they can be comparably organized, but no single paper adequately summarizes the call repertoire of this species. Some parental calls are mentioned by Gross (in Bent 1932), and Lehmann (1941) and various other authors have discussed the sexual and agonistic calls of pinnated grouse. Evidently no special calls in this species announce the presence of enemies; the birds typically freeze or squat silently, not giving their alarm notes until they take flight (Hamerstrom, Berger, and Hamerstrom 1965). Lumsden (1965) reported a possible preflight alarm note in the sharp-tailed grouse but described three silent alarm postures that birds usually assume when they are disturbed.

14. Male display postures of representative North American grouse, including (A) booming by greater prairie chicken, (B) dancing by sharp-tailed grouse, (C) strutting by sage grouse, (D) hooting of blue grouse, (E) strutting by ruffed grouse, and (F) strutting by spruce grouse. Lines indicate reported wild hybrid combinations. Adapted from Johnsgard (1968).

15. Male display postures of representative Eurasian grouse and ptarmigans, including (A) thin-necked upright of capercaillie, (B) rookooing of black grouse, (C) postlanding aggressive posture of rock ptarmigan, (D) territorial calling by red grouse, (E) strutting by sharp-winged grouse, and (F) territorial singing by hazel grouse. Lines indicate reported wild hybrid combinations. After various sources.

TABLE 22

SUMMARY OF ADULT VOCALIZATIONS IN THREE GROUSE SPECIES

Vocalization	Willow Ptarmigan (Red Grouse)[a]	Sharp-tailed Grouse[b]		Pinnated Grouse[c]	
Flock or pair activities (both sexes)					
Flight intention	1				
Social contact	1				
Subtotal	2				
Avoidance of enemies (both sexes)					
Flying predator alarm	1		1? (preflight)		
In-flight alarm	—		1		1
Fleeing (and chase)	1				
Flying predator defense	1				
Hissing defense	1				
Subtotal	4		2		1
Sexual and agonistic					
Song (in flight/on ground) (both sexes)	2	Aggressive cackle (male and female)	1	Long cackle (mainly males)	1
Attack; attack-intention; threat (both sexes)	1–3	Aggressive *Lock-a-lock* (both sexes)	1	Aggressive *Ca-cá-caa* (males)	1
Sexual (both sexes)	1	Cooing (males)	1	Booming (males)	1
		Aggressive whine (males)	1	*Kwier* whine (males)	1
		Squeal & cork calls (males)	2	*Kliee/kwaa/kwah* calls (males)	1[d]
		Chilk and *cha* (males)	2	*Pwiek/pwark/pwk* calls (males)	1[d]
		Pow (male courtship call)	1	*Pwoik* (male courtship call)	1
Subtotal	4–6	9		5–7	
Parental	2[e]	?		2–3[e]	
Totals	12–14	11[e]		8–11	

Source: Adapted from Johnsgard, 1973
[a]Based on Watson and Jenkins 1964; all calls uttered by both sexes.
[b]Based on Lumsden 1965; female parental calls not included in study.
[c]Based on Gross 1928, Lehmann 1941, and personal observations.
[d]Probably variants of whining and *pwoik* calls.
[e]Excluding calls from other categories above.

According to Hjorth (1976) the male display "songs" of all grouse species typically consist of alternating attack-retreat units that are repeated several times, with an increasing emphasis on the retreat component toward the end. The sounds are generated by the syrinx and its associated resonating structures, and in species with low-pitched calls the inflatable esophagus probably helps amplify those frequencies that tend to carry long distances. Further, in the capercaillie the "clicking" phase of the call exhibits repetition, abrupt sonic discontinuities, and a wide frequency spectrum, all characteristics that facilitate location. Hjorth suggested that the lack of the ''cork'' note in the easternmost populations of capercaillies may be related to the acoustically less complex environmental characteristics typical of the species' habitats in these areas.

THE EVOLUTION OF GROUSE MATING SYSTEMS

I have already stated that among grouse the apparent pattern of social evolution and mating systems probably derives from an essentially monogamous one in which the male occupies a fairly large territory, where the female remains during the breeding season and where the brood is reared. The willow ptarmigan is the extant species nearest to this hypothetical ancestral mating system. From that type a variety of promiscuous mating systems has evolved, the culmination of which is the development of lek or arena behavior by males that confine sexual activities to very small contiguous territories and do not participate in any stages of the reproductive cycle after fertilization.

In a review of the grouse family, Wiley (1974) suggested that the evolution of promiscuity in grouse was related to the possibility that natural selection favored sexual bimaturism, or a sexual difference in the onset of breeding in the two sexes. In his view, both sexes in monogamous species of grouse typically breed as yearlings, and in promiscuous species the females breed as yearlings but the males typically do not. This might tend to skew the adult sex ratio toward an excess of sexually active females and force a degree of promiscuity in mating. Wittenberger (1978) has reviewed and criticized this idea, pointing out that not all grouse species follow this stated assumption of correlation between sexual bimaturism and relative promiscuity. He also questioned Wiley's belief that subadult males delay breeding because the risks of survival (associated with predation levels on the display areas) are too great, rather than that they simply are unable to compete effectively for females with older males. Wittenberger believes that instead the major difference between monogamous and nonmonogamous species of grouse lies in their spring behavior, with females of monogamous species nesting and foraging within male territories while those of promiscuous species do not. Monogamous females benefit from and perhaps rely on male vigilance during this vulnerable period, again in contrast to those of promiscuous forms. Wittenberger thus suggests that females of monogamous forms are likely to be limited in their productivity by the availability of food in spring, while those of promiscuous forms are more likely to be limited by innate digestive rates. If this hypothesis is correct, scarce food resources should favor monogamy by males and the defense of large territories through the breeding season. More abundant food should favor promiscuity by males and the holding of smaller or more temporary territories that do not contain significant food resources but instead are favorably situated to attract the maximum number of females.

In a review of black grouse arena behavior, Vos (1979) has taken the position that group living has ecological advantages for males in that it possibly protects them against predators, may provide useful information about the location of limited resources, and might also protect them against conspecific territorial males that control the population's foraging distribution outside display areas. Male clumping during sexual display may likewise be beneficial to females, because the intense male competition may allow the females to make their best choice of a male in a short time. For example, Moss (1980) suggested that female choice in capercaillies might be made on the basis of differential male fighting abilities. Or, as Geist (1977) has suggested, females may choose to mate with the male in whose company they are least molested by other males, namely the dominant cock. This might be based on its position in the lek, irrespective of actual observed fighting behavior. In choosing that male, they minimize the time and energy cost of mating and can invest that energy in more efficient reproduction. Females may also tend to mate with the most conspicuous males in arenas, even though such conspicuousness may be ecologically disadvantageous to such males in terms of predation pressure. In Vos's view, female mating preferences were not initially responsible for the development of arena behavior in males but may have played a role in speeding up the evolution of such mating systems after it began.

In summary, the general trend of grouse social behavior should be away from prolonged monogamous pair bonding whenever the ecological constraints of limited food resources or dangers from predation for the female and young during the breeding season allow for it. The length of time males must remain on their territories and advertise them perhaps is related to interspecific variations in the intensity of intermale competition for mates, as determined in part by the adult sex ratio and certainly in part by the ecological environments varying potential for female attraction and territorial defense. Arena behavior is obviously particularly well suited to an open or steppelike environment. Arena species have relatively large home ranges but defend extremely small territories that they completely control from a single display site. On the other hand, promiscuous forest-dwelling species tend to defend relatively large territories that may be more or less coextensive with their total home ranges and that are impossible to control from a single display site.

According to Vos (1979), the capercaillie occupies an ecological position intermediate between the typical forest species with fairly large, relatively permanent, and scattered territories and the typical arena species that have very small, clustered territories that males usually defend for only a small portion of each 24-hour period, spending most of their time outside them. The capercaillie is somewhat like the arena species in that males often have their ''territory centers'' near territorial boundaries, where they are in visual contact with other males (see Müller 1979), although their total territories are extremely large and comparable to those of typical forest species. Interestingly, Junco Rivera (1975) has reported that in Spain males of the race *aquitanicus* maintain permanent territories where they live with their harems and offspring. This is quite different from the territorial pattern exhibited by capercaillies elsewhere in Europe and is perhaps related to climatic or ecological differences characteristic of this most southerly capercaillie population. Jones (1981) reported that in a Scottish population of capercaillies the males sometimes also defended a harem of females, forming groups that wandered about fairly freely. He believed the relatively wide spacing of capercaillie leks might possibly be related to the availability of cover for predators in forest habitats.

The intermediate condition of the capercaillie's social structure provides some clues to the characteristics of lekking versus nonlekking grouse species and to the probable pattern of evolution of arena behavior. The major characteristics of lek-forming grouse species include (1) display areas that are not part of the male's normal home range but instead are visited and intensely defended for only part of the year; (2) increased sexual dimorphism both in body size and in plumage or soft-part features associated with sexual display; (3) a dominance relationship among males according to their individual ability to attain territorial quality (special position), territorial size, or both; (4) a positive relationship between male territorial dominance and individual reproductive success; and (5) the attraction of several females to specific males for a brief period, apparently only for fertilization. Females normally remain with favored males only long enough to be fertilized and thus form no individual attachments to such males.

In contrast, the characteristics of nonlekking grouse include (1) male mating territories that are simply part of the males' larger overall home ranges, although they too may be only seasonally defended, (2) generally less marked sexual dimorphism in body size, plumage, or development of soft parts (comb, air sac); (3) no obvious dominance hierarchy among males, with each defending a variably sized but typically large and permanent home range or territory, probably with only infrequent boundary disputes rather than intense daily confrontations; (4) no evident correlation between the incidence of territorial behavior and mating success of males; and (5) the individual attraction of females to males or vice versa, perhaps because of overlapping home ranges. At least in some cases (ptarmigans and hazel grouse) individualized and prolonged pair bonds develop between the sexes.

It is highly probable that lek behavior has evolved several times within the grouse assemblage, for the conditions that predispose populations to lek behavior are widespread. One of these might be the evolution of large body size, since the larger the adult bird and the more prolonged its development the greater the probability of marked individual differences among males in holding preferred territories and in winning battles with rivals. Another advantage of social display is that it allows the females to make rapid and efficient choices among congregating males; thus females that are fertilized on social display grounds are more likely to pass on favorable genes than those that are fertilized by lone males displaying out of direct competition with others.

For males, the advantages of arena display must be balanced against the possible attraction of predators to such assemblages. This is especially true in forested areas or other heavy cover. Thus lek behavior will best develop in conjunction with adequate antipredator tactics, such as synchronized periods of activity and silence (sharp-tailed grouse), restriction of most activity to dawn and dusk when few diurnal or nocturnal predators are out, and the evolution of "safe" and easily hidden signals (such as inflatable air sacs and hard-to-locate calls).

Social displays in grouse are thus likely to evolve whenever it is more profitable for females to seek out a group of displaying males and be fertilized by the dominant one than to remain on the territory of a particular male and benefit from the resources and protection he provides to her and her brood. For males, participation in lek display might evolve whenever it is individually more desirable to compete actively for a favored mating territory—even though this might require a year or two of "apprenticeship" with few or no mating opportunities—than it is to display alone, where there may be a lower probability of predation but also a much lower chance of attracting a female. Large and permanent (type "A") territories should occur only in grouse where resources are limited and the female must depend on the male for protection. Mating-station (type "D") territories should occur wherever the female is fully able to nest and raise her brood without need of the male beyond fertilization, and where individual differences in male fitness are reflected in differential social status that females can readily perceive and use in choosing their mates.

With this in mind, the capercaillie's uncertain status as a solitary, territorial species or as a lekking species can be raised again. Evidence supporting the "solitary" position includes the fact that the display area is only a part of the male's overall home range rather than being a special and separate area. Further, the male's display area is relatively large compared with the often miniscule territories of typical lekking grouse. They have strong sexual dimorphism in both plumage and body size, and indeed have the greatest body dimorphism of all grouse. Second, individual males exhibit great differences in their ability to attract and fertilize females, which at least in part is associated with relative male-to-male dominance. Third, intense sexual display occurs over a fairly short period during early morning hours, and females are attracted to these male groups only during the period of maximum female receptivity. Yet, unlike typical lek grouse, females may remain close to the dominant male on his territory for several days, forming a haremlike assemblage. Further, the number of males that normally aggregate on display areas are relatively few in Scotland and Europe. However, this may be a reflection of population density, for in some parts of the USSR as many as 40 to 50 males have been reported on display areas (Pukinsky and Roo 1966).

McNicholl (1978) suggested that the North American blue grouse is also an incipient lekking species. This is indicated by the male's relatively bright "adornments" and elaborate acoustic signals, the "hooting groups" or display assemblages of males, the variable social dominance among males, and the attraction of females to such territorial males, and perhaps only to dominant males. Selection of dominant males by females may perhaps be attained by individual variations in male singing patterns, in McNicholl's view. Another "solitary" species that has also been reported to sometimes form display groups is the ruffed grouse, although this idea has been questioned by Gullion (1976).

After the above discussion was written, Oring (1982) provided a general survey of the occurrence of lek behavior among birds generally, and of the evolution of lek behavior, which according to his summary perhaps exists in as many as forty-three species in eleven families. He considered that at least six grouse species exhibit true lek behavior, but that the capercaillie represents "intermediate clumping." The black-billed capercaillie was not included in his list, though it clearly might have been included in the same category. He hypothesized that lek behavior probably evolves as a result of one or more of five mutually nonexclusive reasons: male stimulus pooling, information sharing, predator avoidance, protection against conspecific territorial males, and various benefits to females such as facilitating the assessment of male quality or increased protection from predators.

9

Aviculture and Propagation

THE PROBLEMS OF KEEPING and breeding grouse and ptarmigans in captivity are distinctly different from and much greater than those of propagating pheasants, quails, and partridges, and as a result relatively few persons have succeeded in keeping and breeding grouse in large numbers or with consistent success. This is largely a reflection of their greater sensitivity to various poultry diseases and parasites that are transmitted by ground contact, forcing the game breeder to keep the birds in wire-bottom cages where they can have no direct contact with the ground or their own droppings. Summaries of the diseases and parasites of galliform birds have appeared elsewhere (e.g., Hofstad 1972; Davis et al. 1971), and only a few specific problems will be mentioned here.

Fleig (unpub. MS.) has summarized the difficulties of keeping grouse (and other galliform birds) on the ground and the treatment or preventive measures for the most commonly encountered diseases and parasites. These include coccidiosis, enteritis, cecal worms, blackhead, and capillaria worms. Coccidiosis is caused by a protozoan parasite (*Eimeria*) that is a serious problem with both quails and grouse, but it can be prevented by adding Amprolium to the diet at the rate of ¾ cup to 25 pounds of feed and can be treated with Sulmet. Intestinal inflammation, or enteritis, can be avoided by adding NF-180 to the feed in the amount of 1 ounce to 25 pounds, though this reduces male fertility and therefore must be discontinued during the breeding season. Cecal worms (*Heterakis*) are probably more serious in grouse than in quail because of the more highly developed ceca of grouse, and a serious infection can be lethal. Hygromix at the rate of 1 ounce to 25 pounds of feed is an effective treatment for these worms as well as most other worm parasites. A related infection is enterohepatitis or blackhead (caused by *Histomonas*), which is often carried by *Heterakis* and affects both the liver and the digestive tract. A preventive measure is Emtryl at the rate of 3 teaspoons per 25 pounds of feed, and higher doses can be used for treatment.

Probably the worst enemy of grouse in captivity is the cropworm (*Capillaria*), which, though not usually a serious threat to wild grouse, may cause severe losses in captive birds. It has been reported in the ruffed grouse, rock ptarmigan, sharp-tailed grouse, and pinnated grouse (Braun and Willers 1967). It is apparently less serious in quail but has been documented (Hobmaier 1932). Flieg (unpub. MS.) reported that 1 ounce of vitamin A premix to 25 pounds of feed may be used to prevent and partially control cropworm. Various persons have found that methyridine is an effective against *Capillaria* in poultry (Davis et al. 1971).

Pullorum disease, a bacterial infection caused by *Salmonella*, and aspergillosis, fungal disease of the respiratory tract, are other serious problems in keeping grouse. Both are difficult to treat, but Flieg reported some success in treating *Salmonella* infections with antibiotics such as Neomycin and Cosa Terramycin. Aspergillosis and similar fungal diseases may be avoided by adding copper sulfate to the drinking water or may be treated with a product of Vineland Poultry Laboratories called Copper-K, a combination of acidified copper sulfate and synthetic vitamin K (Allen 1968). Limited success in treating aspergillosis has been reported with the use of nystatin (Mycostatin-20, Squibb) (Davis et al. 1971).

Many of these problems can be avoided or minimized by keeping the birds on wire, but this poses new problems of providing grit and dusting places for feather maintenance. Also, if the floor is unsteady it may reduce the probability of effective fertilization during copulation. The absence of natural vegetation for hiding and nest building may further inhibit reproductive success in birds maintained in wire-bottom cages.

General principles of breeding game birds, especially quails and partridges, have been summarized well by Greenberg (1949). It is impossible to summarize all his points in the space available here, and so I will mention only a few highlights.

EGG CARE AND INCUBATION

Eggs should not be held longer than a week before being placed in the incubator, and they should be stored at a temperature between 50° and 60° F and at a relative humidity of about 80–90 percent. Keeping the eggs in plastic bags improves their hatchability (Howes 1968; Kealy 1970), and they should be stored pointed end down. Tilting them or turning them daily during the preincubation storage period is also desirable. Incubation may be done in either still air or forced air; the latter is generally preferred although it is considerably more expensive. In either case the eggs should be rotated 90° every 3 to 6 hours, or on a similar regimen, until the last few days of incubation when they are moved to hatching trays. Ideal incubation temperatures differ with the incubator type. Romanoff, Bump, and Holm (1938) stated that the ideal temperature for incubating bobwhite eggs is 103° F in still-air incubators (60–65 percent humidity) during the first 2 weeks, or 99.5° F in forced air incubators (similar relative humidity) during that period. During the last 2 or 3 days of incubation the temperature should be slightly higher (0.25° to 0.5°) for best results, and there should be more fresh air.

In their studies of prairie grouse, McEwen, Knapp, and Hilliard (1969) found that hand-turned incubators were unsatisfactory for grouse eggs and recommended using an incubator with automatic turning and with a temperature for the first 3 weeks of 99.75° F, with a wet-bulb reading of 82–86° F. After the eggs are placed in the hatching incubator, they are held at 99.5° F and a wet-bulb reading of 90–94° F. Moss (1969) reported that ptarmigan eggs could successfully be hatched in a still-air incubator provided the humidity was held as high as possible. Bump et al. (1947) reported that still-air incubators were better than forced-air models for incubating ruffed grouse eggs. They recommended an incubation temperature of 103° F for still-air models and 99.5° F for forced-air machines, and a 60–65 percent relative humidity, with eggs being turned three to four times a day during the first 20 days. During the last few days of incubation the humidity and temperature should be maintained at these same levels.

CHICK CARE

After hatching, chicks must be provided with supplemental heat, either by broody hens as foster mothers or in artificial brooders. With artificial brooders, newly hatched chicks should initially be exposed to a brooder temperature of 95° F, which is gradually reduced so that by the time the birds are about 2 weeks old the temperature is about 70° F. Newly hatched chicks should be provided with a high-protein food such as chick starter, and in addition they may benefit from finely cut fresh green leaves such as lettuce, endive, or dandelion. For many delicate species, live insect food such as mealworms (*Tenebrio*) may be crucial in inducing the young to begin eating. Shoemaker (1961) found that coating the worms with a vitamin-mineral concentrate prevented weakness in the legs (perosis), generally thought to be related to manganese deficiency. Dellinger (1967) indicated that he was able to stimulate feeding in Montezuma quail (*Cyrtonyx montezumae*) chicks by sprinkling Purina Startina mixed with hard-boiled eggs and finely chopped greens on a paper towel, to which he added small live and chopped-up mealworms. For water, he recommended filling jar lids with water and marbles, with ½ teaspoon of Furacin or Terramycin added per quart of water to prevent disease. Coats (1955) dipped mealworms into egg yolk or corn syrup, then dusted them with high-protein starter mash, to initiate chick feeding.

Problems that might be encountered in raising grouse chicks have been discussed by a number of writers, including McEwen, Knapp, and Hilliard (1969), Fay (1963), and Bump et al. (1947). Fay recommended an initial brooder temperature of from 100° to 105° F at chick level. He used various game-bird starter feeds as well as limited amounts of fresh green material. He also added soluble Terramycin to the drinking water at the rate of 1 teaspoon of powder to 2 gallons of water. McEwen, Knapp, and Hilliard (1969) found that water could effectively be provided to young chicks without the danger of drowning by using dripping siphon tubes at the chicks' eye level. The rate of dripping can be controlled by clamps, and the water falls through the mesh floor to be caught below the cage.

Howes (1968) recommended vaccinating young chicks for bronchitis and Newcastle disease if they were kept near other poultry by adding the vaccine to the drinking water. He also advised vaccination against pox.

Cannibalism—chicks pecking one another—is frequently a serious problem, especially where crowding is necessary. Such pecking may be reduced by providing sufficient grit, a source of greens or other roughage at which the birds can peck, and a balanced diet. Trimming the beak may also be necessary to prevent serious damage or even death when pecking becomes a major problem.

CARE AND HOUSING OF ADULTS

Minimum space requirements for grouse are considerably greater than those for quails, both because of the generally larger sizes of the birds and because of their reduced social tolerance. McEwen, Knapp, and Hilliard (1969) recommended at least 30 square feet of floor space per bird to minimize conflicts among prairie grouse. Thus, a 5-by-18- foot pen would accommodate a maximum of 1 male and 2 female grouse, and a 10-by-18-foot pen could hold up to 4 or 5 birds. It is important when keeping grouse to provide natural cover or artificial hiding places for the female to retreat to when the male begins to become highly aggressive during the breeding season (Moss 1969). McEwen, Knapp, and Hilliard (1969) recommended use of dusting boxes (with 5 percent Rotenone powder added) to control external parasites.

Probably the most complete summary of the problems of maintaining grouse in captivity is that provided by Bump et al. (1947) for the ruffed grouse. No doubt many of the techniques described for the ruffed grouse are applicable to other species. They found that breeding pens measuring 6 by 8 feet wide and 3 feet high were adequate for single pairs of grouse, with one end of the pen enclosed and the other end open wire mesh. They noted that up to 20 birds could be maintained in pens measuring 8 by 32 feet, especially if 10 inch-high cross-boards were placed at 4-foot intervals to help establish territorial boundaries. A wintering flight pen measuring 25 by 110 feet was judged able to hold up to 300 full-winged grouse and was constructed around a service room that facilitated feeding and watering the birds.

RECORDS OF INITIAL PROPAGATION OF GROUSE AND PTARMIGANS

Records of successful propagation of grouse are relatively few. Perhaps the earliest record of any North American grouse being successfully maintained in captivity is that of W. L. Bishop (quoted by Bendire 1892), who kept spruce grouse in captivity for some time. At present, very few spruce grouse are in captivity, and the only recent rearing success was reported by Pendergast and Boag (1971a).

Blue grouse are seen in captivity almost as infrequently as spruce grouse, and the earliest report of successful rearing of this species I am aware of is that of Simpson (1935). Smith and Buss (1963) hatched and reared 4 blue grouse through their juvenile stages, and Zwickel and Lance (1966) hatched and reared 27 chicks from eggs taken in the wild.

99

A few records of sage grouse propagation exist, including those of Batterson and Morse (1948) and Pyrah (1963, 1964), who hatched and raised birds from eggs taken in the wild.

Ruffed grouse have probably been raised in captivity more frequently than any other grouse species. Edminster (1947) reviewed the history of this species' propagation in captivity and noted that the first instance of rearing birds from eggs taken in the wild came in 1903 but that A. A. Allen developed the basic techniques needed for successful propagation during the 1920s. Later work by the state game biologists of New York resulted in the rearing of nearly 2,000 grouse, including birds of the tenth generation.

Success in rearing and propagating ptarmigans has been quite limited. Seth-Smith (1929) indicated that willow ptarmigan and the related red grouse were successfully reared in England during the early 1900s, but that the rock ptarmigan had only rarely been kept in captivity. Additionally, the pinnated grouse, sharp-tailed grouse, and ruffed grouse have been maintained in captivity in recent years (Carr 1969), rock ptarmigan have been reared from eggs to maturity, and willow and rock ptarmigans have survived well in captivity after being caught as adults in the wild. Moss (1969) has described techniques for hatching and rearing ptarmigans from eggs taken in the wild. He reported success in breeding captive stock over several years, so that breeders four or more generations removed from wild birds have been obtained. White-tailed ptarmigans have recently been reared by Nakata (1973).

One of the first persons to propagate pinnated grouse in captivity was J. J. Audubon, who obtained 60 wild-caught birds in Kentucky. He indicated that many of these birds laid eggs, and a number of young were produced. The history of recent attempts to propagate prairie grouse has been summarized by McEwen, Knapp, and Hilliard (1969), who noted that it is only recently that any real success has been attained with pinnated grouse and sharp-tailed grouse. They have maintained individual greater prairie chickens and sharp-tailed grouse in captivity for many years, with one male sharptail at least 7 years old still vigorous and breeding, and one male pinnated grouse attaining 6 years of age. From more than 4,400 eggs laid by captive birds, they reared 375 pinnated and sharp-tailed grouse. Perhaps the greatest success in rearing prairie grouse in captivity has been attained by Lemburg (1962). He has been rearing sharp-tailed grouse since 1960 and greater prairie chickens since 1965, and he began raising lesser prairie chickens in 1966. During the past decade he raised an average of 60 to 70 prairie grouse per year and in some years raised as many as 100 birds.

Based on the summary by Hopkinson (1926), pinnated grouse were initially bred in the Paris Jardin d'Acclimatation in 1873, which is certainly one of the earliest records of successful breeding of grouse in Europe. By the 1860s there were at least four records of captive breeding of the red grouse, but the willow grouse was not bred until 1909 at the Regents Park Zoo in London, according to Hopkinson. He believed that only one doubtful record of captive breeding of capercaillie then existed, and there were no records of black grouse breeding under these conditions. However, only a few years later, Moody (1932) reported that female black grouse nest ''readily'' in captivity and that female capercaillie have on several occasions nested under such conditions. During the late 1920s and early

1930s, Heinroth and Heinroth (1928–31) raised capercaillies, black grouse, and even a hazel grouse at the Berlin Zoo, the last-named species perhaps for the first time in captivity. Later Kratzig (1939) hatched and raised 3 hazel grouse from eggs obtained in the wild, but it was not until the late 1970s that Aschenbrenner, Bergmann, and Müller (1978) were able to induce captive birds to display, lay in captivity, and raise young successfully. Similarly, Höglund (1955) had considerable success in breeding capercaillies in captivity, and Aschenbrenner (1982) has recently raised a variety of Eurasian grouse with high success.

There are no records of the Caucasian black grouse or the sharp-winged grouse having been bred in captivity (Flint 1978), and likewise there is no evidence that the black-breasted hazel grouse has ever been brought into captivity.

10

Hunting, Recreation, and Conservation

IT IS PROBABLY SAFE to state that the grouse and ptarmigans of the world have provided sportsmen with some of the most exciting and memorable upland game hunting imaginable. In central Europe, hunting capercaillie is essentially restricted to the most privileged ranks. There the species is so rare that a hunter may consider himself lucky to kill a single trophy bird in a lifetime. In eastern North America, where the ruffed grouse represents the epitome of upland game hunting, a hunter may also be content with but a single bird for a day's hunt, or even for a season. Thus in Indiana, for example, during 1977 an estimated 11,700 ruffed grouse hunters harvested an estimated 8,300 birds in the entire 1977 season (Derek Major, in litt.), or only about one bird per 12.5 hours of hunting time. Clearly, the regard for hunting such a species must be much greater than the weight or number of the birds brought home.

Nevertheless, one obvious way to evaluate the sporting value of each species is to try to estimate the annual hunter kill for all the states and provinces in which it is legal game. Such estimates are regularly made by most but not all state and provincial game agencies, but since the techniques used for these estimates vary greatly, their accuracy varies as well. Nevertheless, in the belief that an inexact estimate is better than none at all, I have attempted to gather annual hunter-kill estimates for all the North American species concerned (table 23). In some cases these were derived from annual reports of the game agencies or from technical or semitechnical periodic publications of these agencies, while in others they represent unpublished estimates that are normally used for management purposes or other functions. Because of the diversity of origins of the data, these sources are not indicated in the table, and clearly the estimates should be regarded only as general ones, even though they are not usually rounded off to the nearest thousand. Wherever possible, I have used and averaged figures from a several-year period rather than listing the most recently available single-year data, since there tend to be major yearly variations in hunter success. A summary of the harvest or population data for individual states, Canadian provinces and territories, and European countries is provided in Appendix 3.

TABLE 23

ESTIMATED RECENT HARVESTS OF GROUSE AND PTARMIGANS IN THE UNITED STATES AND CANADA

| Species | Estimated Total Number of Birds Taken by Hunters Annually | | | | | |
| | Late 1960s[a] | | | Late 1970s | | |
	United States	Canada	Total	United States	Canada	Total
Sage grouse	250,000	Few	250,000	272,000	1,200	273,200
Blue grouse	240,000	130,000	370,000	386,000	134,000	520,000
Spruce grouse	140,000	300,000	440,000	188,000	360,000	548,000
Ptarmigans	100,000	200,000	300,000	100,000	200,000	300,000
Ruffed grouse	2,700,000	1,000,000	3,700,000	3,575,000	2,366,000	5,941,000
Greater prairie chicken	84,000	0	84,000	61,000	0	61,000
Lesser prairie chicken	ca. 1,000	0	ca. 1,000	5,700	0	5,700
Sharp-tailed grouse	255,000	200,000	455,000	439,000	260,000	704,000

[a]Data from Johnsgard 1973.

While some of these figures, such as those for the ptarmigans, are little more than educated guesses, those for the sage grouse, greater prairie chicken, and lesser prairie chicken are relatively accurate. It is apparent that for most species hunting pressure on grouse is probably greater than a decade ago, and only in the case of the greater prairie chicken have harvests moved substantially downward. It is in such species that hunting can sometimes play a significant role in the species' population dynamics; Hamerstrom and Hamerstrom (1973) estimated that a limited hunting season on greater prairie chickens in Wisconsin increased annual mortality by about 25 percent during the single year in which it was held. Persons charged with decisions for such seasons must make hard choices between generating income from hunting licenses that might be used to buy additional habitat or otherwise help conserve the species and placing added mortality risks on already marginal populations. Such considerations perhaps account for the recently increased hunting levels on lesser prairie chickens, which are either declining or at most barely stable throughout their limited range. The current harvests of this form probably average about 12 percent of the fall population and, except in areas of poorest habitat, are perhaps compensatory (Crawford 1980).

It is of course impossible to place a dollar value on any living creature; and the grouselike birds offer a special aesthetic quality for lovers of nature. Leopold (1949) beautifully stated this view as follows: "Everybody knows that the autumn landscape in the north woods is the land, plus a red maple, plus a ruffed grouse. In terms of conventional physics the grouse represents only a millionth of either the mass or the energy of an acre. Yet, subtract the grouse and the whole thing is dead."

Thus, the value of our "game birds" to bird watchers is real; indeed, to this group perhaps the birds are at least as valuable as they might seem to hunters. To many people the first bobwhite whistle not only is the harbinger of spring, it is the spring. To others the muffled drum roll of ruffed grouse in a distant glade is anticipated as eagerly as the earliest hepatica blossom, and on the midwestern prairies the vernal predawn booming of prairie

TABLE 24

GROUSE AND PTARMIGANS REPORTED ON AUDUBON CHRISTMAS COUNTS, 1957–79

| | 1957–79 | | | 1957–62 | |
Species	Years Reported	Average High Count	Highest Count	Average Count per Station[a]	Average Number of Stations
Sage grouse	17/23	62.3	303	10.3	1.3
Blue grouse	21/23	6.7	17	3.4	2.3
Spruce grouse	18/23	5.4	37	3.0	7.3
Ruffed grouse	23/23	58.5	93	5.4	145.0
Willow ptarmigan	17/23	21.3	68	3.0	0.2
White-tailed ptarmigan	18/23	6.0	28	2.0	0.8
Rock ptarmigan	9/23	17.4	49	—	—
Sharp-tailed grouse	23/23	165.9	436	20.8	11.6
Greater prairie chicken	21/23	57.9	138	17.2	4.8
Lesser prairie chicken	14/23	80.4	443	48.0	0.2

[a]Excluding stations not reporting species.

chickens at their ancestral leks is as rich a heritage as the big bluestem and Indian grass that lie golden in the swales.

To be individually appreciated by humans, a grouse must first be seen. This is not to say that a white-tailed ptarmigan on an inaccessible mountain peak that has yet to be climbed is of no value. To many, in fact, a ptarmigan on a mountain meadow, surrounded by dwarf alpine flowers and framed by a glacial cirque, is the very essence of the American wilderness and represents an aesthetic value beyond measure. But for the average American, tied to a city job during the week and enclosed by a concrete jungle of maddening noise and confusion, there is a special attraction in driving a few miles into the country in the hope of catching a glimpse of the local wildlife. To obtain some measurement of this relative accessibility of the grouse to American bird watchers, I have extracted data from the annual Audubon Society Christmas counts, whose distribution reflects in some measure the distribution of people in the country and their relative bird-watching opportunities. For the twenty-three years from 1957 through 1979 I have tabulated (table 24) the number of years the various species of grouse and ptarmigans have been reported by at least one party; the average number of birds seen on the highest yearly counts (excluding years when the species was not seen at all); and the highest individual count during the entire twenty-three-year period. For the years 1957–62 (later summaries of this nature were not compiled), the average count of each species seen in all stations where the bird was reported at all has been calculated, providing a rough index of the population density and perhaps also the relative sociality of each species. From these summaries it may be seen that the ruffed grouse is the species most likely to be encountered over much of North America during winter months, though it is typically seen in rather small numbers. On the other hand, the sharp-tailed grouse is typically seen in rather large numbers during the winter, and in nonforested areas especially it is probably the species of native grouse most commonly encountered. Most of the other species were observed by relatively few Christmas count groups and generally were found only in small numbers. These species have a kind of "rarity appeal" that adds to their attractiveness for winter bird watchers, and their appearance on a daily checklist provides ample testimony to the effort expended in locating the birds.

Both hunters and nonhunting nature lovers can wholeheartedly agree to the need for conserving our irreplaceable grouse and ptarmigans. Perhaps too often the nonhunter might accuse the upland game sportsman of "killing off our birds," whatever the species concerned might be. With the present controls on hunting this is, of course, utter nonsense; every species included in this book has a relatively high reproductive rate associated with a comparable mortality rate, and in most circumstances hunting cannot measurably alter the mortality rate of the species. Far more important than the number of birds shot during the fall hunting season is the amount of winter food and cover available to support the survivors until the following breeding season. Except in rare circumstances, it is the simple presence or absence of adequate cover to fill the species' daily and annual needs that will determine whether or not a wild species can survive and prosper in an area. Unlike the situation with our migratory waterfowl, we cannot blame people living somewhere else when our upland game populations diminish; the local environment is the critical factor in their success. In most cases this does not necessarily require the retention of large wilderness areas. What is serious, however, is widespread habitat disturbance or destruction during the nesting or brooding season or the reduction of adequate winter food and cover so that the birds are forced into marginal habitats and increasingly exposed to the elements and to predators.

We have only recently become fully aware of another threat to our wild populations that is unrelated to cover, hunting pressure, or any other of the classic concerns of game biologists. This is the threat of pesticides and their insidious ability to permeate the natural environment before we are really aware of the enormous damage they might do. These particularly include the "hard" or persistent insecticides such as the chlorinated hydrocarbons, which can remain in soils and living tissues for great lengths of time, becoming increasingly concentrated as they are passed up the food chain. Since the grouse and quails feed primarily on plant material as well as some insects, they do not suffer from this "biological magnification" to the degree that is true of various predators, fish-eating species, and the like. However, they do store materials such as DDT in their body fat, and not only might these materials cause physiological damage during times of fat utilization, but they may also be passed on to human comsumers or to predators. So far, DDT levels high enough to affect eggshell thickness and thus reduce hatchability have not been detected in either the grouse or the quails of North America. We need not compliment ourselves on this, however; enough damage has been done by DDT to our fish-eating birds and to other avian predators such as the falcons to justifiably indict the pesticide industry and its apologists for a disaster of unprecedented magnitude.

Environmental pollutants of greater immediate threat to upland game birds are the organic mercury fungicides used to treat seed wheat and other grains. Since small grains are a major food source for prairie grouse and other upland game birds in the Great Plains states, the birds are likely to ingest considerable amounts of the fungicide when they eat treated grain. In the fall of 1969 the Alberta Department of Lands and Forests found it necessary to close the hunting season on pheasants and gray partridges because of the concentrations of mercury found in these birds, and the Montana Fish and Game Department similarly found sufficient concentrations of mercury to cause them to caution hunters against eating the birds. Mercury causes poisoning in far smaller concentrations than DDT, and it operates directly on the central nervous system. The physiological effects of DDT on vertebrates are far less localized, and a wide variety of organ systems and processes are disrupted. The first case of closing a game bird season because of dangerous DDT levels in a game bird species occurred in 1970, when New Brunswick closed its season on woodcock. Since then federal restrictions on pesticide use have improved the situation, but even in 1982 it has been found that endrin levels in Montana game birds have been found to be more than ten times those allowed in domestic poultry.

This is a sad period in the history of North America for lovers of wildlife and the outdoors. We are witnessing the progressive extirpation of the greater prairie chicken from one state after another, and we must soon face the possibility that both the Attwater prairie chicken and the lesser prairie chicken will join the heath hen in the shadow of extinction. It also seems unlikely that the magnificent sage grouse will be able to withstand indefinitely the combined onslaughts of sage clearing and sage destruction through herbicide spraying, and it will be fortunate to survive the rest of this century. A few short-term advances have been made and are properly rejoiced in, such as the establishment of several grassland refuges for prairie chickens, but at the same time the tide of increasing population and its associated degradation of our natural environment silently inches ever higher and begins to threaten our own survival.

We are not separate from our environment; each species we destroy and each habitat we ravage, whether by bulldozer or pesticides, represents one more bridge we have burned in our own ultimate battle for survival. It is a melancholy thought that, after its compatriots had disappeared, the last surviving male heath hen in North America faithfully returned each spring to his traditional mating ground on Martha's Vineyard, Massachusetts, where he displayed alone to an unhearing and unseeing world. Finally, in the spring of 1932 he too disappeared. With him died the unique genes that reflected the sum total of the species' history, from Pleistocene times or earlier through uncounted generations of successful survival to the very last, when inbreeding, habitat disruption, fire, and disease inexorably tipped the balance of survival a final time. No one knows exactly how or when that last survivor died, and no bells tolled to mourn his passing. Indeed, only by the absence of his dirgelike booming on a March morning of 1932 was the heath hen's extinction finally established, and the bird that had been as much a part of our New England history as the Pilgrims was irrevocably lost.

Accounts of
Individual Species

Sage Grouse

Centrocercus urophasianus (Bonaparte) 1827

Other Vernacular Names

Sage hen, spiny-tailed pheasant, sage cock, sage chicken; tétras des armoises (French); Beifusshuhn (German).

Range

From central Washington, southern Idaho, Montana, southeastern Alberta, southern Saskatchewan, and western North Dakota south to eastern California, Nevada, Utah, western Colorado, and southeastern Wyoming. Reintroduced into northern New Mexico and extirpated from British Columbia and Oklahoma.

Subspecies

C. u. urophasianus (Bonaparte): Eastern sage grouse. Resident from southern Idaho, eastern Montana, southeastern Alberta, southern Saskatchewan, and western North Dakota south to eastern California, south-central Nevada, Utah, western Colorado, and southeastern Wyoming.

C. u. phaios (Aldrich): Western sage grouse. Resident from central and eastern Washington south to southeastern Oregon; formerly north to southern British Columbia. Aldrich and Duvall (1955) consider most California birds intermediate forms, but the validity of *phaios* might be questioned.

MEASUREMENTS

Folded wing: Males 282–323 mm; females 248–79 mm. Using flattened wings, females range from 240 to 285 mm and males from 288 to 334 mm, with 290 mm a calculated best division point (Crunden 1963).

Tail: Males 297–332 mm; females 188–213 mm.

1. Current distribution of eastern (E) and western (W) races of the sage grouse. The known historic distribution is indicated by stippling, and an area of intergradation is shown by crosshatching.

1. Sage grouse, male. Photo by Alan Nelson.

2. Sage grouse, males. Photo by author.

3. Sage grouse, male and females. Photo by Ken Fink.

4. Sage grouse, male and female. Photo by author.

5. Sage grouse, male and female. Photo by author.

6. Sooty blue grouse, male. Photo by Ken Fink.

7. Sooty blue grouse, female. Photo by Ken Fink.

8. Dusky blue grouse, male. Photo by author.

9. Sharp-winged grouse, male. Photo by Boris Verprintsey.

10. Richardson blue grouse, male. Photo by Ken Fink.

11. Canada spruce grouse, male. Photo by Hans Aschenbrenner.

12. Canada spruce grouse, male. Photo by Ken Fink.

13. Franklin spruce grouse, male. Photo by Ken Fink.

14. Canada spruce grouse, male and female. Photo by author.

15. Canada spruce grouse, male. Photo by author.

16. Norwegian willow ptarmigan, male. Photo by author.

17. Newfoundland willow ptarmigan, male. Photo by author.

18. Attu rock ptarmigan, male. Photo by Ken Fink.

19. Keewatin willow ptarmigan, female. Photo by author.

20. Male red grouse calling, Scotland. Photo courtesy Nicholas Picozzi.

21. Newfoundland rock ptarmigan, male and female. Photo by author.

22. Newfoundland rock ptarmigan, male. Photo by author.

23. Northern rock ptarmigan, female. Photo by Ken Fink.

24. Southern white-tailed ptarmigan, male. Photo by author.

25. Southern white-tailed ptarmigan, male and female. Photo by author.

26. Caucasian black grouse, males and female. From J. Gould's *Birds of Asia*.

IDENTIFICATION

Adults, 19–23 inches long (females), 26–30 inches long (males). The large size and sagebrush habitat of this species make it unique among grouse. Both sexes have narrow, pointed tails, feathering to the base of the toes, and a variegated pattern of grayish brown, buffy, and black on the upperparts, with paler flanks but a diffuse black abdominal pattern. In addition, males have blackish brown throats, narrowly separated by white from a dark v-shaped pattern on the neck, and white breast feathers concealing the two large, frontally directed gular sacs of olive green skin. Behind the margins of the gular sacs are a group of short white feathers with stiffened shafts that grade into longer and softer white feathers and finally into a number of long, black hairlike feathers that are erected during display. Males also have rather inconspicuous yellow eyecombs that are enlarged during display. Females lack all these specialized structures but otherwise generally resemble males. Their throats are buffy with blackish markings, and their lower throats and breasts are barred with blackish brown.

FIELD MARKS

The combination of sage habitat, large body size, pointed tail, and black abdomen is adequate for certain identification. Males take flight with some difficulty and fly with their bodies held horizontally; females take off more readily, and while in flight their bodies dip alternately from side to side. When the bird is in flight the white under wing coverts contrast strongly with the blackish abdomen.

AGE AND SEX CRITERIA

Females may readily be separated from adult males by their weights and measurements (see above), by the absence of black on the upper throat, and by the fact that the white tips of the under tail coverts extend part way down the feather rachis (Pyrah 1963). Crunden (1963) provides a sex and age key based on primary measurements.

Immatures (under one year old) resemble females but are paler, the outer primaries are more pointed and mottled than the others, the outer wing coverts are narrowly pointed instead of being unmottled dark gray and are marked with brown and white and have white tips (Petrides 1942). Immatures also have light yellowish green toes, unlike the dark green toes of adults. Males do not usually achieve full breeding condition their first year; subadult males have narrower white breast bands than do adults. The tail feathers of immature males are also blunter and are tipped with white. During their first fall immature birds have bursa depths of more than 10 mm (averaging 18.9 mm in October), whereas adults have maximum bursa depths of 7 mm and average depths of 1.6 mm (Eng 1955).

Juveniles have conspicuous shaft streaks on their upper body feathers and tail feathers, with white central shafts that spread out into narrow terminal white fringes (Ridgway and Friedmann 1946).

Downy young of sage grouse have a distinctive "salt and pepper" pattern dorsally that consists of a mottled combination of black, brown, buff, and white, devoid of striping. The head is whitish, spotted with brown and black in a fashion similar to blue grouse downies, and the underparts vary from grayish white to buff and brownish on the chest region, where a brown-bordered buff band is usually evident. The malar and nostril spots of this species are unique (Short 1967), and a definite loral spot is also present. The broken pattern of dark markings on the forehead and crown in this species probably corresponds to the black border that occurs around the brown crown patch in most other grouse (Short 1967).

DISTRIBUTION AND HABITAT

At one time this species was found virtually wherever sagebrush (*Artemisia*, especially *A. tridentata*) occurred, throughout many of the western and intermountain states. In early times it occurred in fourteen or fifteen states and was the principal upland game species in nine (Rasmussen and Griner 1938). However, overgrazing and drought contributed greatly to the species' demise. By the early 1930s it was a major upland game species in only four states (Montana, Wyoming, Idaho, and Nevada), and by 1937 only Montana retained a regular open season. Restricted hunting was then still permitted in Nevada and Idaho, but all other states had closed seasons (Rasmussen and Griner 1938). After 1943 Montana also established a closed season that lasted nine years. The species became completely extirpated from British Columbia and New Mexico, although New Mexico has repeatedly attempted to reestablish the bird, and British Columbia attempted the same (Hamerstrom and Hamerstrom 1961). There are no recent specimen records from Nebraska, although a few birds may occasionally stray across the Wyoming state line. There are no Oklahoma records since 1920 (Sutton 1967), when it was found in Cimarron County. Currently in New Mexico a few birds may still survive in the Taos area.

A low ebb in sage grouse populations in the western states occurred in the middle to late 1940s. Idaho reported an upturn in population after 1947 and, after four years of protection, reopened hunting in 1948. Nevada reestablished limited hunting in 1949, followed by Washington in 1950. Permit-only seasons were established by Wyoming in 1948 after eleven years of protection and by Utah in 1950. California opened one county (Mono) to hunting in 1950 after five years of protection. Judging from figures presented by Patterson (1952), the total United States kill in 1951 was fewer than 75,000 sage grouse.

Except for two years (1944 and 1945), Colorado maintained a closed season from 1937 until 1953, and in 1952 Montana held its first season since 1943. South Dakota began to permit hunting sage grouse again in 1955 after nineteen years of protection, and in 1964 North Dakota held its first season since 1922. Alberta initiated a highly restricted season in 1967.

In recent years the sage grouse has recovered sufficiently to be a major game species again in about five states. Hamerstrom and Hamerstrom (1961) reported estimated hunter-kill figures for 1959 of about 44,000 birds in Wyoming, 23,000 each in Idaho and Montana, 15,000 in Colorado, and 12,000 in Nevada, plus approximately 2,000 each in California, Washington, and Utah, totaling more than 100,000 for the country as a whole.

The most recently available hunter-kill estimates indicate that the sage grouse is at least maintaining its population sufficiently to be a major game species in six states and of secondary importance in four more states and one province. The estimated 1979 kill in Idaho was 92,600 birds, and in the same year 94,426 birds were harvested in Wyoming and 66,398 in Montana. The 1979 estimated kill in Utah was 28,280 birds, and the 1978 estimates for Nevada and Colorado were 17,693 and 11,724 birds, respectively. Considerably smaller harvests occur in California (3,980 in 1978), Washington (740 in 1979), South Dakota (no data, a three-day season in 1979), Alberta (no data, a six-day season in 1979), and North Dakota (closed season in 1979, usually fewer than 300 birds killed in prior years). Populations were too small to open seasons in Oregon, Saskatchewan, and New Mexico. The total yearly harvest is thus currently about 280,000 birds, or not much different from that of a decade ago, but substantially more than that of the late 1950s. These larger harvests do not reflect so much a recent increase in grouse populations as increased hunting pressure and a recognition that limited harvests are not a controlling factor in the security of sage grouse.

Patterson (1952) estimated that some 90 million acres of preferred sagebrush-grassland habitat existed in the early 1950s and that an additional 40 million acres of desert scrub habitat was also available to sage grouse. If the 90-million-acre figure is assumed to be currently representative, this would total about 140,000 square miles of preferred habitat. If an average population density of 10 birds per square mile can be assumed, the total sage grouse population might be roughly estimated at 1,500,000 birds. The present yearly harvest of 280,000 would then represent 17 percent of the total, which does not seem exorbitant.

In spite of this seemingly comfortable number of birds, it is difficult to be optimistic about the long-term future of the sage grouse in North America. The continued clearing of extensive areas of sage for irrigated farming, as has been widely done in central Washington, and the expanded use of herbicides to improve grazing conditions are likely to further reduce sage grouse habitat and populations in future years. Schneegas (1967) estimated that 5 to 6 million acres of sagebrush have been removed in the past thirty years, a portent of things to come.

Only a few studies of the effects on sage grouse population have been made of spraying or removing sagebrush. In an 12,000-acre area of sagebrush in Wyoming, spraying over a five-year period resulted in the loss of an entire wintering population of 1,000 birds, although a few began to return about five years after the spraying (Higby 1969).

In Montana, which currently supports sage grouse on about 11 million acres, approximately 10 percent of the native range has already been destroyed, primarily as a result of sagebrush control programs. Locally, as in one area in Meagher County, nearly 50 percent of the sagebrush habitat has been converted to cropland. The effects on grouse of spraying or mechanical elimination of sagebrush seem to vary greatly according to the size of the areas affected, whether strip or block removal is utilized, and perhaps other factors. Certainly large-scale removal of sagebrush often has marked effects on winter grouse use, on strutting ground distribution and male occupancy, and on total grouse populations, which may be either rather temporary or long-lasting in nature (Wallestad 1975).

POPULATION DENSITY

Patterson (1952) estimated sage grouse densities by determining strutting ground sizes and numbers in two study areas that totaled 250 square miles. He reported an average of one strutting ground per 5.7 square miles and a density of 12.5 males per square mile. This, of course, excluded from consideration all females and probably some immature males. Edminster (1954) thus calculated that the total spring population of sage grouse might be from 30 to 50 birds per square mile, or 13 to 21 acres per individual. Rogers (1964) likewise reported that certain counties of Colorado support 10 to 30 birds per square mile in some sections, while the remaining habitat supports 1 to 10 birds per square mile.

HABITAT REQUIREMENTS

Wintering Requirements

During winter, sagebrush not only provides nearly 100 percent of the food for sage grouse but also furnishes important escape cover. Edminster (1954) pointed out that during winter sagebrush has the important attributes of being evergreen, tall enough to stand above snow, and highly nutritious. Rogers (1964) indicated that the best wintering areas in Colorado are at the lowest elevations, where sagebrush is available all winter. Local topography may influence the availability of sagebrush, because of snow cover, but sage grouse may be expected to occur wherever exposed sagebrush is found through the winter. Dalke et al. (1963) reported that wintering concentrations of sage grouse in Idaho usually existed where snow accumulations were less than 6 inches, which occurred in areas some 30 to 50 miles from the habitats used during fall and spring. Black sage (*Artemisia nova*) is the preferred winter food in eastern Idaho but is often covered by snow.

In a Colorado study, Beck (1977) reported that nearly 80 percent of winter use of an area of 1,252 square kilometers dominated by sagebrush was on less than 7 percent of the total. This was primarily the result of differential availability of sagebrush above snow, as associated with snow depth, steepness of slope, and sagebrush disturbance. Flocks typically used south- to west-facing slopes of less than 5 percent gradient, which were usually kept fairly snow free and provided forage even where the sagebrush was very short. Similarly, in Montana it was found that prime wintering areas for sage grouse have large expanses of dense sagebrush with little if any slope, both characteristics that make the areas prime targets for sagebrush control programs. Winter ranges of five females monitored by radio tracking varied from 2,615 to 7,760 acres during two different years (Eng and Schladweiler 1972).

Spring Habitat Requirements

In late winter, male sage grouse begin to leave their wintering areas and return to their traditional strutting grounds. Of 45 strutting grounds classified by type of land area, Patterson (1952) found that 11 were on windswept ridges and exposed knolls, 10 were on flat sagebrush areas with no openings, 7 were on bare openings on relatively level lands, and the remaining 17 occurred in seven other habitat types. Relatively open, rather than dense, sage cover is clearly the preferred habitat for strutting grounds, as indicated by a number of writers such as Scott (1942) and Dalke et al. (1963). The latter study reported that new strutting grounds could readily be established by clearing areas of ¼ to ½ acre in dense stands of sage.

When not on their strutting grounds, males spend most of their time on locations with a canopy coverage of sagebrush ranging from 20 to 50 percent, and among plants that are usually no more than a foot tall (Wallestad 1975).

Nesting and Brooding Requirements

Patterson (1952) reported that 92 percent of the nests he found were under sagebrush plants, usually in cover from 10 to 20 inches tall, and in drier sites where the shrub cover was less than 50 percent. In Utah, Rasmussen and Griner (1938) found that a related species, silver sage (*A. cana*), provided preferred nest cover, with plants of this species from 14 to 25 inches tall providing cover for 33 percent of 161 nests, while the more common big sage (*A. tridentata*) of the same height category accounted for 24 percent of the nests. The highest nesting densities (up to 23 nests on 160 acres, or 1 nest per 6.95 acres) were in dense second-growth sagebrush. Klebenow (1969) found that 91 percent of 87 nests or nest remains were associated with three-tip sage (*A. tripartita*). In nesting habitats he noted that the sagebrush averaged only 8 inches tall but that taller plants were preferred for nest sites. No nests occurred where the shrub cover exceeded 35 percent. In the best nesting areas, nest densities of up to 1 nest per 10 acres were found. Similarly, Montana studies indicate that female sage grouse tend to seek out taller sagebrush plants when looking for nest sites, and that successful nests usually had significantly greater sagebrush cover within 24 inches of the nest than did unsuccessful ones (Wallestad 1975).

Brooding habitat requirements are evidently slightly different from sage grouse nesting requirements. Klebenow (1969) reported that 83 percent of the broods he observed were in big sagebrush but not in dense stands. All but 3 of 98 broods recorded were seen in areas of less than 31 percent shrub cover. As the summer progressed, broods moved into moister areas that still contained green plant material, until by late August they had gathered near permanent water sites. However, green vegetation, rain, and dew evidently provide adequate moisture for sage grouse.

Observations of Martin (1970) in Montana indicated that in 158 locations young broods used areas having less plant density and lower crown cover (9 to 15 inches high) than did older broods or adults (7 to 25 inches high). Rogers (1964) also reports that low sage (7 to 15 inches high) is preferred for feeding, nesting, and roosting cover, while taller plants serve for nesting, shade, and escape cover. Spraying with the herbicide 2,4-D in Montana greatly reduced summer use by sage grouse, apparently by altering vegetational composition, particularly of favored food plants (Martin 1970). Similarly, Peterson (1970) concluded that components of brood habitat for sage grouse include a diversity of forms and a density of sage ranging from 1 to 20 percent.

During summer and early fall, males tend to remain segregated from broods and hen flocks but are typically found in flocks within 2 to 3 miles of strutting grounds (Wallestad 1975).

FOOD AND FORAGING BEHAVIOR

The importance of sagebrush as food for adult sage grouse is impossible to overestimate. Martin, Zim, and Nelson (1951) reported that sage made up 71 percent of the diet in 203 samples and that use of animal material ranged from 9 percent in summer to 2 percent in spring and fall. Apart from sagebrush, vegetable food consists largely of the leaves of herbaceous legumes and weeds (collectively called forbs) and grasses, which are utilized primarily in late spring and summer (Edminster 1954). Patterson (1952) reported that sage constituted 77 percent (of a total of 95.7 percent plant material) of foods found in 49 samples from adult sage grouse in Wyoming and 47 percent (of a total of 89 percent plant material) from 45 juvenile sage grouse analyzed.

A study of year-round food intake of sage grouse, based on 299 crops from Montana, indicated that 97 percent of the material eaten by adult birds is vegetable matter. Sagebrush constituted 62 percent of the total food volume throughout the year, and between December and February it was the only food item found in all crops. Only between June and September did sagebrush contribute less than 60 percent of the diet (Wallestad 1975).

During early life, young sage grouse feed heavily on ants, beetles, and weevils and later add grasshoppers to their food intake (Patterson 1952), although the total animal content of the diet drops from as much as 75 percent to less than 10 percent. The study of Klebenow and Gray (1968) indicates that insects predominate in the diet only during the first week of life; thereafter forbs become the predominant food, with shrubs only gradually assuming primary importance. The importance of forbs is also indicated by Trueblood (1954), who found that this food category comprised from 54 to 60 percent of the major food items consumed by juvenile sage grouse in Utah and from 39 to 47 percent of those eaten by adults. On lands partially reseeded to grass, he found that adults persisted in their preference for shrubs, while juveniles preferred forbs and had a strong aversion to grasses.

Martin's study (1970) has provided additional evidence of the value of a variety of forbs as summer food for sage grouse. He found that, in a sample of 35 sage grouse collected from July to September, sagebrush totaled 34 percent of the food, while dandelion (*Taraxacum*) constituted 45 percent. Collectively, these plants plus two additional forb genera (*Trifolium* and *Astragalus*) contributed more than 90 percent of the food material. Two California studies (Leach and Hensley 1954; Leach and Browning 1958) also indicate that weedy forbs such as prickly lettuce (*Lactuca*) and cultivated herbaceous broad-leaved plants such as clover and alfalfa are important early fall food sources for sage grouse.

One of the most complete studies available on juvenile food requirements is the recent study of Peterson (1970), who analyzed the food of 127 birds up to 12 weeks of age. During that period, forbs constituted 75 percent of the diet, and two genera (*Taraxacum* and *Tragopogon*) together made up 40 percent of the food consumed. Insect use declined from a high of 60 percent in the first week to only 5 percent by the twelfth week, and sagebrush was used very little by chicks before the age of 11 weeks.

MOBILITY AND MOVEMENTS

Seasonal Movements

The most complete study on seasonal movements of sage grouse so far available is that of Dalke et al. (1963). Patterson (1952) had previously summarized the literature on possible migratory movements of these birds, noting that in Oregon a winter migration to lower elevations was followed after nesting by a migration to summer ranges at 8,000-foot elevations. Possible winter movements of Wyoming and Montana birds into South Dakota were discussed by Patterson, and he mentioned a male that was banded in Wyoming and recovered the following fall still in Wyoming but some 75 air miles from the point of banding.

In the mountainous country, wintering grounds of sage grouse are often some distance from spring and summer habitats, at considerably lower elevations. With the gradual regression of snow, male grouse on their wintering grounds begin working toward the strutting areas. Dalke et al. (1963) reported that these birds move in small flocks, flying short distances, during this migration. Many such birds in Idaho may move from 50 to 100 miles along established routes before reaching their strutting grounds. Adult females evidently reach the strutting grounds about the same time as adult males or somewhat later. Patterson (1952) noted that males began to arrive on Wyoming strutting grounds as early as February and were followed in 1 or 2 weeks by females. Dalke et al. (1963) found that males and even females occupied grounds in late March or early April that were not yet free of snow. A rapid buildup of adult males occurred in early April, subadult females arrived about a week after adult females, and subadult males did not appear in numbers until most of the females had already left the grounds in late April.

Movement of birds between strutting grounds is evidently fairly rare, both within one season and from year to year. Dalke et al. (1963) noted that, of 78 adult males banded in 1959 and 1960, a total of 14 (18 percent) were observed later on grounds other than those where they had been banded. During the same two years 107 females were banded, and 6 of these were subsequently observed visiting other strutting grounds. Males moving between strutting grounds covered distances of from 550 yards to 4.3 miles. Dominant males were only rarely involved in these movements, suggesting that the movements are the result of attempts by subordinate males to establish territories in various locations. Earlier, Dalke et al. (1960) had reported that 70 percent of banded sage grouse that were again observed on strutting grounds in the first three years were seen on their original strutting grounds and no others. Some master cocks occupied nearly identical territories in successive years, while others lost their territorial positions.

It is not well known how far the females move from strutting grounds to build their nests, but current evidence suggests that it is usually not very far. Klebenow (1969) noted that on one area of three-tip sage (a favored nesting cover) more than a mile from the nearest strutting ground no nests were found and only one very young brood was seen. In each of two areas of big sage, nests were found within a half mile and at only slightly lower elevations. However, unpublished Colorado studies indicate that females regularly move 3 or 4 miles from a display ground to a nest site and may travel as far as 7 miles. On the other hand, a Montana study indicated that 68 percent of 22 nests made by radio-tagged females were constructed within 1.5 miles of the strutting ground on which they had been captured, and only 2 of the nests were more than 3 miles away (Wallestad and Pyrah 1974).

After nesting, females gradually move their broods to places where food supplies are plentiful, usually moist areas such as hay meadows, river bottom lands, irrigated areas, and the like. Patterson (1952) estimated that family units break up and juveniles become relatively independent at about 10 to 12 weeks, when they have completed their molt into juvenal plumage.

Spring dissolution of the strutting grounds by males is a gradual process, and some subadult males may remain after most adult birds have left for summer ranges (Dalke et al. 1960). However, Eng (1963) found that adult males were the last to leave the strutting area. These are usually at higher elevations, but the birds may move down into alfalfa fields near irrigated valleys. Schlatterer (1960) reported that the sequence of arrival of birds on the summering areas in Idaho was males, unproductive females, and productive females. In southern Idaho the summer brood range may be from 13 to 27 miles from the nesting grounds, a considerable movement for these recently fledged birds.

Fall movement toward wintering areas is likewise a gradual process, and the rate probably varies according to weather conditions. Pyrah (1954) reported that immature females were the first to leave for wintering areas, followed by mature females, then adult males. Immature males associated with immature and mature females. Dalke et al. (1963) reported that birds collected in flocks near water holes as freezing temperatures began and

118

that movements were quite noticeable by the time the daily minimum dropped to 20° F. Birds usually remained in a single place for several days, then moved out in groups. By the time the first snow fell, flocks were usually composed of between 50 and 300 birds in loose associations. During severe weather, flocks of up to 1,000 birds could be seen, but in midwinter they normally consisted of fewer than 50 individuals, with old males often in groups of fewer than 12.

Daily Movements

Daily movements and activity patterns of sage grouse have yet to be carefully documented, but some work with banded birds is of interest. Lumsden (1968) noted the daily locations of several individually marked males on a strutting ground and confirmed that individual males returned daily to their specific territories. However, their territorial boundaries were rather ill defined and exhibited considerable overlap. On one occasion, when a cluster of hens formed about 55 meters from Lumsden's blind, 6 males left their territories and moved toward the hens, apparently maintaining their positions relative to one another. Of 27 individually marked hens, 16 were observed later on the same display ground. Four were seen to visit the ground on three mornings, 1 was seen twice, and 8 only once. Seven were observed mating, in each case only once, and none of these birds was seen again.

Males arrive on the strutting grounds long before dawn, and early in the season they may remain all night. Hens arrive before dawn and usually leave shortly after sunrise. After daybreak, immature males are the first to leave the grounds, followed by successively more dominant males and finally the master cock. The birds normally walk to feeding areas that may be within a half mile of the strutting grounds (Pyrah 1954). Hens rarely return to the strutting grounds in the afternoon.

Observations on nesting hens by Girard (1937) and Nelson (1955) indicate that they normally leave their nests twice a day during incubation. Girard reported that these foraging periods occurred between 9:30 and 11:30 A.M. and between 2:00 and 3:00 P.M., whereas Nelson reported earlier morning and later afternoon periods. The feeding periods usually lasted between 15 and 25 minutes, according to Nelson.

In the late summer, sage grouse roost until about 6:00 A.M., forage until about 10:00 A.M., rest until about 3:00 P.M., forage again until 8:00 P.M., and finally go to roost again about 9:00 P.M. (Girard 1937). Unlike the prairie grouse, sage grouse exhibit no fall display activities. During winter, daily movements of sage grouse have no definite pattern, and, apart from foraging, the birds spend much time resting and preening. Roosting occurs on rocky outcrops (Crawford 1960; Dalke et al. 1963).

REPRODUCTIVE BEHAVIOR

Prenesting Behavior

In a sense, the sage grouse may be regarded as the classic lek-forming species of North

119

American grouse. Not only are the lek sizes the largest in terms of average numbers of males participating, but also the degree of segregation according to dominance classes is the most evident. Further, although Scott (1942) was by no means the first to describe the social strutting behavior of sage grouse, his study first recognized the complex social hierarchy of males and designated the most dominant males master cocks. This term has been applied to most other lek-forming grouse, such as pinnated grouse and sharp-tailed grouse.

As soon as traditional display grounds are relatively free of snow, male sage grouse begin to occupy them. In different years conditions may vary, but in the northern United States the birds are usually on their strutting grounds by late February or March. Most studies indicated that the first birds to occupy the grounds are the adult males, which may return to virtually the same territorial sites they occupied in previous years.

It might be assumed that the male behavior patterns exhibited on the strutting grounds perform two separate functions: proclamation and defense of territory on the one hand, and attraction and fertilization of females on the other. Although natural selection thus operates through the differential successes of individual males in attracting females, it is of interest that apparently in all grouse the behavior patterns serving to attract females are derived directly from hostile behavior patterns associated with the establishment and defense of territory. As a result, relatively few of the displays performed by male grouse in lek situations serve strictly as male-to-female displays; rather, those postures and calls that function in territorial establishment are for the most part utilized in sexual situations as well. It is therefore generally impractical to fully separate signals associated with attack and escape (agonistic displays) from those that function sexually to attract females (epigamic displays). The resulting close relation between relative individual success in performing territorial behavior (achieving male-male social dominance) and relative individual reproductive success (fertilization of females) provides a basic key to understanding social behavior in lek-forming grouse. This contrasts with the situation in socially displaying duck species, in which agonistic and sexually oriented displays are much more separable, probably because of the absence or insignificance of territoriality during pair-formation in waterfowl.

That most male displays performed by lek-forming grouse are derived from hostile responses further complicates their dual role as sexual attractants. Female grouse not only must be attracted to these signals, but must in turn identify themselves as females in order to avoid attack by territorial males. This is usually achieved by submissive postures that in general are associated with inconspicuousness through slimmed plumage, silent movements, and general lack of malelike signals. Thus a kind of paradox may be seen in lek-forming grouse. Whereas in non-lek-forming species of grouse (e.g., ptarmigans) the females may perform fairly elaborate and often malelike displays, in the social species the degree of development of female display is perhaps inversely proportional to the relative development of male displays and other male signals. The role of the female in lek-forming grouse is therefore reduced to simply appearing on the lek, being attracted to particular males, and allowing copulation to occur. This last point is achieved by a precopulatory squatting display with wings partially spread, which is virtually identical in all grouse so far studied. In the sage grouse, where hens often cluster in groups around specific males (master cocks), fighting between hens may sometimes occur, but it is not likely that this happens in other species, except for capercaillies.

Male Territorial Advertisement Behavior

Although strutting has been described by many writers, the accounts by Lumsden (1968) and Hjorth (1970) are by far the most complete and accurate. The following summary is therefore in large measure based on their descriptions. Lumsden and Hjorth have confirmed the basic findings of Scott (1942), who discovered the relation of social dominance to sexual sucess, with the master cock maintaining a central territory that is selectively sought out by females for copulation. It is important to note, however, that the strutting of master cocks differs in no obvious way from that of birds occupying lower social ranks, such as the secondary-status "subcocks" and "guard cocks" or the peripheral attendant males. But strutting by nonterritorial yearling males is poorly developed and may readily be distinguished from that of older birds. Such immature birds probably represent the "heteroclite" males described by Scott.

Overt fighting between males is largely but not entirely limited to the edges of territories. Fighting males typically stand 10 to 20 inches apart, head to tail and nearly parallel with heads upright and feathers usually lowered. The tail may be raised or lowered and is sometimes shaken rapidly, producing a rattling sound that perhaps corresponds to the tail-rattling display of sharp-tailed grouse. Periodically the males attempt to strike each other with their nearer wing, but, unlike the prairie grouse, they do not fly into the air and strike with their feet. The associated calls are *kerr* sounds, often in a series of eight to twelve repeated notes.

Overt fighting is less common in sage grouse than is ritual fighting, in which the same parallel position is assumed but the birds remain virtually motionless. At times the birds may actually close their eyes as if sleeping in this posture, which Lumsden interprets as "displacement sleeping." When threatening, male grouse draw up the skin on the sides of the neck, thus erecting the filoplumes and increasing the exposed areas of white feathers. The tail may also be cocked and spread and the body held more upright when in such a threat posture. In general, the amount of white feathers a male exhibits is a relative index of his aggressive tendencies. It is thus of interest that female grouse lack white areas and that the white neck area of yearling males is smaller than that of adults. When charging, the adult male assumes a posture strongly similar to that held during the strutting display. This would suggest that strutting represents a ritualized form of charging, in which the forward body movement has been almost entirely lost.

When on territory and between strutting sequences, the male is usually in an "upright" posture (Hjorth 1970), with tail cocked and spread, wings slightly drooped, neck feathers ruffled, and the esophageal pouch partly inflated and pendulous. In this posture he may jerk his head upward and utter a soft snoring note that is apparently associated with the inhaling of air (Hjorth 1970).

The strutting display ("ventro-forward" of Hjorth 1970) is a complex sequence of stereotyped movements and sounds that lasts about 3 seconds, which Lumsden has divided into ten stages. In the first stage the male stands erect with the tail fanned and held slightly behind the vertical, lowers his folded wings, and takes a step forward. The back is gradually raised, so that by stage two it is held at a 45° angle to the ground. The anterior neck feathers then suddenly part, exposing two olive green skin patches. The third stage begins as the bird opens his beak and apparently takes a breath. The pendant esophageal bag is then lifted and the skin patches disappear, another step forward is taken, and the folded wings are quickly drawn across the stiffened feathers at the sides of the neck as it is jerked upward ("first vertical jerk" of Hjorth), producing a brushing sound. In the fourth stage the beak is shut, the wings are moved forward again, and the esophageal bag is lowered. In stage five the neck again swells, the oval skin patches are exposed a second time but again are not greatly inflated, and a second although silent backward stroke of the wings is performed. In stage six a third step forward is taken, the wings are again moved forward, the skin patches are somewhat more fully expanded, and the esophageal bag begins to move upward again. In stage seven the neck is diagonally extended ("second vertical jerk" of Hjorth) as the esophageal bag is strongly raised, nearly hiding the head, and the wings are again rubbed against the breast feathers as they make their third backward stroke. In stage eight the head is withdrawn into the erected neck feathers, the esophageal bag bounces downward, and the inflated bare skin patches form large oval bulges ("first forward thrust" of Hjorth) while the wings move forward and back a fourth time. In stage nine the head is quickly withdrawn into the neck feathers so that it becomes completely concealed, compressing the esophageal bag so greatly that the skin patches bulge strongly outward in the shape of hemispheres ("second forward thrust" of Hjorth), and the wings complete a fifth backward stroke. Pressure on the trapped air in the esophagus is now suddenly released, causing the skin to collapse with two plopping sounds, and the head is moved upward toward a normal position. In the tenth and final stage the head returns to the original starting position, the white neck feathers close over the bare skin areas, and the body returns to the stance assumed at the beginning of the display.

The major motor elements of the entire display sequence thus consist of several forward steps (Hjorth reported four to seven), five rotary wing movements, two brushing sounds of the wings against the sides of the breast and neck, and four increasingly greater inflations of the esophagus, with associated expansions of the colored skin patches. The predominant nonvocal sound is a "resonant squeaking, swishing" noise (Lumsden 1968) followed by two plopping sounds. However, a call is also uttered, which Lumsden described as sounding lke *wa-um-poo*, only the last part of which can be heard at any distance. Hjorth (1970) determined that there are actually four vocal notes produced, of which the second is the loudest.

122

16. Male displays of the sage grouse, showing lateral view of strutting sequence. Numbers indicate elapsed time (in tenths of seconds) since onset of display. Also shown (*bottom row*) are rear and front view of strutting male, and copulation posture. Primarily after Hjorth (1970).

Studies by Wiley (1973*a*) on the strutting display indicated that this behavior is highly stereotyped, with duration typically having a coefficient of variation of less than 4 percent and some measurements varying even less than 1 percent. Wiley observed no geographic variation in display characteristics, nor did he find that individual differences were related to reproductive success. Finally, the displays of young males were no more variable than those of adults.

The sage grouse lacks much of the pivoting action of the pinnated grouse's booming, but as Lumsden pointed out strutting is not a specifically frontal display. Although visually impressive when seen from the front, the long and colorful under tail coverts are also conspicuous signals when seen from behind. Lumsden found no strong tendency of males to face hens when performing strutting displays, and often they faced directly away from them.

In contrast to nearly every other North American grouse (the ruffed grouse is the only other case), the sage grouse lacks a flight display. Lumsden is probably correct in explaining this on the basis of the male's large size and poor agility, plus the fact that needs for territorial advertisement are reduced in sage grouse because of the large number of males usually present and the conspicuous nature of individual birds. Lumsden also believes that ''call flights'' by hens advertise the location of the strutting ground. Such ''quacking'' calls are uttered by hens when flying toward the ground or when flying from one part of the ground to another, and occasionally when the hen flies away from the strutting ground. Lumsden also described a ''wing-bar signal'' display, which he states may be performed by females in flight before landing, perhaps functioning as a landing-intention signal. This display is sometimes, but not always, associated with a call flight and is produced by drawing the white under wing coverts up over the leading edge of the wing to they are visible from above and behind the bird. A somewhat similar ''shoulder-spot display'' occurs in both sexes of sage grouse while on the ground. Lumsden regards this display as an expression of conflict, with fear as one of the components.

Calls of male sage grouse include the strutting call, grunt, and fighting call already mentioned, as well as a high-pitched and repeated *wut* note used as an alarm call (Lumsden 1968). Males, especially yearlings, may also utter a squawking note, perhaps as a flight-intention signal. Hens also have well-developed fighting notes, as well as whining notes in agonistic situations. Both sexes may also hiss when being handled, according to Lumsden.

Apart from the fighting call and the call uttered during strutting, only one other male call has been reported for sage grouse. Lumsden noted a deep grunting sound that males produced both in threat situations and when near hens and often as a prelude to fighting. The same call was occasionally heard from hens. Hjorth (1970) called this vocalization a ''grunting chatter.''

The strutting behavior of males when hens are present is not noticeably different from when they are absent, except perhaps for the greater frequency of displays. Sage grouse hens typically gather together in tight groups near master cocks; from 50 to 70 hens have been seen in single clusters in large leks. Lumsden noted that, although hens clustered at twenty different locations during his observations, the groups nearly always formed near the dominant male. Thus, hens are clearly attracted to specific males rather than to specific mating spots on a lek. Clusters of hens evidently serve as a sexual stimulus for females, and precopulatory squatting by one apparently stimulates others to behave similarly. Usually males quickly mount any soliciting female, and copulation lasts only a few moments. Unlike other grouse, the male does not normally grasp the female's nape in his beak while mounted, perhaps because of the considerable disparity in size between the sexes.

Most studies indicate that the majority of copulations are achieved by only one, or at most two, males in any center of mating activity. Scott (1942) found that master cocks performed 74 percent of 174 observed copulations, Patterson (1952) found mating success similarly restricted to a few males, and Lumsden (1968) found that two males accounted for more than half of the 51 copulations he observed. However, Hjorth (1970) found that four males took part in the matings he observed on one lek.

Although Wiley (1973a, 1978) agreed that at least 90 percent of the copulations at any particular lek may be performed by no more than about 10 percent of the males, he believed that this was not the result of the females' "choice" of preferred males, but rather the tendency of the females to gather at the center of the lek, thus favoring the males that have their territories in that area. He believed that, simply by surviving long enough, a male would gradually shift toward and eventually reach the mating center, and thus age rather than "beauty" might be the key to reproductive success for individual male sage grouse.

After copulation, the female usually runs a short distance forward, shaking her wings and tail for several seconds before starting to preen. Usually females leave the strutting grounds within a few minutes after copulation. Males usually squat motionless for several seconds after copulation, which Lumsden regards as a ritualized display posture that he believes may reduce disruption of the hen cluster.

Nesting Behavior

Once fertilization has been accomplished, the hen apparently leaves the strutting ground for nesting. There is no present evidence that a hen requires more than one successful copulation to complete her clutch. Patterson (1952) believed that females begin laying within a few days after mating, although Girard (1937) indicated that from 7 to 12 days may be taken up in locating a nest site and in nest construction. This kind of delay does not seem to be normal, and Dalke et al. (1963) found a good correlation between actual and calculated hatching periods by assuming that 10½ days would be required to lay an average clutch of 8 eggs, and that 26½ days more would be required for incubation, for a total time of 37 days between mating and hatching.

Estimates of average clutch size usually range from 7 to 8 eggs. Patterson (1952) reported an average clutch size of 7.26 eggs in 80 nests during one year, and 7.53 eggs in 74 nests the following year. Griner (1939) reported an average clutch size of 6.8 eggs in Utah, Nelson (1955) reported 7.13 in Oregon, and Keller, Shepherd, and Randall (1941) reported 7.5 in Colorado. Patterson (1952) believed that very limited renesting might occur, judging from smaller late clutches and the presence of new nests near destroyed or deserted nests. Although Eng (1963) found a second peak of females on strutting grounds in late May, this was not reflected in a second late hatching peak, and he concluded that reduced male fertility late in the season prevents effective renesting.

Patterson's estimate (1952) of a 25- to 27-day incubation period for sage grouse has generally been supported by later workers such as Pyrah (1963), who utilized data from captive grouse. This contrasts with various earlier estimates of 20 to 24 days. Sage grouse appear to have a high rate of both nest destruction and nest desertion. Gill (1966) summarized data on fates of nests from eight different studies, which ranged in hatching success from 23.7 to 60.3 percent. Predators were responsible for a large part of the nesting losses, accounting for 26 to 76 percent of the lost nests in six of the studies. Of a total of 503 nests represented, 47.7 percent were destroyed by predators. Coyotes, ground squirrels, and badgers are evidently among the more important mammalian predators, while magpies and ravens may also be significant.

EVOLUTIONARY RELATIONSHIPS

For reasons that have never been evident, taxonomists have traditionally regarded the sage grouse as closely related to the true "prairie grouse," namely the pinnated grouse and the sharp-tailed grouse. Not until the analysis by Hudson, Lanzillotti, and Edward (1964) was it proposed that the sage grouse may have its nearest affinities with the "forest grouse" instead. Short (1967), using various lines of evidence, supported the view that *Centrocercus* probably evolved from an ancestral type similar to *Dendragapus* and that *D. obscurus* represents the nearest living relative of the sage grouse. Lumsdan's analysis of behavior (1968) also presented this view, and he pointed out that the male sage grouse shares with the blue and spruce grouse the characteristic of having a white V on the throat that apparently has signal value at least in the sage grouse. Lumsden suggested that the sage grouse and blue grouse diverged from a common ancestral type that was a forest-dwelling bird, to which the spruce grouse and Siberian sharp-winged grouse are the nearest modern equivalents. In contrast, Short suggested that the ancestral grouse was a woodland-edge species, of which the earliest offshoot was a woodland form ancestral to *Tympanuchus*, followed later by separation of pre-*Dendragapus* and pre-*Centrocercus* types.

I believe that both adult and downy plumage characteristics strongly favor the view that *Dendragapus* and *Centrocercus* are closely related, and that the male displays of sage grouse and blue grouse have much in common. The evolution of lek behavior by the sage grouse produced some convergent similarities to the social displays of prairie grouse, but these should not be regarded as evidence for close common ancestry.

126

Blue Grouse

Dendragapus obscurus (Say) 1823

Other Vernacular Names

Dusky grouse, fool hen, gray grouse, hooter, mountain grouse, pine grouse, pine hen, Richardson grouse, sooty grouse; tétras sombre (French); Felsengebirgshuhn (German).

Range

From southeastern Alaska, southern Yukon, southwestern Mackenzie, and western Alberta southward along the offshore islands to Vancouver and along the coast to northern California, and in the mountains to southern California, northern and eastern Arizona, and west-central New Mexico (A.O.U. *Check-list*, 1957).

Subspecies (ex A.O.U. *Check-list*).

D. o. obscurus (Say): Dusky blue grouse. Resident in the mountains from central Wyoming south through eastern Utah and Colorado to northern and eastern Arizona and New Mexico.

D. o. sitkensis Swarth: Sitkan blue grouse. Resident in southeastern Alaska south through the coastal islands to Calvert Island and the Queen Charlotte Islands, British Columbia.

D. o. fuliginosus (Ridgway): Sooty blue grouse. Resident from the boundary between Yukon and Alaska south through the mainland of southeastern Alaska, coastal British Columbia including Vancouver Island, western Washington, and western Oregon to northwestern California.

D. o. sierrae Chapman: Sierra blue grouse. Resident on the eastern slope of the Cascade Mountains of central Washington south into California and from southern Oregon south along the Sierra Nevada into California and Nevada.

D. o. oreinus Behle and Selander: Great Basin blue grouse. Resident in mountain ranges of Nevada and Utah.

2. Current distribution of dusky (D), Great Basin (G), Mount Pinos (P), Oregon (O), Richardson (R), Sierra (Si), Sitkan (S), and Sooty (So) races of the blue grouse.

D. o. howardi Dickey and van Rossem: Mount Pinos blue grouse. Resident on the southern Sierra Nevada from about latitude 37° N to the Tehachapi range and west to Mount Pinos, where it may be extirpated.

D. o. richardsonii (Douglas): Richardson blue grouse. Resident from the southern Yukon and Alaska south through interior British Columbia to the Okanagan valley and western Alberta to Idaho, western Montana, and northwestern Wyoming.

D. o. pallidus Swarth: Oregon blue grouse. Resident from south-central British Columbia south through eastern Washington to northeastern Oregon.

N.b.: Three of the forms listed (*sitkensis, fuliginosus,* and *sierrae*) are sometimes specifically separated from the remaining ones and tend to have 18 rather than 20 rectrices, yellowish rather than grayish downy young, and certain other minor structural differences.

MEASUREMENTS

Folded wing (unflattened): Adult males 196–248 mm; adult females 178–235 mm (adult males of all races average over 217 mm; females, under 216 mm).

Tail (to insertion): Adult males 131–201 mm; adult females 111–59 mm (adult males average over 150 mm; females, under 150 mm).

IDENTIFICATION

Adults, 17.2–18.8 inches long (females), 18.5–22.5 inches long (males). This is the largest of the coniferous-forest grouse of the western states and provinces. Sexes differ somewhat in coloration, but both have long, squared, ane relatively unbarred tails (pale grayish tips usually occur in both sexes of all races except *richardsonii* and *pallidus,* which sometimes have suggestions of a pale tip). Upperparts of males are mostly grayish or slate-colored, extensively vermiculated and mottled with brown and black, and the upper wing surfaces are more distinctly brown. White markings are present on the flanks and under tail coverts, and feathering extends to the base of the middle toe. The bare skin over the eyes of males is yellow to yellow orange, and the bare neck skin exposed during sexual display varies from deep yellow and deeply caruncled (in the *fuliginosus* group) to purplish and somewhat smoother (in the *obscurus* group). Females have smaller areas of bare skin and are generally browner overall, with barring or mottling on the head, scapulars, chest, and flanks.

FIELD MARKS

Blue grouse are likely to be confused only with the similar but smaller spruce grouse; the ranges of the two overlap in the Pacific northwest. Male blue grouse lack the definite black breast patch of male spruce grouse. Female blue grouse have relatively unbarred, grayish underparts compared with the spruce grouse's white underparts with conspicuous blackish barring. A series of five to seven low, hooting notes is frequently uttered by territorial males in spring.

AGE AND SEX CRITERIA

Females may be recognized by barring on the top of the head, nape, and interscapulars that is lacking in adult males (Ridgway and Friedmann 1946) and by the bases of the neck feathers around the air sacs, which are grayish brown rather than white. The sex of adults may be determined from the wings alone: females have a more extensively mottled brownish pattern on their marginal upper wing coverts; in males these feathers are gray, with little or no mottling (Mussehl and Leik 1963).

Immatures (in first-winter plumage) may be recognized by one or more of the following criteria: the outer two primaries (retained from the juvenal plumage) are relatively frayed and more pointed (van Rossem 1925) as well as being lighter and more spotted than the inner ones, the outer tail feathers are narrow and more rounded (up to ⅞ inch wide at ½ inch below tip, as opposed to at least 1¼ inch wide in adults), and the tail is shorter than in adults (the maximum length of plucked feathers of juvenile males is 152 mm, of juvenile females, 134 mm, compared with 162 and 138 mm in adult male and female *fuliginosus* (according to Bendell 1955*b*). Immatures of both sexes generally resemble adult females but may usually be recognized by their pale buffy or white breasts, the absence of a gray area on the belly, and (except in *richardsonii* and *pallidus*) the absence of a gray bar at the end of the tail (Taber, in Mosby 1963).

Juveniles (in juvenal plumage) may be distinguished by the conspicuous white (tinged with tawny) shaft streaks of the upperparts, wings, and tail and the brown rectrices, which may be mottled or barred and lack a gray tip (Ridgway and Friedmann 1946). The juvenal plumage is carried only a very short time in this species, as in other grouse, and the juvenal tail feathers are molted almost as soon as they are fully grown. Zwickel and Lance (1966) provided a method of age determination for juveniles, which was later slightly modified by Redfield and Zwickel (1976). Growth curves for wild chicks have been constructed by Redfield (1978). A key for aging and sexing by wing characteristics was provided by Bunnell et al. (1977), who noted that juveniles at least 12 weeks old could be sexed by their wing length (females less than 228 mm from wrist joint to tip of seventh primary, males at least 228 mm), and adults could likewise be sexed either by wing length (237 mm to the division point) or by wing color. Braun (1971*a*) found that birds over 6 weeks old could be aged and sexed by wing color.

Downy young of blue grouse exhibit considerable variation in down coloration among the numerous races (Moffitt 1938). Downy blue grouse lack the chestnut crown patch of spruce grouse, instead exhibiting irregular black spotting over the crown and sides of the head and a conspicuous black ear patch. The black head marking in young blue grouse also includes a central crown mark that connects with frontal spotting, two indefinite lateral stripes, and a faint brownish area posteriorly that is bordered by slightly darker markings (Short 1967). This species is thus intermediate between the extreme type of head markings found in the sage grouse and the more *Lagopus*-like markings typical of the spruce grouse.

RANGE AND HABITAT

The overall North American range of the blue grouse is closely associated with the distribution patterns of true fir (*Abies*) and Douglas fir (*Pseudotsuga*) in the western states (Beer 1943). Its range conforms more closely with that of the Douglas fir than any other conifer tree species, but this is probably a reflection of both species' being closely adapted to a common climate and community type rather than of any likelihood that the blue grouse is closely dependent on Douglas fir. The species actually occupies a fairly broad vertical range in the western mountains, breeding at lower elevations, sometimes as low as the foothills, and spending the fall and winter near timberline or even above it. Rogers (1968) reports that in Colorado the birds are usually to be found at between 7,000 and 10,000 feet but have been seen at elevations as low as 6,100 feet and as high as 12,400 feet, averaging about 9,000 feet. At least in the moist Pacific northwest, lumbering and fire produce a more open forest that improves the breeding habitat of blue grouse by opening the forest cover, but heavy grazing on lower slopes can be deleterious (Hamerstrom and Hamerstrom 1961).

In contrast to several grouse species, the blue grouse has had no major range changes of importance in historical times (Aldrich 1963). In none of the states and provinces where the species occurs is it in danger of extirpation, although the southern populations in New Mexico and Arizona are relatively sparse and scattered.

Although the blue grouse depends heavily on coniferous cover for wintering, its preferred habitat also includes a number of deciduous tree species, shrubs, and forbs. Foremost among broadleaf trees are aspens (*Populus*), and a variety of shrubs provide food and escape cover. Rogers (1968) summarized records of dominant trees, shrubs, forbs and grasses associated with blue grouse observations in Colorado over several years. In all years, aspen was the dominant tree, snowberry (*Symphoricarpos*) was the dominant shrub, bromegrass (*Bromus*) was the dominant grass, and groundsel (*Senecio*) and vetch (*Astragalus*) were the dominant forbs. Trees recorded less frequently were juniper (*Juniperus*), spruce (*Picea*), Douglas fir, and ponderosa pine (*Pinus ponderosa*). Although hens and broods were sometimes seen in piñon pine (*Pinus edulis*) and juniper cover, summer concentrations of males were usually in open coniferous stands of spruce and fir. Rarely were blue grouse seen more than a mile from trees or shrubs, and females with broods were usually not far from water.

Similar observations on blue grouse habitat have been made in southern Idaho by Marshall (1946). There the vertical range used by the species extends from less than 5,000 feet in ponderosa pine–Douglas fir forest, which is infrequently used by blue grouse, to subalpine forests reaching over 8,000 feet, which provide wintering areas for both sexes and summering habitats for males. In these higher ridges they use the conifers, especially Douglas fir, for both food and cover. In all but 8 of 25 cases, the grouse were observed to land in conifers on being flushed, while the rest landed on the ground. Of 159 observations, 87 birds took cover in Douglas fir, 41 in subalpine cover, 25 on banks of streams, and the remaining 6 in grass or brush.

A study by Fowle (1960) on Vancouver Island provides comparable data for the coastal population of blue grouse. Summer habitat there consists of second-growth cover produced by fire and logging of Douglas fir forests. About 45 percent of the sample areas had no vegetation at all, while in the rest mosses, lichens, ferns, and grasses, as well as a variety of shrubs and forbs, made up most of the cover. Except near water, where alders (*Alnus*), willows (*Salix*), and dogwood (*Cornus*) occurred, trees grew only in scattered groups. About 20 percent of the area was covered with important grouse foods, including bracken fern (*Pteridium*), willow, Oregon grape (*Mahonia*), blackberry (*Rubus*), huckleberry (*Vaccinium*), salal (*Gaultheria*), and cat's-ear (*Hypochaeris*). These plants made up a total of more than 90 percent of adult food samples and more than 80 percent of juvenile food samples.

By the end of September the birds move up to higher slopes, and they winter in the coniferous zone (Bendell 1955*c*), where they are found primarily in subalpine forests. Zwickel, Buss, and Brigham (1968) point out that winter habitat is probably determined more by cover type than by altitude per se and in Washington may occur as low as 4,000 feet, between the ponderosa pine and Douglas fir zones, with the critical factor apparently being the presence of interspersed Douglas fir and true firs.

POPULATION DENSITIES

Blue grouse population densities are difficult to estimate because of the cover inhabited and the generally solitary nature of the species. Rogers (1968) summarized results of grouse surveys from vehicles; over a three-year period in two study areas they averaged 1 grouse per 26.07 survey miles, ranging from 10.3 to 38.72 miles in various years.

Using a strip-count census method, Fowle (1960) counted adult grouse on Vancouver Island during two summers. In four areas totaling 272 acres he determined a density in 1943 of 2.6 acres per bird. Later work in the same area by Bendell and Elliott (1967) indicated that the density of territorial males in dense and sparse populations respectively was approximately 0.44 and 0.13 or fewer males per acre, or from about 2.3 to 7.7 acres per territorial male. Similar counts of territorial male blue grouse were made by Mussehl and Schladweiler (1969) in Montana on six study areas that were in part exposed to insecticide spraying. Numbers of territorial males on sprayed and unsprayed areas did not appear to differ and averaged about 1 male per 18 acres, ranging from 12 to 24 acres per male.

On an isolated population of grouse on Prevost Island, in the Gulf Islands of British Columbia, the density of males on territories in spring was estimated at 4.4 males per 100 acres, or about 23 acres per male (Donaldson and Bergerud 1974). This was fairly low, at least compared with three areas on Vancouver Island that had been burned over between seven and sixteen years previously, and supported from 5 to 35 males per 100 acres, or 2.8 to 20 acres per bird. The data of Zwickel and Bendell (1972), also for Vancouver Island, suggest that under ideal conditions the density of adult male grouse may at times reach 90 birds per square kilometer (2.76 acres per bird) or approach the size of the average territory estimated by Martinka (1972) and Harju (1974).

Whether the blue grouse is subject to population "cycles" is perhaps questionable, but at least major population fluctuations and corresponding changes in density evidently do occur. Fowle (1960) and Hoffmann (1956) summarized historical data on grouse populations during the 1900s, but neither attempted to explain these fluctuations. Zwickel and Bendell (1967) hypothesized that population fluctuations in the species are related to the nutritional condition of females, as determined by the summer range conditions, which might affect chick survival and in turn determine subsequent autumn population densities. However, no relation was found between the number of young in autumn and the breeding density in the following year. They suggest that the death rate or dispersal of juveniles between autumn and early spring is the single most important factor regulating breeding densities.

Later, Zwickel and Bendell (1972) could find no relation between grouse populations and such habitat factors as vegetation, soil fertility, or food, nor any obvious controlling effects of weather, predation, or disease. Blue grouse populations are apparently controlled by variations in juvenile mortality between autumn and spring. Boag (*Proceedings of the Fifteenth International Ornithological Congress*, p. 178) suggested that perhaps the survival of juvenile grouse may be related to their varying abilities to shift successfully from summer foods to winter foods.

HABITAT REQUIREMENTS

Wintering Requirements

Primary wintering needs for the blue grouse appear to be enough trees for roosting and escape cover and a supply of needles from trees of the genera *Abies, Tsuga*, or *Pseudotsuga* as food. Beer (1943) reports that adult blue grouse subsist almost entirely on needles from November through March. Needles, buds, twigs, and seeds of Douglas fir may all be eaten in winter, and needles, buds, and pollen cones of true firs are also used. Where both *Abies* and *Pseudotsuga* are present, the former appears to be preferred. Larch (*Larix*) may be used until its needles are shed, and various species of pines are used for their buds, pollen cones, and seeds. Marshall (1946) noted that 99 percent of the food contents of 9 birds killed during winter in Idaho consisted of needles and buds of Douglas fir. Grit is evidently retained in the gizzard through the winter in spite of the deep snow cover. Hoffmann (1956) reported that white fir (*Abies concolor*) provides favored winter roosts in California.

Spring Habitat Requirements

As the winter ends, both sexes begin to move downward from the coniferous zones, and males seek out areas suitable for territories. Bendell and Elliott (1966) analyzed the habitats used by both sexes of blue grouse on Vancouver Island from spring through August, classifying cover as "very open" (40 percent tree, log, stump, and salal cover) or "very dense" (100 percent woody cover). The relative grouse use in two types was 115 in very open cover compared with 18 in very dense cover. The use of the very dense cover was limited to some territorial males that apparently had established territories there before it became so heavily overgrown. The authors concluded that the blue grouse is better adapted to a dry habitat than is the ruffed grouse and may indeed have evolved from a grassland species. Supporting this view was their finding that young captive blue grouse required only about half as much water as captive ruffed grouse. They concluded that the breeding habitat of blue grouse might be defined as open and dry, with shrubs and herbs interspersed with bare ground.

Martinka (1972) analyzed habitat characteristics of blue grouse territories in Montana and found that territories averaged 2.0 acres in area and were primarily composed of thickets of coniferous trees. These thickets varied in number and size but provided relatively constant amounts of edge, averaging about 675 feet. The average thicket size was 0.2 acre, and most trees in the thicket were from 10 to 60 years old. Thickets probably support territories for some 40 to 50 years before they become unsuitable and are abandoned. Harju (1974), working in Wyoming, similarly found that territories there averaged between 0.7 and 1.0 hectare (1.7 to 2.5 acres) per bird and that breeding habitats were usually dominated by lodgepole pine (*Pinus contorta*) and aspen (*Populus tremuloides*). Donaldson and Bergerud (1974), studying an insular population of grouse off the coast of British Columbia, concluded that the preferred habitat of males consists of forest of irregular heights, with patchy shrub layer and a discontinuous crown canopy, all factors associated with visibility for displaying males. In both areas, preferred territorial sites provide a maximum amount of edge. Lewis (1981) found that, judging from the environmental characteristics he measured, persistent and transient territorial sites chosen by males had virtually identical vegetational and structural characteristics, differing only in height or elevation of the display area relative to that of surrounding areas, with persistent sites elevated locations.

In California, Hoffmann (1956) found that the persistence of snow cover determined the onset of hooting in spring and the transition to spring behavior in a study area where there was virtually no seasonal migration. Blackford's studies (1958, 1963) in Montana provide additional information on territorial requirements for an interior population (*obscurus*) of this species. In this area, hooting occurred either at ground level or in trees during strutting. Strutting areas were in forest-edge habitats with combined grassy, open-forest border and a dense coniferous stand. They contained occasional rocky outcrops, and old logs were present on the forest floor. Blackford's observations established that earlier, widely reported differences in territorial defense and strutting behavior between coastal and inland populations of blue grouse are not absolute.

Yearling males may migrate downward to the breeding areas or may remain on the wintering areas through the summer. Bendell and Elliott (1967) estimated that about half of the yearling males move to the summer range their first year. There they are silent, move about widely, and may be attracted to hooting territorial males. These authors observed two cases of territorial yearling males. Females may return to the same general area of the summer range in subsequent years but are not nearly so localized in this respect as are males (Bendell 1955c). Unlike males, females are not particularly aggressive to one another, and their home ranges may overlap. However, Stirling (1968) suggested that during the squatting and egg-laying periods females do become somewhat aggressive, and this behavior tends to scatter them and perhaps promotes spacing of nests.

Nesting and Brooding Requirements

Surprisingly little has been written on specific nesting needs for blue grouse, perhaps because their nests are rather difficult to locate. Usually nests are near logs or under low tree branches and are fairly well concealed. Bendire (1892) stated that most nests are under old logs or among roots of fallen trees and are generally to be found in more open timber along the outskirts of the forest. He found one nest beside a creek in rye grass some 2 miles away from timber and another in an alpine meadow under a small fir tree, with no other trees within 30 yards. Bowles (in Bent 1932) noted that nests are usually in very dry, well-wooded sites, and they are often at the bases of trees or under fallen branches or some other shelter. However, they may be up to 100 yards from trees, with little or no concealment. Lance (1970) found that nests were usually fairly near territorial males but well separated from the nests of other females.

Brooding habitat for blue grouse appears to be that which provides ample opportunities for the young to feed on insects and other invertebrates. Beer (1943) suggested that blue grouse usually nest in open situations where there will be abundant insects for the newly hatched birds. For the first 10 days, the young feed almost exclusively on animal material, especially ants, beetles, and orthopterans, according to Beer. As they grow older, they seek out berries, such as currants (*Ribes*) and Juneberries (*Amelanchier*), and the young birds and adults gradually move upward as they follow the ripening berry crop.

Wing, Beer, and Tidyman (1944) reported that broods occupy home ranges characterized by semiopen vegetation and available water. Relatively open areas are used by newly hatched chicks, while older broods move into more densely vegetated areas. Mussehl (1963) found that brood cover in Montana is consistently low (averaging 7 to 8 inches), has little bare ground (8 to 20 percent), and is predominantly herbaceous in nature, with grasses next in importance, followed by low shrubs and forbs. Woody growth increases in importance as food and escape cover as the birds mature.

Similarly, Donaldson and Bergerud (1974) observed that females with broods favored grassy openings in old logged areas, especially in meadows surrounded by forest. They believed that an extensive herb layer and proximity to cover were the most important components of brood habitat.

FOOD AND FORAGING BEHAVIOR

In spite of the rather broad geographic range of the blue grouse, its food requirements appear to be fairly consistent. Martin, Zim, and Nelson (1951) report that Douglas fir was the most important food in 158 samples from the northern Rocky Mountains, and in 154 samples from the Pacific northwest Douglas fir and true firs provided the major food items. They also list a variety of herbaceous plants and sources of berries that are used in summer and fall. Judd (1905) indicated that winter blue grouse foods include ponderosa pine, Douglas fir, true firs (*Abies concolor* and *A. magnifica*), and hemlocks (*Tsuga heterophylla* and *T. mertensiana*).

Beer (1943) analyzed more than 100 crops and gizzards of blue grouse, mostly from Washington and Oregon, and noted that adult foods were 98 percent plant materials, with conifer needles constituting 63.8 percent, berries 17 percent, miscellaneous plant materials 17.2 percent, and animal material 1.7 percent, of the specimens examined. Beer noted that the grouse reach the peak of their morning feeding by 7 a.m. and stop by 9 a.m. Later feeding periods are just before noon, during later afternoon, and particularly toward evening, when the most intensive foraging of the day occurs. Growing young feed more continuously than adults, but birds of all ages forage most heavily during the last 3 hours of daylight. Similar observations were made by Fowle (1960), who noted that, although grouse fed throughout the day, they ate the most food after 6 p.m. Males often alternate feeding with hooting, but females with young evidently restrict their foraging to the evening. Fowle never saw wild grouse drink water and believed it might not be improtant when berries or other succulent foods are available.

Hoffmann (1961) noted that blue grouse in California rely during the winter almost entirely on needles of white fir (*Abies concolor*), which he analyzed for protein content. He found that needles from high in the tree had a higher protein content than those from lower branches but that there were no apparent yearly differences over a three-year period during which the grouse population suffered a major decline.

MOBILITY AND MOVEMENTS

Seasonal movements

Altitudinal movements of blue grouse to coniferous wintering areas have been reported for most areas, the exception being Hoffman's study in California (1956). Doubtless the horizontal distances involved in movements between summering and wintering areas differ greatly in various regions, but relatively little detailed information is available. One banding study by Zwickel, Buss, and Brigham (1968) in north-central Washington indicates that autumn migrations of blue grouse may be fairly long. The longest movement recorded by a banded bird was 31 miles, in less than 2 months. Of 30 birds recovered, 50 percent had moved more than 5 miles, and 30 percent were recovered more than 10 miles from where they had been banded. In contrast, Mussehl (1960) reported a maximum fall movement of 3.4 miles in Montana, while Bendell and Elliott (1967) found a maximum fall movement of 10 miles on Vancouver Island. Zwickel, Buss, and Brigham (1968) speculated that at least some breeding females leave their broods behind and return to their previous wintering areas, which stimulates wandering by young birds and possible colonization of new wintering areas.

Daily Movements

Evidently relatively little daily movement is performed by adult male blue grouse from the time they arrive on the summer range and establish territories until they begin their fall movement back to the wintering areas. Males probably establish territories as soon as weather permits, and maintenance activities such as foraging, dusting, and sleeping are all carried out within the territorial boundaries (Bendell and Elliott 1967). Territorial size presumably varies inversely with population density. In dense populations of about 0.44 male to the acre, Bendell and Elliott estimated that territorial sizes averaged about 1.5 acre. In sparse populations of about 0.13 male to the acre, territories were at least 5 acres.

Similarly, female grouse probably exhibit little daily movement, at least after fertilization. Until then they presumably move about through the territories of males until sufficiently stimulated to permit mating. Various studies of marked broods (Mussehl 1960; Mussehl and Schladweiler 1969) indicate that before dispersal the broods move about relatively little, and individual brood ranges may overlap considerably.

In a study in Washington, Zwickel (1973) reported that females leading broods tend to remain dispersed while the broods are very young but that later in the summer both broodless females and those that have raised broods successfully tend to aggregate. He suggested that such spacing behavior of reproductively active females may have significance as a population regulation mechanism in promiscuous species such as the blue grouse.

REPRODUCTIVE BEHAVIOR

Territorial Establishment

Male blue grouse evidently become territorial immediately after they arrive on the breeding range (Blackford 1963) or as soon as snow cover permits (Hoffmann 1956). Territorial site requirements are somewhat ill defined and may vary locally or with subspecies. In Colorado, Rogers (1968) states that display sites may be in aspen–ponderosa pine, mixed fir and aspen, open and dense aspen, mixed shrubs, sagebrush, wheat fields, and on roadbeds, but preference is shown for fairly open stands of trees or shrubs. Physical features include earth mounds, rocks, logs, cutbanks, and occasionally tree limbs. Preference is generally given to flat, open ground, although steep slopes are also used. Display sites may be near heavy cover, but this is normally used for escape rather than for display. Two observations were made of birds displaying at more than 20 feet, but ground display is typical of interior populations of blue grouse.

In contrast, Hoffmann (1956) found that in a California population (*fuliginosus*) the males normally hooted from the tops of white fir or sometimes from Jeffrey pine (*Pinus jeffreyi*) or lodgepole pine (*P. contorta*). Bendell and Elliott (1966, 1967), studying the same subspecies on Vancouver Island, found that many hooting sites were elevated areas on the ground and that territories included diverse cover types, with males hooting from virtually all types of cover within their territory. In dense cover with small openings, territories are related to the location of openings. Thickets within territories are used for resting and concealment. This combination of open areas for display and shelter in the form of fir clumps, logs, or stumps used for hiding and as observation posts fulfills the basic territorial requirements. Several display sites may be used within a single territory; Rogers (1968) noted that from two to eleven hooting sites for one bird have been recorded.

Territorial Advertisement

Male blue grouse proclaim their territories by a combination of postures, vocalizations, and movements that are collectively called hooting. In spite of reported differences in hooting behavior among different populations, current evidence indicates that actual differences are few and tend to be quantitative rather than qualitative. Thus, the interior populations (dusky grouse) have much weaker hooting calls that are barely audible more than 50 yards away, whereas the coastal populations (sooty grouse) have strong hooting notes that carry several hundred yards. The former typically call from the ground but may use trees, while the latter more often call from tree limbs. The gular sac of dusky grouse males is generally purplish, while that of sooty grouse is more heavily wrinkled and yellowish. The eye combs of dusky grouse are large and vary from yellow to a bright red under maximum stimulation; those of sooty grouse are smaller and usually are lemon yellow but sometimes become vivid red (Bendell and Elliott 1967).

During hooting the male partially raises and spreads his tail and opens the feathers of his neck to expose an oval gular sac that is surrounded by white-based neck feathers, forming a "rosette" pattern. Both wings are slightly drooped toward the ground. In this posture (called the "oblique" by Hjorth 1970) the gular sac is partially inflated in a pulsing manner as up to seven but usually five (in the dusky grouse) or six (in the sooty grouse) *hoot* sounds are uttered in fairly rapid succession. These are repeated at frequent intervals. Bent (1932) reported intervals of 12 to 36 seconds between call sequences of *fuliginosus*, Steward (1967) determined a mean interval of 24.2 seconds in *sitkensis*, and Rogers (1968) noted intervals of from 6 to 23 seconds for *obscurus*. Such hooting is uttered at various times during the day but is most prevalent in early morning and again in late evening, primarily between 3 and 5 A.M. and between 7 and 10 P.M. (Bendell

1955c). Hjorth (1970) noted that although in both subspecies groups the call sequence lasts about 3 seconds, the fundamental frequencies of dusky grouse calls (95 to 100 Hz) are lower than those of sooty grouse (100 to 150 Hz) and have much less amplitude. Males may periodically move about between hooting sites, and while walking they keep the head low and the tail cocked and spread, exposing the spotted under tail coverts ("display walking" of Hjorth 1970).

Strutting Displays

When in the presence of another grouse, the male stands erect with his tail tilted toward the other bird ("upright cum tail tilting" of Hjorth 1970), the eye combs enlarged, and the wing away from the intruder drooped in proportion to the amount of tail tilting. In this posture the male may perform vertical head jerking, with the gular sac nearer the intruder expanding in synchrony with these head movements (Hjorth 1970). Hjorth also reported that these downward head movements ("bowing cum asymmetric apteria display") may be greater in the dusky grouse group than in sooty grouse.

In this erect and tilted-tail posture, the male typically advances toward the intruder. Bendell and Elliott (1967) stated that in the sooty grouse the head and neck are held broadside to the other bird in such a way as to be framed against the background of the dark tail. Rogers (1968) has a photograph of the comparable posture of a Colorado dusky grouse. The approach display is climaxed by a quick, arcing dash toward the other bird ("rush cum single hoot" of Hjorth 1970) that is associated with maximal tail cocking and spreading, extreme engorgement of the eyecombs, and a drooping of the wings so that they drag on the ground. In this posture the male jerks his head several times, then lowers it and runs forward with short, fast steps, terminating the run with a deep *oop* or *whoot* note. Rogers (1968) noted that this sound could be heard as far as 510 feet away, in contrast to the hooting series in Colorado grouse, which could not be heard beyond 105 feet. Bendell and Elliott (1967), as well as Hjorth (1970), observed that it is actually a double note, with a short squeal or whistle following the deeper sound. Hjorth (1967, 1970) noted that during the forward dash the male deflates his neck, turns his tail toward the other bird, and holds his neck in such a way that the cervical rosette is maximally exposed. The head is held low, the tail is twisted to provide maximum surface exposure, and the wing on the far side is increasingly drooped as the tail is twisted. After uttering the call the bird gradually resumes a normal posture.

If the other bird is a receptive female she may remain in place, and the male then displays about her, raising and lowering his body and jerking his head, always keeping the neck rosette and nearer eyecomb in full view of the female. After 2 to 3 minutes of such display, the male moves behind the female and attempts to mount her. During treading the male grasps the hen's nape in his beak and holds her body against his lowered wings as she squats. After treading the male again assumes his upright display posture (Hjorth 1970).

Flight Displays

The other primary aspects of display by male blue grouse involve fluttering or flying

17. Male displays of the blue grouse, showing sequence of rush display. Elapsed time (in seconds) since beginning of sequence is indicated. After Hjorth (1970).

movements that have been variably ritualized to produce sound and advertise the male's presence. They are difficult to classify, since they have been described differently by various observers. Blackford (1958, 1963) attempted to classify these aerial displays based on his observations in Montana, which may be summarized as follows:

"Wing fluttering" is a brief flapping of wings as the bird rises about 8 or 10 inches in the air, producing relatively little noise. It may be performed by either sex, both on the ground and in trees.

"Wing drumming" is the typical male display flight, or flutter-jump. It is a short vertical leap into the air as the bird beats his wings strongly a few times before descending. Often one wing is beat much more strongly than the other, producing a rotary movement ("rotational drumming") and causing the bird to make an incomplete turn before landing.

"Wing clapping," so far noted only by Blackford, is an upward leap associated with a single, very loud wing note.

Blackford distinguished "drumming flight" from normal wing drumming by a circular flight some 10 to 12 feet in diameter, made before the bird lands again near the takeoff point.

Blackford (1963) noted several other possible wing signals, including a "double wing flutter," a "perching signal," an "explosive flush," and an "aerial signal." Since they have not been well studied or described by others, they need not be given further consideration here.

Vocal Signals

Male vocalizations other than the *hoot* and *oop* calls are relatively few, judging from most accounts. Rogers (1968) reported a "gobbling" sound uttered by a male after he made a clapping, wing-beating flight to a branch. This was followed by regular hooting sounds until he uttered a single two-note *ca-caw* about 18 minutes later.

Female vocalizations reported by Blackford (1958) include an in-flight alarm call, *kut-kut-kut*, a low warning note uttered before flight, *kr-r-r*, and an "excitement" call, *kutter-r-r-r*, that fluctuates greatly in pitch. Rogers (1968) noted that the in-flight alram call of females was the note most commonly heard. Female blue grouse also produce a "whinny" call that is highly effective in stimulating males to begin hooting and to move toward the source of the sound. Using tape recordings of such calls is an effective method of censusing blue grouse (Stirling and Bendell 1966). Likewise, recorded chick distress calls evoke clucking responses from broody hens.

Stirling and Bendell (1970) have recently reviewed the behavior and vocalizations of adult blue grouse. They described and presented sonagrams of three male calls, including the hooting call, the *whoot* call associated with the rush display, and a growling *gugugugug* associated with attack. Females were believed to have two calls related to reproduction: the "whinny," related to copulation readiness, and the "quaver call" or *qua-qua* that consists of a pulsed series of notes produced by breeding females just before males reach maximal reproductive development, thus possibly synchronizing breeding cycles. Females also utter a "hard cluck" or *bruck-duck* call, which apparently serves as a threat signal.

141

Collective Display

Although the blue grouse is regarded as a species that normally defends farily large territories and displays in a solitary fashion, several observations of collective display have been made. Bendell and Elliott (1967) noted that, of 420 territorial males studied, the average distance between nearest territorial neighbors in open cover was approximately 600 feet. In 5 percent of the 1,000-foot circular areas they studied there were 7 or 8 hooting males, which were usually 200 to 500 feet apart and formed a "hooting group" that usually called in chorus. They regard such hooting groups as indicating a habitat favorable for territories rather than as a variant of lek behavior, since, they point out, blue grouse remain on their territories through the breeding season, in contrast to typical lekking grouse. However, Blackford's observations (1958, 1963) of collective display indicate that males will leave their territorial sites and cross over adjacent territories to perform in a "communal court." In one case he noted that at least two males, two females, and one bird of unknown sex converged on the territory of another male, where collective display occurred. This kind of temporary establishment of collective display areas by males, which perhaps follow females into the territory of an unusually effective resident male, might provide the evolutionary basis for typical lek behavior, provided such "hooting groups" are more efficient in attracting females than are individual males displaying in a solitary fashion.

Nesting and Brooding Behavior

Since the male plays no role in nest defense, incubation, or brooding, the female undertakes these duties alone. Evidently nearly all females, including yearlings, attempt to nest (Zwickel and Bendell 1967). Further, most hens that fail to produce a brood do so because of nest destruction rather than nest desertion. Zwickel and Carvath (1978) reported that of 164 nests found, 94 hatched, or 57.3 percent. Of 113 found during incubation, 76 hatched, or 67 percent, the difference resulting from differential nest desertion rates. How much renesting might occur after nest destruction or desertion is still uncertain, but Zwickel and Lance (1965) reported two definite instances indicating that renesting might occur even when the first nest is destroyed late in incubation and that a second clutch can be started within about 14 days after such nest destruction.

Zwickel (1975) reported that 118 nests on Vancouver Island had an average clutch size of 6.37 eggs. Zwickel and Lance (1965) indicate that the laying rate for blue grouse is 1.5 days per egg, and that the incubation period is 26 days.

Upon hatching, blue grouse chicks soon become fairly independent of the female. Zwickel (1967) found that chicks begin to eat plant materials at 1 day of age, can fly at 6 to 7 days, and by 2 weeks of age can fly up to 60 meters. No chicks older than 11 days were observed being brooded by the hen, and few over 7 days old were seen being brooded. Contrary to other writers, Zwickel (1967) doubts that chilling by rain or cold days normally plays an important role in chick survival. Zwickel noted several calls of brooding females. When the chicks wailed loudly with their distress note, the females uttered a low brood call, *cu-cu-cu*. While foraging, hens produced a similar but less audible series of notes that Zwickel terms a contact call. When calling the brood together, the female sometimes produced a high-pitched *kwa-kwa-kwa* call to which the chicks responded by wailing. When the hen returned to her brood after a considerable absence she would cluck loudly or produce a high-pitched *kweer-kweer-kweer* that was audible for up to ¼ mile under favorable conditions. Zwickel concluded that vocal signals were highly important in maintaining brood organization and exhibited considerable plasticity to meet varying needs.

Evidently most chick losses occur during the first 2 weeks, according to Zwickel and Bendell (1967). These authors present data indicating that brood sizes for chicks up to 14 days old average from 3.3 to 4.4 young, while brood sizes for chicks estimated to be older than 42 days average 2.9 to 3.7 young. Mussehl's study in Montana (1960) indicated that the movements of 8 marked broods for periods of 19 to 47 days were restricted to areas having maximum diameters of 440 to 1,320 yards. During early July these broods primarily used a mixed grass-forb cover, but with gradual drying of the prairie forbs they moved into deciduous thickets for the remainder of their brooding period. Little use of montane coniferous forest was noted. By the end of August most of the brooding range had been abandoned, and broods began to disperse. Juveniles then moved singly or in small groups, with individual birds making lateral movements of up to 2.1 miles as they worked their way up toward the wintering ranges.

In a recent analysis of Arizona blue grouse populations, Brown and Smith (1980) have found a positive relation between precipitation during winter and spring and blue grouse density, reproductive success, and subsequent survival rates. They suggest that annual variations in mortality rates may be influenced by variations in the spring growth of herbaceous vegetation and by grouse density. Such variations in survival may be more effective in determining population levels than are variations in reproductive success, according to these authors.

EVOLUTIONARY RELATIONSHIPS

The blue grouse presumably had its evolutionary origin in western North America, either in coniferous forest or in a forest-grassland edge habitat. Jehl (1969) concluded that two species of *Dendragapus* occurred in western North America in the late Pleistocene, one of which presumably directly gave rise to the modern blue grouse. I believe that the ancestral blue grouse probably originated in North America, whereas the ancestral spruce grouse may have had its origins in eastern Asia, only later coming into contact with the blue grouse.

It seems probable that the sage grouse also had its origin in the western part of North America and may be much more closely related to the blue grouse than adult plumage patterns suggest. The surprising similarities of the downy young support this view, and the strutting behavior patterns of the two species are not greatly different. To a much greater extent than is usually appreciated, the breeding habitat of the blue grouse is relatively arid and open, and the bird is in no sense a climax coniferous forest species.

I would suggest that North America was invaded relatively early from Asia by a *Tetrao*-like ancestral type, which as it moved southward produced the more montane-dwelling blue grouse ancestor, and also the intermontane or valley-dwelling sage grouse ancestor. A second invasion brought the spruce grouse into North America, possibly as recently as late Pleistocene times.

Spruce Grouse

Dendragapus canadensis (Linnaeus) 1758
(*Canachites canadensis* in AOU *Check-list*, 1957)

Other Vernacular Names

Black partridge, Canada grouse, cedar partridge, fool hen, Franklin grouse, heath hen, mountain grouse, spotted grouse, spruce partridge, swamp partridge, Tyee grouse, wood grouse; tétras du Canada, tétras des savanes (French); Tannenwaldhuhn (German).

Range

From central Alaska, Yukon, Mackenzie, northern Alberta, Saskatchewan, Manitoba, Ontario, Quebec, Labrador, and Cape Breton Island south to northeastern Oregon, central Idaho, western Montana, northwestern Wyoming, Manitoba, northern Minnesota, northern Wisconsin, Michigan, southern Ontario, northern New York, northern Vermont, northern New Hampshire, Maine, New Brunswick, and Nova Scotia (A.O.U. *Check-list*). Introduced in Newfoundland.

Subspecies (ex A.O.U. *Check-list*)

D. c. canadensis (Linnaeus): Hudsonian spruce grouse. Resident in east-central British Columbia, central Alberta, central Saskatchewan, southwestern Keewatin, northern Manitoba, northern Ontario, northern Quebec, and Labrador south to central Manitoba, central Ontario, and central Quebec. Introduced into Newfoundland in 1964; now well established and expanding in range.

D. c. franklinii (Douglas): Franklin spruce grouse. Resident from southeastern Alaska, central British Columbia, and west-central Alberta south through the interior of Washington, central Idaho, western Montana, and northwestern Wyoming.

D c. canace (Linnaeus): Canada spruce grouse. Resident from southern Ontario,

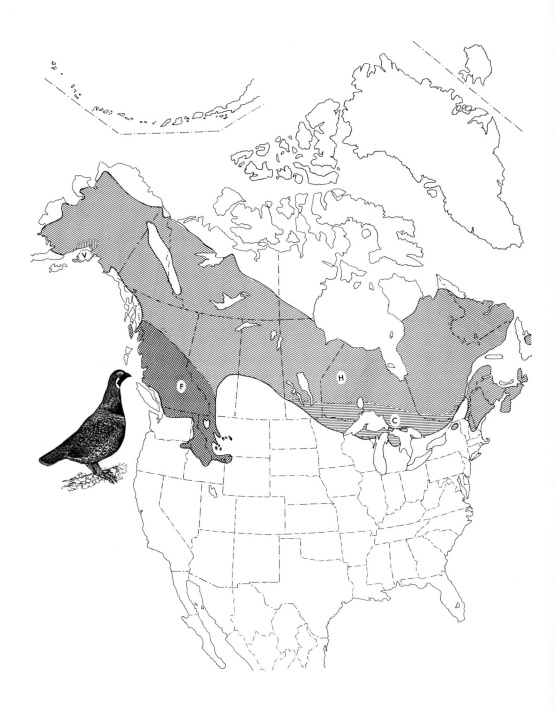

3. Current distribution of Canada (C), Franklin (F), Hudsonian (H), and Valdez (V) races of the spruce grouse.

southern Quebec, New Brunswick, and Cape Breton Island south to northern Minnesota, northern Wisconsin, Michigan, northern New York, northern New Hampshire, northern Vermont, northern and eastern Maine, New Brunswick, and Nova Scotia.

D. c. atratus (Grinnell): Valdez spruce grouse. Resident in the coast region of southern Alaska from Bristol Bay to Cook Inlet, Prince William Sound, and perhaps Kodiak Island (no recent records).

MEASUREMENTS

Folded wing: Males 161–92 mm; females 159–91 mm (males average 2 mm longer).

Tail: Males 107–44 mm; females 94–119 mm (adult males of all races average over 120 mm; females, under 110 mm).

IDENTIFICATION

Adults, 15–17 inches long. The species is associated with coniferous forest throughout its range. The sexes are quite different in coloration, but both have brown or blackish tail feathers that are unbarred and are narrowly tipped with white (*franklinii*) or have a broad pale brownish terminal band. The upper tail coverts are relatively long (extending to about half the length of the exposed tail) and are either broadly tipped with white (in *franklinii*) or tipped more narrowly with grayish white. The under tail coverts of both sexes are likewise black with white tips (males) or barred (females). Feathering extends to the base of the toes. Males are generally marked with gray and black above, with a black throat and a well-defined black breast patch bordered with white-tipped feathers. The abdomen is mostly blackish, tipped with tawny (laterally) to white markings that become more conspicuous toward the tail. The bare skin above the eyes of males is scarlet; no bare skin is present on the neck. The females are extensively barred on the head and underparts with black, gray, and ochraceous buff in varying proportions; the sides are predominantly ochraceous, and the underparts are mostly white.

FIELD MARKS

In the eastern states and provinces spruce grouse are likely to be confused only with the ruffed grouse, from which the spruce grouse can be readily separated by its shorter tail with a lighter tip rather than a darker band toward the tip. The conspicuous black-and-white markings of the underparts of males will distinguish spruce grouse from blue grouse, and the prodominantly white underparts of females will help distinguish them from the generally similar female blue grouse.

AGE AND SEX CRITERIA

Females may be distinguished from adult males by their tawny to whitish throats and breasts, barred with dark brown (these areas are black or black tipped with white in males). Accurate determination of sex in most races is possible from either the breast feathers (males' breast feathers are black tipped with white, females' are barred with brown) or the tail feathers (males have black rectrices, tipped and lightly flecked with brown; females' are black or fuscous, heavily barred with brown). In *franklinii* the breast condition is the same, but the tails of females are barred or flecked with buffy or cinnamon brown, while the males have uniformly black tails or black tails flecked with gray (Zwickel and Martinsen 1967).

Immatures resemble adults of their sex, but the two outer juvenal primaries are more pointed than the others and (at least in *franklinii*) are narrowly marked with buff rather than whitish on the outer webs (Ridgway and Friedmann 1946). Ellison (1968*a*) also reported that the tip of the ninth primary in immature Alaskan spruce grouse is mottled and edged with brown, while in adults it is only narrowly edged with brown.

Juveniles resemble adult females but have white or buffy markings at the tips of the upper wing coverts as well as on their primaries and secondaries. Their tail feathers are dark brown, barred, speckled, and vermiculated with lighter markings (Ridgway and Friedmann 1946). McCourt and Keppie (1975) provided a technique for estimating the age of juveniles up to 75–80 days of age, using the seventh and ninth juvenal primaries and the ninth postjuvenal primary. After 35–40 days of age the sex of juveniles can be determined by color differences in the ventral feathers, the upper tail coverts, and the rectrices, as in adults.

Downy young of this species more closely resemble those of *Lagopus* than those of blue grouse and have a discrete chestnut brown crown patch margined with black. Downy spruce grouse, however, lack the feathered toes of ptarmigans, are more generally rufous dorsally, and have less definite patterning on the back. Their dorsal patterning is evidently almost identical to that of the Siberian sharp-winged grouse (Fjeldså 1977).

DISTRIBUTION AND HABITAT

The overall geographic distribution of the spruce grouse is a transcontinental band largely conforming to that of the boreal coniferous forest (Aldrich 1963). East of the Rocky Mountains, the species' range generally conforms with that of the balsam fir (*Abies balsamea*) and also the black and white spruces (*Picea mariana* and *P. glauca*). In the Rocky and Cascade ranges the bird's southern limit occurs well north of the limits of montane and subalpine coniferous forest, suggesting that other limiting factors are influential in that area. What role competition with blue grouse might play in limiting the western range of the spruce grouse is unknown.

Probably only in the southeastern limits of its range have the populations of spruce grouse undergone serious reduction. In Michigan, where the species was once common to abundant, it had become noticeably reduced as early as 1912 (Ammann 1963*a*). It is now uncommon on the Upper Peninsula and rare in six counties of the Lower Peninsula, and hunting was last permitted in 1914. In Michigan spruce grouse are more often found associated with jack pines (*Pinus banksiana*) than with spruces.

In Minnesota the spruce grouse was fairly abundant in coniferous forests as late as 1880 but almost completely disappeared with the cutting of this forest (Stenlund and Magnus 1951). Roberts (1932) believed the species was doomed to be extirpated from the state "before many years have passed." However, by 1940 the second-growth forest that followed lumbering began to develop an understory of conifers (especially black and white spruce) and jack pine, and the spruce grouse again became common in several northern areas (Stenlund and Magnus 1951). In recent observations reported by these authors, associated cover type was most commonly jack pine, followed in order by black spruce, balsam fir (*Abies balsamea*), and tamarack (*Larix laricina*). Of 79 observations, 44 percent were made in cover that was completely evergreen, and 72 percent were in upland cover rather than in lowland or swamp cover. Shrader (1944) has also noted recent population gains in the spruce grouse in Minnesota after its near extinction.

The situation in Wisconsin for spruce grouse is apparently still extremely unfavorable. Scott (1943, 1947) has documented the historical changes in spruce grouse populations of that state. His map indicates that the species probably originally extended across northern Wisconsin from Polk County to Marinette County, but as of 1942 was limited to about ten counties, with an estimated population of 500 to 800 birds.

Finally, in southern Ontario, spruce grouse have nearly disappeared from the area of Lake Nipissing (Hamerstrom and Hamerstrom 1961).

Lumsden and Weeden (1963) pointed out that in the early 1960s spruce grouse had sufficiently high populations to be hunted in Maine, Montana, Washington, Idaho, Alaska, and all the Canadian provinces and territories except Nova Scotia (where it is protected) and Prince Albert Island (where it has been extirpated). In 1970 Minnesota also allowed the hunting of spruce grouse, and since then a small number have been harvested there annually. However the species has remained protected in Wisconsin, Michigan, New York, Vermont, and New Hampshire, and Maine also has had no open season since the late 1960s. Robinson (1980) estimated the total annual kill of spruce grouse in the late 1970s at between 340,000 and 450,000 birds, or not greatly different from my earlier (Johnsgard 1973) estimate of 440,000 for the late 1960s.

POPULATION DENSITY

Several estimates of population densities in spruce grouse are now available. Robinson (1980) estimated an average spring density of 8.8 males and 9.2 females per square mile on a 2.5 square mile study area in northern Michigan. This is similar to a spring density of about 7 to 10 males per square mile, or 20 to 30 birds per square mile, in the Kenai peninsula of Alaska (Ellison 1968b, 1974). Other estimates include one of 12 to 13 birds per square mile in southern Alberta (McCourt 1969), and about 7 males per square mile in Montana (Stoneberg 1967). Ellison (1975) reported that an area of Alaskan forest that had an August preburn grouse density of 40 adults per square kilometer (104 per square mile) temporarily increased to 97 per square kilometer after an intense forest fire that forced many birds into nearby unburned habitats. Preburn July densities of adults were about 1 male per 25 hectares (62 acres) and 1 female per 16 hectares (39.5 acres). The five-year average of males in spring was 3.9 males per square kilometer, while the summer density of females was 10.4 (Ellison 1974). Boag et al. (1979) found spring densities of grouse to vary from 10.5 to 19.3 per 100 hectares (12.7–23 acres per bird) over a ten-year period.

HABITAT REQUIREMENTS

A careful analysis of all the habitat needs of the spruce grouse remains to be done, but Robinson (1969) provides a valuable analysis of summer habitat needs. By analyzing tree composition, as well as that of shrubs and low herbs, and comparing locations of spruce grouse sightings, he obtained a useful indication of habitat selection. Of 430 trees where spruce grouse were seen, 32 percent were spruces, although spruces (*Picea mariana* and *P. glauca*) made up only 3 percent of the tree cover. On the other hand, jack pines made up 91 percent of the tree composition but accounted for only 51 percent of the sightings. Pure stands of either jack pine or spruce were not used as much as mixed stands. In the shrub layer, young black spruces accounted for a large proportion of spruce grouse sightings than would be expected from their relative abundance, while jack pines again provided a smaller proportion of sightings. Balsam firs were more than seven times as abundant at sighting points as at random sites. As to low vegetation, blueberry (*Vaccinium*), trailing arbutus (*Epigaea*), black spruce, and logs and stumps all were associated with higher than expected sightings of spruce grouse. In general, mature stands of either jack pine or spruce were not favored, apparently because of the lack of concealing cover at ground level. Robinson found that molting males used the same habitat in late summer as did females with broods, and indeed they were often seen accompanying broods. Robinson concluded that populations of spruce grouse in Michigan were highest in areas of boreal forest and jack pine forest. In one such area the grouse selected habitats that had a mixture of spruce and jack pine, had a prevalence of young spruces in the shrub layer, and had a varied ground cover that included blueberries, trailing arbutus, and scattered stumps and logs.

In a comparable study of Alaskan spruce grouse, Ellison (1968*b*) noted that hilltops covered with white spruce, birch (*Betula*), and species of *Populus* were not a preferred habitat, although where an understory of alder was present some brood use and use by molting adults occurred in late summer. Two upland cover types provided preferred habitat. These were a white spruce and birch community with understories of grasses, spiraea, blueberry, and cranberry, and a black spruce community with a blueberry, cranberry, and lichen understory. Grouse sometimes also used dense lowland stands of black spruce, and broods were often found in stunted black spruce borders at the edges of bogs.

MacDonald (1968) noted that the habitat of the Franklin race of spruce grouse in Alberta consisted of lodgepole pine forests with some clumps of aspen and poplar. Somewhat open stands of pines, 20 to 30 feet tall, were evidently preferred areas for display by territorial males.

Winter habitat needs of the spruce grouse, to judge from their known food habits, consist simply of coniferous trees of various species that provide both food and cover.

150

Robinson (1980) observed that winter habitat in Michigan seems to have a somewhat lower proportion of spruce and a higher proportion of jack pine, probably because the more arboreal behavior of the birds in winter makes living in lower branches of spruces relatively less important. Further, the diet then is nearly 100 percent jack pine needles. Estimated tree density in winter habitat was also somewhat greater than in that used during summer, averaging about 336 trees per acre compared with 134 per acre during summer.

Generally, spruce grouse prefer younger stands of pine and spruce, or sparser stands of older pines that include young spruces, where there are trees with living branches that still reach the ground. These branches provide excellent concealment and in sparse stands still allow adequate room for flight. Broods also apparently avoid low and dense cover where there is very limited visibility and thus are often found on relatively high and dry ground (Robinson 1980).

FOOD AND FORAGING BEHAVIOR

The survey by Martin, Zim, and Nelson (1951) indicated that spruce grouse in Canada and the northwest feed extensively on the needles of jack pine, white spruce, and larch and on the leaves and fruit of blueberries. A small fall and winter sample from British Columbia included a diverse array of berry species as well as lodgepole pine and spruce needles.

Jonkel and Greer (1963) analyzed crop contents during September and October in Montana and noted that western larch (*Larix occidentalis*) was an important early fall food but that it declined in use during October. Other important foods were needles of pine, spruce, and juniper, clover leaves, the fruits of huckleberry (*Vaccinium*), snowberry (*Symphoricarpos*), and white mandarin (*Streptopus*), and grasshoppers. A study by Crichton (1963) indicated that before snowfall in central Ontario spruce grouse fed mostly on needles of jack pine and tamarack (*Larix laricina*) and the leaves of blueberries. After the shedding of the tamarack needles and the fall of snow, jack pine needles became almost the sole source of food in spite of a high availability of black spruce.

A seasonal analysis of spruce grouse foods in Alberta by Pendergast and Boag (1970) indicated that during winter lodgepole pine needles (*Pinus contorta*) made up nearly 100 percent of the food. In spring, the ratio of spruce needles to pine needles increased. The summer diet of adults was mostly ground vegetation, such as *Vaccinium* berries. In the fall the adults returned to feeding on conifers, but berries remained important. In contrast, chicks under a week old apparently subsisted entirely on arthropods. Later they began to eat *Vaccinium* berries, but arthropods remained an important source of food through August. By October the juveniles were starting to eat needles, and by November both the adults and the young were using needles as a major food.

A study in Alaska by Ellison (1966) yielded generally similar conclusions, except that the winter diet consisted primarily of needles of both black and white spruce. With spring, spruce was taken in decreasing amounts, and blueberry leaves, buds, and old cranberries were taken, as well as unripe crowberries (*Empetrum*). Summer food consisted largely of berries (crowberry, blueberry, and cranberry), and berry consumption continued into fall, as spruce needles again began to appear in the diet. Ellison reported that the protein content of spruce needles ranged from 5.7 to 6.3 percent, or about the same protein content as has been reported for Douglas fir and white fir.

MOBILITY AND MOVEMENTS

Spring Movements

Among the most detailed information on spruce grouse movements so far available is that provided by Ellison (1968b), who used radio transmitters to obtain movement data. He found that all adult males but only some yearling males established territories and became relatively sedentary. Those birds that were considered territorial remained localized on from 3 to 21 acres of forest during late April and most of May. Immature males considered nonterritorial occupied "activity centers" of from 6 to 16 acres during this time but also made fairly long trips of up to 1.25 miles from these centers, frequently entering the territories of other males in the process, evidently attracted to them by displaying males. Ellison noted that in each year of the study, juvenile males tended to establish territories on the periphery of those held by especially active territorial males, a tendency reminiscent of "hooting groups" of blue grouse, and this has also been noted in ruffed grouse (Gullion 1967a). The actual estimated territorial size of 4 adult males ranged from 4.6 to 8.9 acres and averaged 6.9 acres. After May 21 these same males occupied larger home ranges of from 4.5 to 29.6 acres, averaging 20.1 acres. Considering 4 immature and territorial males as well, the maximum sizes of the home ranges of all 8 males was 61 acres, while 3 of 5 nonterritorial males moved about over areas of 270 to 556 acres. In a later study, Ellison (1973) reported that 9 males remained within their territories of from 1.2 to 8.5 hectares (3–21 acres) during their period of intense display activity, but that 5 males later had appreciably larger molting ranges of from 3.6 to 20 hectares. The minimum home ranges of 3 females during the preincubation period ranged from 6.5 to 21 hectares (16–52 acres); the females apparently spent about 90 percent of their time outside the males' territories. According to Herzog and Boag (1978), at that time of the year females also defend territories that overlap only slightly with male territories and tend to be uniformly spaced around the periphery of the aggregated male territories.

Summer and Fall Movements

Females evidently become considerably more mobile after hatching their broods. Hass (1974) noted that the estimated minimum home range of females averaged 45.6 acres in the prenesting period, compared with 75.5 acres for hens with broods and 23.3 acres for unsuccessful females. The total prenesting and postnesting home ranges for hens that hatched broods averaged 80.2 acres. Similarly, Ellison (1973) observed that most broods utilized a range of about 50 hectares (123.5 acres). Fall and winter home ranges and movements are evidently quite variable. Ellison observed that 7 grouse (3 males, 4 females) had fall home ranges ranging from 5.7 to 159 hectares (14–393 acres), while the winter home ranges of 10 birds varied from 3.2 to 112.7 hectares (7.9–278 acres).

REPRODUCTIVE BEHAVIOR

Territorial Establishment

Ellison (1968b) reported that spruce grouse males established their territories and activity centers in stands of fairly dense spruce or in stands of spruce and birch with trees 40 to 60 feet tall. Stands of trees up to 80 feet tall, with dense undercover, were sometimes used by nonterritorial males but apparently were not suitable for territorial purposes. MacDonald (1968) indicated that pines from 20 to 30 feet tall that were not too closely spaced were preferred display sites. Stoneberg (1967) stated that of 4 males he studied 3 displayed in small openings in dense forest and 1 was in less dense forest. He estimated that the 4 marked males he studied had home ranges of 10 to 15 acres. Two remained in very localized sites during the display period, while one of the others used several display sites within a 25-yard radius and the last moved about extensively and used no specific sites. However, this last bird was the only one that had no female on his territory at the time. MacDonald thought that males have favored display sites within their home ranges but that the latter are too large to have definite boundaries except in areas of contact with adjacent males.

Although various workers have described the territorial pattern of spruce grouse as dispersed, a study by Herzog and Boag (1978) suggests that the territories are actually clumped. Adult male territories, which are uniformly spaced within the aggregate of territories, tend to be fairly small. Yearling males sometimes occupy large areas peripheral to territories of established males and may overlap with the areas of adult females, which also establish territories that tend to be spaced around the periphery of the territories of the adult males. Some young females also establish territories in spring, while others, along with many young males, leave the area.

Territorial Advertisement

Several detailed accounts of strutting behavior are now available. Displays of the Franklin race of spruce grouse have been described by Stoneberg (1967) and MacDonald (1968), and those of the nominate race by a number of writers, including Bishop (in Bendire 1892), Breckenridge (in Roberts 1932), Harper (1958), and Lumsden (1961a). Only a few differences appear to be present in the two forms, as will be noted below.

The basic male advertisement or "strutting" display is performed in a standing position ("upright" of Hjorth 1970). In this posture the tail is cocked at an angle of from about 70° to 90°, exposing the white-tipped under tail coverts that are held out at varying angles, the neck is fairly erect, the wings droop slightly, and the crimson eyecombs are engorged. The throat feathers are lowered to form a slight "beard," and the lateral black neck feathers are lifted, as are the lower white-tipped feathers at the sides of the neck and the upper breast. No bare skin is exposed, but the pattern of feather erection is much like that of the male blue grouse. Lumsden (1961a) has noted that the esophagus is evidently slightly inflated as well, but no hooting sound is normally heard. However, an extremely low-pitched sound (ca. 85–90 Hz) may be produced by male spruce grouse (Stoneberg 1967; Greenewalt 1968). Stoneberg heard series ranging from one to four such notes, and I have heard similar sounds coming from boxes containing several recently trapped males and females. MacDonald (1968) likewise heard hooting sounds apparently produced by a male when it rushed toward a female. However, Hjorth (1970) questioned on anatomical grounds whether male spruce grouse can produce such low-pitched sounds, believing that reports of such calling were the result of confusion with blue grouse hooting.

When in the strutting posture, the male usually walks forward with deliberate paces, typically spreading the rectrices on the opposite side as he raises each foot, making the spread tail asymmetrical ("display walking cum tail swaying" of Hjorth 1970). This lateral tail movement, which produces a soft rustling sound, may also occur when the bird is not walking, as has been noted by Stoneberg as well as by me. A similar display is tail fanning, in which the rectrices of both sides are quickly fanned and shut again. This also produces a rustling sound and may occur during walking or when the bird is standing still, often alternating with tail flicking. On one occasion I saw a male performing tail fanning before a female as it uttered a series of low hissing notes that started slowly and gradually sped up, with a fan of the tail accompanying each note. Lumsden (1961a) observed this when a male saw his reflection in a mirror. Michael Flieg (pers. comm.) informed me that a similar tail fanning during calling is typical of the capercaillie.

When approaching a female in the strutting posture, the male may perform several displays that have been given different names by various writers. One is a vertical head bobbing that may grade into or alternate with ground pecking (Harper 1958; Lumsden 1961a; Stoneberg 1967; MacDonald 1968). During the pecking movements the male faces the female and often tilts his head to the side, thus exposing both combs to her view. Wing flicking may likewise occur at this time (Stoneberg); Harper also noticed what appeared to be wing beating movements suggestive of the ruffed grouse's drumming.

Two other major male displays occur when a male comes close to a female. These are the "neck jerk" display described by Lumsden, which MacDonald preferred to call the "squatting" display; and the "tail flick" described by Lumsden, but which Stoneberg calls the "head-on rush."

The tail flicking, or head-on rush, display (called the "rush cum momentary tail fanning" by Hjorth 1970) is apparently homologous to the short forward rush of the male blue grouse. It begins with the male's making several short, rapid steps toward the female, stopping a few inches away, partially lowering his head, and suddenly snapping his tail open with a swishing sound. The wings are simultaneously lowered to the ground, and a hissing vocalization is uttered, followed by a high-pitched squeak. The wings are then withdrawn leaving the alulae exposed, the tail is closed, and the head is tipped downward with the neck still extended diagonally. In this rigid posture the tail is fanned a second time and is held open longer. During this display the male is usually oriented so that his head faces the female, exposing to her view the visual effect of the eyecombs, fanned tail, and contrasting breast coloration. In the Franklin race the white-tipped upper tail coverts are made conspicuous by the tail movements, but they are not evident in the nominate race.

154

18. Male displays of the Franklin spruce grouse (after MacDonald 1968). Shown are wing clapping (1–2) and stages in the tail swishing display (3–6).

MacDonald noted that during this display (which he described under the general tail swishing display) a single, soft hooting noise could be heard at very close range.

The squatting display is performed by the male as a possible precopulatory signal, according to Lunsden, and MacDonald agrees with this interpretation but notes that it is sometimes omitted from the sequence. As the male approaches the female, the head-on rushes (or arcing rushes, since MacDonald indicates that the male may move in arcs in front of the hen) increase in frequency until he is quite close to her. After watching her intently for several seconds, he sinks to the ground in a squatting position, with neck stretched, head nearly parallel to the ground, and tail held vertical and partially spread, while the wings are slightly spread and lowered. I have observed this display only once, and to me it closely resembled the ''nuptial bow'' of pinnated grouse, which serves as a precopulatory display in that species. Hjorth (1967) illustrates the posture and agrees that it is homologous to the nuptial bow of prairie grouse. He believes it is stimulated when the male's displays elicit neither attack nor pairing behavior.

Squatting as described by MacDonald probably does not correspond to the typical head jerk described by Lumsden and Stoneberg, since MacDonald mentions no actual head jerking movements I likewise noted none during one observation of the squatting display. Lumsden mentions seeing repeated sudden upward movements of the head, first to one side then to the other, as well as occasional circular head movements. With each upward movement the tail was fanned open and again shut, producing the usual rustling sound. Stoneberg noted two types of head jerking movements, one of which was a rapid tossing of the head from on side to the other for up to 3 seconds, repeated after a pause, with the tail kept vertical and the head near the ground. A slower type of head jerking was associated with strutting, when the bird would stop, facing the female, and jerk his head from one side to the other while fanning or flicking his tail.

Aggressive male displays of the spruce grouse consist of at least two postures. MacDonald reports that when two males meet at a distance the resident territorial male sleeks his plumage, raises his tail, and flashes his lateral rectrices and upper tail coverts while uttering a series of gutteral notes. These notes no doubt correspond to the calls I heard from a male when I interrupted his strutting, which Lumsden describes as harsh hissing sounds. Stoneberg describes the rapid notes as ''throaty kuks.'' The male then runs toward the opponent with head low, neck extended, and tail down (Lumsden's ''head and tail down'' display posture), with the wings held slightly away from the flanks. MacDonald found that such behavior was enough to cause a trespasser to fly away or at least to fly into a tree. A mounted male or a mirror may elicit actual attack behavior. Stoneberg found that by placing bright red pieces of felt on a male skin he was able to elicit strong attack behavior. The male approached the skin with plumage sleeked except for the chin feathers, paused, then leaped at the skin, beating his wings and pecking at the head and breast. In a second attack the male succeeded in removing the combs as well as the feathers and skin from the neck and upper breast.

So far, only one description of copulatory behavior exists. Harju (1971) observed that strutting, tail swishing, and tail flicking all occurred during a precopulatory situation.

Additionally, the male uttered a "challenge call" that was coordinated with his neck snapping and tail flicking. As the male approached the female he performed foot stamping and repeated lateral head jerking. While thus jerking his neck, he swished his tail and flicked his wings laterally, after which he moved around behind the female and mounted. Postcopulatory strutting was vigorous and resembled the precopulatory displays.

Aerial Displays

In contrast to the terrestrial displays of spruce grouse, some population variation may occur in the aerial displays of males. Lumsden (1961*a*) has summarized the observations of aerial display by the nominate subspecies, which apparently includes several variations. One of these is a short, vertical flight from a few to about 14 feet in the air, drumming on suspended wings, and fluttering back to the ground. This behavior is closest to the typical flutter-jump of prairie grouse. More commonly, however, the male flies either vertically upward or horizontally toward a tree perch, checks his flight, and either lands on the perch or drops back to earth. If he lands on the elevated perch he may stay there for varying lengths of time; Lumsden reports periods as short as 10 seconds and as long as 4 minutes. The flight back down is always performed in the same manner, by dropping steeply downward to about 4 to 6 feet from the ground, then swinging the body into a nearly vertical position and descending on strongly beating wings. Although the drumming sound produced by the wing beats can be heard as far as 200 yards away, neither Lumsden nor Ellison (1968*b*) reported any wing clapping sounds by males of this race, nor have earlier observers. Apparently no vocal calls are uttered during the flight.

Descriptions of the aerial display flights of the Franklin race are somewhat at variance with this general situation. Stoneberg (1967) states that the downward phase of the flight is as Lumsden described except that during the final drop to the ground two loud sounds are produced, apparently by clapping the wings together. Once Stoneberg heard wing clapping before the bird landed in a tree, and in 2 of 45 cases only one clapping sound was produced rather than two. The wing clapping display was most commonly heard near sunrise and sunset but often could be heard during the middle of the day as well. Stoneberg believed that cool temperatures favored the display.

MacDonald's observations of wing clapping are unusually complete, and he regarded the display as an advertisement of the location of territorial males. He noted that the wing clapping flight was never started from the ground but always began from some elevated site. Flying out from a branch some 10 to 20 feet high, the male moves on shallow wing beats through the trees, with tail spread and tail coverts conspicuous. On reaching the edge of a clearing, he rises slightly, makes a deep wing stroke, and brings the wings together above the back, producing a loud cracking sound. A second clap follows as the bird drops vertically toward the ground. The male soon selects another branch overhead and begins the sequence again. MacDonald noted that a resident male wing clapped in the presence of an intruder and, after he had driven it away, began a sequence of vigorous displays and wing clapping.

According to MacDonald, the vertical flight to a perch may be followed by display on the perch before launching into the wing clapping display. He reported that after alighting on a branch and before the wing clapping flight, the male may perform either or both of two displays. These include a short rush along the branch followed by a spreading of the wings and tail, closing them, and again spreading the tail, apparently a variant of the tail-flicking display. A second display consists of three or four shallow wingstrokes, like the drumming of a ruffed grouse, producing a similar thumping sound.

Vocal Signals

Two distinct vocal signals of males have been mentioned; one of these is the low-pitched *hoot* of a male in a sexual situation. These calls may be uttered as single notes or may occur in a series of notes roughly half a second apart (Greenewalt 1968). They are notable for their extremely low frequency, less than 100 Hz.

Males also utter a series of rather gutteral notes in situations of aggression. When I placed an adult and an immature male in a box together, both birds produced such calls. These usually consisted of two preliminary low, growling *kwerr* notes, followed by from two to eight more rapidly repeated *kut* notes. Occasionally the two types of calls were uttered independently of one another. The younger male gave the calls at a noticeably higher pitch than the adult male.

Caged female spruce grouse produced at least three different types of notes. The loudest and highest pitched was a repeated squealing or whining *keee'rrr* call that resembled the distress call of various quail species. Females also uttered a softer series of *pit, pit, pit* notes when disturbed and a fairly low-pitched gutteral *kwerr,* which presumably correspond to the two types of agonistic male notes mentioned above. When in a tree looking down on a human or other potential enemy, females utter a series of clucking sounds that quickly reveal their presence. Bent (1932) described these as *kruk, kruk, kruk* sounds, and a *krrrruk* that no doubt corresponds to the *kwerr* note mentioned above. In-flight alarm notes have not been reported.

Nesting and Brooding Behavior

There is no evidence that the male spruce grouse participates in nest or brood defense, although males may often be seen with females and well-grown broods in early fall. I observed this in southern Ontario during September 1970, when at least four males were seen associated with females and broods. However, the males made no attempt to defend the broods; instead they simply appeared intent on displaying to the adult females.

Nests of the spruce grouse are usually in well-concealed locations, often under low branches, in brush, or in deep moss in or near spruce thickets. Ellison (Alaska Department of Fish and Game, *Game Bird Reports*, vols. 7–9, 1966–68) reported on 19 nest locations, 14 of which were in open, mature white spruce, birch, or spruce-birch-alder ecotones, while 2 were in open black spruce, 2 were in moderately dense black spruce, and 1 was in a mixture of alder and grass. Of 67 nests found by Keppie and Hertzog (1978), 55 were close to the base of a tree or between two or three closely adjoining trees. The most frequent site (55 percent) was the base of a single pine tree.

Clutch sizes in the spruce grouse are surprisingly small. Ellison (1974) reported that 26

Alaskan nests averaged 7.54 eggs, and Tufts (1961) found an average clutch size of 5.8 eggs in 39 Nova Scotia nests. Keppie (1975) reported an average of 4.9 eggs among 21 Alberta nests, Robinson (1980) an average of 5.7 eggs in 7 Michigan nests, and Hass (1974) an average of 4.7 eggs in 7 Minnesota nests. Apparently nearly all females attempt to nest. Ellison (1972) estimated that 90 percent of the females in his study area nested and that 80 percent raised broods. Ellison reported finding a single renesting effort following abandonment of a complete clutch. Keppie (1975) also found a case of renesting in a female that had lost its first nest (which had contained a single egg).

Although the incubation period was reported by Pendergast and Boag (1971a) to be only 21 days for a captive bird, more recent observations (McCourt et al. 1973) suggest that it may be somewhat longer (23.5 days) in the wild. Ellison (1974) estimated hatching time at between 22 and 25 days for 8 nests. Evidently female spruce grouse are extremely attentive incubators, probably spending more than 90 percent of the daylight hours on the nest and taking only brief periods for feeding (McCourt et al. 1973; Herzog 1978).

Ellison (1974) estimated hatching success at 81 percent (21 of 26 nests), and he also found a 91 percent hatchability among the eggs of the successful nests. Boag et al. (1979) estimated that between 32 and 44 percent of the females they studied over a four-year period hatched clutches, and they further believed that only 60 to 70 percent of the yearling females attempted to breed. They estimated an average brood size of 3.15 young at the time of independence, compared with a potential brood size of 4.9 during incubation, or a reduction of about 36 percent. Redmond et al. (1982) recently reported a 48 percent nesting success (100 nests) with adults more successful than yearling females, and females of *canace* more successful than those of *franklini*.

Ellison (1974) found no evidence of high juvenile mortality rates between hatching and August, but Haas (1974) observed summer juvenile mortality rates of between 36 and 64 percent, averaging 55 percent over three years, or about the same as reported for several other species of grouse and ptarmigans.

It is generally believed that further losses of young during fall and winter are primarily responsible for changes in grouse breeding populations the following year. However, Ellison (1974) suggested that, if juveniles do indeed suffer higher losses than adults, these must occur during late winter and spring rather than during fall and early winter. Keppie (1979) estimated that autumn and winter losses in juveniles averaged only 12 percent in his study and that spring dispersal was an important factor in changing the numbers of potential recruits. Similarly, Boag et al. (1979) suggest that spring densities are largely regulated by territorial adults' forcing yearling birds to disperse. They believed that most adult mortality occurs during spring and summer and that the number of juveniles surviving to independence is inversely related to the density of adults and yearling birds present, which tends to dampen population fluctuations.

EVOLUTIONARY RELATIONSHIPS

Short's recommendation (1967) that *Canachites* be merged with *Dendragapus* appears to me to be fully warranted, for reasons he outlined. It seems that the nearest living relative to

159

the spruce grouse is *Dendragapus* (*"Falcipennis"*) *falcipennis*, the Siberian sharp-winged grouse, since it not only occupies a very similar habitat but evidently has nearly identical courtship displays (Short 1967; Hjorth 1970). Some similarities in courtship characteristics between the spruce grouse and the blue grouse are also evident, including the short run toward the female followed by a single-note call, the very low-pitched hooting sounds, the tail fanning displays, and the drumming flight behavior. Some interesting features of the male spruce grouse display also suggest affinities with the capercaillie. These include the general posture, erection of the chin feathers to form a "beard," and calling with simultaneous tail fanning. The general plumage appearance of both sexes is also very similar in these two species and the Siberian sharp-winged grouse. Similarities between the display of the capercaillie and the Siberian sharp-winged grouse have also been noted (Kaplanov, in Dementiev and Gladkov 1967).

It seems probable that the evolutionary origin of the spruce grouse was in eastern Asia, where separation into two populations gave rise to the Siberian sharp-winged grouse and the North American spruce grouse, the latter gradually moving southward and eastward through boreal forest and western coniferous forests. Contacts in the west with early blue grouse stock may have provided the selective pressure favoring the evolution of conspicuous upper tail covert patterning and wing clapping during aerial display as sources of reinforcement of isolating mechanism differences between these two related types. There is apparently no fossil record of either *"Canachites"* or *"Falcipennis"* except for a late Pleistocene specimen from Virginia, whereas typical *Dendragapus* fossil remains are known from several localities in the western states (Jehl 1969).

Sharp-winged Grouse

Dendragapus falcipennis (Hartlaub) 1855

Other Vernacular Names

Siberian spruce grouse, sicklewing grouse, spiny-winged grouse, Ochotsian grouse; Sichelhuhn (German); tétras de Sibérie (French); dikush (Russian).

Range

Resident in the USSR in the eastern parts of Transbaikalia (from the Olekma basin in the middle reaches of the Shilka), the southern parts of Yakutia (upper and middle reaches of the Aldan and the basins of its tributaries), the Amur basin (the upper reaches of the rivers Oldaya and Zei, the southern part of the Selemdyhi basin, and the upper parts of the Bureya and Kur basins), the shore of the Gulf of Okhotsk, the Sikhote Alin area (from the valley of the lower parts of the Amur and the Gulf of De Kastri in the south to the heights of Great Ussurka and Dalnegorsk) and on Sakhalin. Currently declining and listed in the *Red Data Book for the USSR* (Flint 1978). Present occurrence within the Chinese boundaries is doubtful. See distribution map 9.

Subspecies

None recognized.

MEASUREMENTS

Folded wing: Males 185–200 mm, average of 5, 191.6 mm; females 175–95 mm, average of 5, 191.6 mm.

Tail: Males 105–37 mm, average of 3, 117.3 mm; females 88–95 mm, average of 4, 92.0 mm.

IDENTIFICATION

Length 15–17 inches. Adult males are generally dark chestnut dorsally, the feathers strewn with white shaft streaks and tips and barred with pale ochre and russet, producing a variegated pattern on the upperparts. The nape and crown are black, with faint grayish bars, becoming lighter at the base of the beak. The anterior back is similar, but the feathers are marked with light transverse edging, and the rest of the upperparts are lighter, with barring and longitudinal shaft streaks, especially on the scapulars and rump. The sides of the head and neck are spotted with white and light ochre, and the throat and upper breast are blackish with a whitish line extending from the eyes backward and downward to the junction of the throat and upper breast, enclosing the black throat. The sides, flanks, and underparts are barred and spotted with black and white. The remiges are blackish brown. The rectrices are mostly black, with white tips, often barred or vermiculated basally, but the central pair are brown and vermiculated. The under tail coverts are also tipped with white. The bill is black, and there is a reddish area of skin above the eye. Females are much lighter than males, because of their extensive light spots and streakings both above and below, and have strong russet tones, especially on the head and neck. The front of the neck is black or barred with black and white; all of the upperparts are more russet than in males, and the breast is darkly barred with ochre (Dementiev and Gladkov 1967; Short 1967).

FIELD MARKS

This species is found in coniferous forest, especially areas rich in berries, and is reported to be remarkably tame. No similar grouse is present within its range, and it apparently closely resembles the North American spruce grouse in posture, behavior, and plumage. The birds frequently perch in trees, and when alarmed the female utters rather quiet, coarse notes similar to those of the female black grouse (Dementiev and Gladkov 1967).

AGE AND SEX CRITERIA

Adult males can easily be separated from females by their black throats and generally darker coloration.

First-winter males are reported to have their upperparts spotted with russet brown and have light ochre shaft streaks (Dementiev and Gladkov 1967). The condition of the outer two primaries has not been described but may be helpful in estimating the age of females.

Juveniles are undescribed, but probably are femalelike, with extensive shaft streaking, as in *canadensis*.

Downy young are apparently almost identical to those of *canadensis* and have a black-bordered crown cap, upperparts that are yellow or buff and brown, and yellow underparts (Short 1967; Fjeldså 1977). A description of the downy plumage has been provided by Kirpichev (1972) but does not provide any basis for distinction from *canadensis*. Jon Fjeldså (in litt.) studied one specimen, but was not able to establish any definite diagnostic traits on the basis of this single individual. A recent photo of a chick by Yuri Pukinski is provided in plate 18.

DISTRIBUTION AND HABITAT

Although this species has a fairly small geographic range, its habitats have not been specifically defined. On the western slopes of the Sikhote Alin range, it occurs 700 to 1,500 meters above sea level, occurring at higher elevations in summer than in winter. In general it is limited to the zone of east Asian taiga forest, or ''Okhotsk vegetation,'' and its range is coextensive with this vegetation type (Dementiev and Gladkov 1967). This particular vegetation is centered on the Sea of Okhotsk and the lower reaches of the Amur River and is characterized by Yeddo spruce (*Picea jezoensis*), Khingan fir (*Abies nephrolepis*), and some Dahurian larch (*Larix dahurica*). In the Amur basin the sharp-winged grouse is especially characteristic of Yeddo spruce forests (Berg 1950). In general the species occurs below 1,600 meters elevation, in groves of open and mature coniferous forest having well-developed ground cover.

The breeding habitat consists of spruce-fir forests, spruce woods, and stands of larches, mingled with a combination of moist and shady areas and clearings, and with a dense carpet of mosses rich in berry-bearing plants such as blueberries (*Vaccinium*), crowberries (*Empetrum*), cloudberries (*Rubus chamaemorus*), wild currants (*Ribes*) and the like. On the Sikhote Alin area the nesting habitats include alpine tundra above timberline, montane spruce coppices, burned-over areas covered with aspen and birch regrowth, and craggy areas dominated by oak woods mixed with Dahurian birch (*Betula davurica*). Such areas are rich in bilberries, elfin woodlands of stonebirch, stunted forests of Japanese stone pine (*Pinus pumila*), rhododendrons (*Rhododendron chrysanthum*), bergenia (*Bergenia*) and other high mountain vegetation (Dementiev and Gladkov 1967).

In the lower Amur basin the birds inhabit forests, predominantly those of various firs (''remote''and ''silver'') on the slopes of mountain ridges, and more rarely occur in groves of ''mountain'' fir and larch. Outside the breeding season they occur on treeless knolls, in thickets of cedars and alders, in mountain tundra, and in sphagnum swamps (Nechayev, in Kirikov and Shubinkova 1968).

POPULATION DENSITY

Few specific estimates of population density are available. Between 1960 and 1965 the population in the Slemdyhi basin averaged 0.25 birds per square kilometer. In a survey along 121 kilometers of Sakhalin's River Rukutan, only 3 birds were observed, and likewise in the Vazi River valley, of similar length, only a single bird was seen during winter (Vshivtseu and Vononor, in Kirikov and Shubinkova 1968). In many areas the species is declining because of forest fires and cutting of the dense coniferous forest in which it is most abundant (Flint 1978).

HABITAT REQUIREMENTS

Nothing specific has been reported, but apparently the species is associated with coniferous trees throughout the year, depending on the needles during winter. In winter the birds spend much of their time perched in the tops of conifers and feed primarily on needles of spruce and ''silver fir'' (Flint 1978). The use of snow holes during winter has not been reported. A supply of berry-bearing plants during the summer and fall is probably also an important habitat component.

FOOD AND FORAGING BEHAVIOR

As already noted, these birds evidently rely heavily on the needles of conifers throughout the year. In autumn they also eat bilberries (*Vaccinium*), crowberries (*Empetrum*), and other berries (Nechayev, in Kirikov and Shubinkova 1968). Eleven samples obtained in late August contained almost exclusively vegetable material. This consisted mainly of larch needles (*Larix dahurica*), followed by whortleberries (*Rubus saxatalis*) and spikelets of sedges (*Carex lasicarpa*). Two of the stomachs included insect remains (grasshopper and ant), and all contained considerable grit. This material ranged in weight from 2.5 to 13 grams per bird, usually between 8 and 12 grams. A male shot on the upper Iman in late June contained mostly the leaves of foxberry, as well as Khingan fir (*Abies nephrolepis*) needles, while the stomach contents included larch needles, beetle remnants, one caterpillar, and an ant. Another obtained in August had been eating primarily larch needles, together with some cloudberries and cranberries. Three birds obtained during summer in the Sikhote Alin areas had mainly eaten foxberries and some needles of Khingan fir. Winter specimens from the lower Amur basin contained fir needles exclusively; in one case the crop alone contained 56 grams of this material (Dementiev and Gladkov 1967).

Juvenile grouse feed primarily on insects, gradually changing to berries as subadults. However, young birds may begin to eat fir needles as early as late summer, indicating the importance of that food source (Dementiev and Gladkov 1967).

MOBILITY AND MOVEMENTS

There is no specific information on mobility and movements. The birds in the lower Amur region apparently move in autumn from the coniferous forests to the berry fields of the

164

forest zone and move back to the heavier forests during the winter (Nechayev, in Kirikov and Shubinkova 1968). Likewise, on the upper Tetyukhe and Iodzykh they descend lower in the mountains during fall to concentrate in berry-rich areas (Dementiev and Gladkov 1967). However, the birds are otherwise quite sedentary, and it is unlikely that real migratory movements occur in any part of their range. Even in the fairly mountainous parts of its range, such as in Sakhalin, the birds are nonmigratory. Yamashina and Yamada (1935) observed flocking and groups of 5 to 10 individuals in autumn and winter but saw no indication of migration. However, the birds do move down into river valleys and the lower slopes of mountains of Sakhalin during winter (Gizenko, cited by Hjorth 1970).

REPRODUCTIVE BEHAVIOR

Territorial Advertisement

Compared with material on the North American spruce grouse, observations on this species are extremely limited, but in general they suggest that it is very similar to the North American species in its behavior. An early observation by Kaplanov (cited by Dementiev and Gladkov 1967) is fairly complete: ''The male perches in a tree with his plumage distended and struck the attitude of a courting black capercaillie, elevating his tail fanwise and erecting his head (though this was held up less steeply than in the courting capercaillie), all the while perching soundlessly except for a slight rustling of the tail feathers on being raised or fanned; the cock then dropped to the ground, held out his feathers, issued a prolonged vibrating sound resembling 'u-u-u-r-r-r,' and once more rose some 20–30 cm in the air while uttering a twice-repeated clicking sound resembling that of a capercaillie. This procedure was repeated a number of times in succession at equal intervals of some 2–3 minutes. First came the protracted voice, then a hop with the concurrent double click. Not far off in a tree perched a female.'' This observation was in mid-May, in the Sikhote Alin mountains, at 1,500 meters above sea level and on a burned-over alpine tundra habitat.

Similar observations by Middendorf (cited by Dementiev and Gladkov 1967) were made of a male strutting ''like a turkey'' in the middle of a small forest glade. He ''batted'' his eyes, clicked, and occasionally whinnied. According to Take-Tsukaza (cited by Dementiev and Gladkov 1967), the male displays on a stump, extending his wings and uttering a sort of *loo* or *kuu*. Likewise, Yamashina and Yamada (1934, 1935) photographed a male displaying on a large larch stump, with his head and neck feathers erected and fanning and elevating his tail. After flapping his wings strongly, the male jumped up in the air about a meter and uttered a loud screaming note. According to Flint (1978), the display call can be heard from the second half of April until the end of May, and eggs are laid during the second half of May.

All the observers reported that tail raising and tail fanning are important display elements, as well as feather erection, especially in the head and neck region. However, descriptions of the vocalization vary greatly, from soft cooing notes to clicking sounds and

19. Comparison of male displays and plumage features of the sharp-winged grouse and spruce grouse. Flutter-jumping and crouched forward postures (1), and display walking (2) of sharp-winged grouse are compared with normal postures (3) of male and female. Corresponding display posture of spruce grouse, together with outline of eighth primary and entire male wing are also shown. After various sources.

even loud screaming. Some of these discrepancies have been clarified by the recent description of Potapov (1969), who observed a male displaying for about an hour in a spruce-fir forest on the upper Tym River. The male moved in circles, sometimes pausing to assume a thin-necked upright posture (terminology of Hjorth 1970). The same display sequence was repeated monotonously for an entire hour and began with an accelerated walk, the feathers fluffed and the neck held at an oblique angle. The tail was raised and the wings were lowered, though they did not drag on the ground. The tail was occasionally flicked open and shut, producing a swishing noise. Then the male stopped and assumed a wide-necked upright posture with his wings dragging. At the same time he uttered a ''whistling-cooing'' note that rose in pitch and lasted nearly 3 seconds but was not audible beyond about 40 meters. Shortly after uttering this call the male jumped into the air twice in succession, each time rising about 30 centimeters and advancing a short distance. At the start of the first leap the male produced a clicking or sometimes a double clicking sound, and during the secnd leap he uttered two double clicks. These sounds were relatively loud and could be heard for about 100 meters. As the male landed the second time, he crouched forward for a moment, with the tail raised nearly to the vertical and the wings partly extended (translation of Hjorth 1970).

Nesting and Brooding Behavior

According to Dementiev and Gladkov (1967), the breeding season occurs between May and July. The nests are on the ground at the base of trees. One nest found in southern Sakhalin contained 8 incubated eggs on June 8. It was in dense spruce vegetation and in moss cover surrounded by *Vaccinium* bushes. The nest was a simple depression, lined with twiglets of conifer and *V. chamissois*. Flint (1978) stated that the clutch size ranges from 8 to 15 eggs.

The incubation period is still unreported, but females with small chicks have been observed in late June in the Sikhote Alin range, and a brood of 8 flying juveniles was seen at Taba Bay in early August. There is no indication that the male participates in brood care, but by late August males may again be attending females and performing courtship display (Dementiev and Gladkov 1967).

EVOLUTIONARY RELATIONSHIPS

In his review of grouse genera, Short (1967) considered the criteria for the genus *Falcipennis*, which was erected solely on the basis of the attenuated outer primaries. Short hypothesized that this might produce a rattling or whistling sound during courtship display but believed it was certainly not a generically significant character. He also noted the close similarities of courtship displays between the sharp-winged grouse and the North American spruce grouse, as well as resemblances in adult plumage, in eggs, and in downy young. Short considered *falcipennis* most like the presumed ancestral *Dendragapus* species, though he imagined the genus to have evolved in North America. He tentatively concluded that *falcipennis* should be considered a full species closely related to *canadensis* but that future studies might indicate that the two forms should be regarded as subspecies.

In another review, Stepanyan (1962) made a close comparison of the adult plumage patterns of *falcipennis* and *canadensis*. He too agreed that the two species should be ongeneric, but he thought they should be in the genus *Falcipennis* and included between *Tetrao* and *Tetrastes* in the list of Palaearctic grouse species. He did not, however, take *Dendragapus obscurus* into account in his discussion, and clearly this form is more closely related to the two "*Canachites*" species than is either of the genera just mentioned.

I support Short's position on this question and believe that *falcipennis* and *canadensis* represent a superspecies having closest affinities with *obscurus* and are best included in the genus *Dendragapus*. The name *Canachites* (which has priority over *Falcipennis*) may be retained for subgeneric distinction.

Willow Ptarmigan

Lagopus lagopus (Linnaeus) 1758

Other Vernacular Names

Alaskan ptarmigan, arctic grouse, red grouse (Scotland form), British grouse, white grouse, white-shafted ptarmigan, willow grouse, willow partridge; Moorschneehuhn (German); lagopède des saules (French); belaya kuropatka (Russian).

Range

Circumpoler. Breeds in North America from northern Alaska, Banks Island, Melville Island, Victoria Island, Boothia Peninsula, Southampton Island, Baffin Island, and central Greenland south to the Alaska Peninsula, southeastern Alaska, central British Columbia, Alberta, Saskatchewan, Manitoba, central Ontario, central Quebec, and Newfoundland. Also breeds in the British Isles and across northern Eurasia, south to about 55° N latitude in the west, about 49° in the Kirghiz steppes, about 47° in Mongolia and to Amurland, northern Ussuriland, and Sakhalin. Some populations make limited southward migrations during winter.

Subspecies

L. l. albus (Gmelin): Keewatin willow ptarmigan. Breeds from northern Yukon, northwestern and central Mackenzie, northeastern Manitoba, northern Ontario, and south-central Quebec south to central British Columbia, northern Alberta and northern Saskatchewan, and the Gulf of Saint Lawrence in Quebec. Wanders farther south in winter.

L. l. alascensis Swarth: Alaska willow ptarmigan. Breeds from northern Alaska south through most of Alaska. Winters in southern part of breeding range. May extend east to Old Crow, Yukon (*Arctic* 23:240–53).

L. l. muriei Gabrielson and Lincoln: Aleutian willow ptarmigan. Resident in the Aleutian Islands from Atka to Unimak, the Shumagin Islands, and Kodiak. Now possibly extending no farther west than Unimak.

L. l. alexandrae Grinnell: Alexander willow ptarmigan. Resident on the Alaska Peninsula south to northwestern British Columbia.

L. l. ungavus Riley: Ungava willow ptarmigan. Resident in northern Quebec and northern Labrador south to central Ungava.

L. l. leucopterus Taverner: Baffin Island willow ptarmigan. Resident from southern Banks Island and adjacent mainland to Southampton Island and southern Baffin Island; wanders farther south in winter.

L. l. alleni Stejneger: Newfoundland willow ptarmigan. Resident in Newfoundland.

L. l. scoticus (Latham): Scottish willow ptarmigan ("red grouse"). Resident in Great Britain from Wales to Yorkshire and north to Scotland, Ireland, the Orkneys, and the Hebrides. Introduced in Exmoor and eastern Belgium.

L. l. variegatus Salomonsen: Norwegian willow ptarmigan. Resident on coastal Norway, where it is restricted to the islands off Trondheim Fjord.

L. l. lagopus Linnaeus: Scandinavian willow ptarmigan. Breeds in Norway, Sweden, Finland, and northern Russia. Probably subject to some winter movements.

L. l. rossicus Serebrovsky: Russian willow ptarmigan. Breeds in Russia south of *lagopus* from the vicinity of Leningrad east to the Kama basin and south to the Baltic countries. Probably subject to some winter movements.

L. l. koreni Thayer and Bangs: Siberian willow ptarmigan. Breeds in Siberia from the Urals and Yamal Peninsula east to the Chukotski Peninsula, and south to about 56° N latitude in the west and to the northern coast of the Sea of Okhotsk, Anadyrland, Koryakland south to Kamchatka, and the northern Kuriles. Probably undergoes some winter movements.

L. l. maior Lorenz: Steppe willow ptarmigan. Resident in the steppes of southwestern Siberia and northern Kazakhstan.

L. l. brevirostris Hesse: Oriental willow ptarmigan. Resident in the Altai and the Sayans south to Tannu Tuva and extreme western Mongolia.

L. l. kozlowae Portenko: Mongolian willow ptarmigan. Resident in northwestern Mongolia.

L. l. okadai Momiyama: Baikal willow ptarmigan. Resident in eastern Siberia from the mountains west of Lake Baikal east to the Sea of Okhotsk, south to the mountains south of Lake Baikal, eastern Mongolia, eastern Transbaikalia, the mountains of northern Manchuria (probably), Amurland, northern Ussuriland, and Sakhalin.

MEASUREMENTS

Folded wing: Adult males 182–235 mm; adult females 168–222 mm (males average 10–15 mm longer than females).

170

Tail: Adult males 108–35 mm; adult females 94–139 mm (males average 10–20 mm longer than females).

IDENTIFICATION

Adults, 14–17 inches long. All ptarmigans differ from other grouse in that (except during molt) their feet are feathered to the tips (winter) or bases of their toes (midsummer) and their upper tail coverts extend to the tips of their tails. The primaries and secondaries of the North American and continental Eurasian populations of this species are white in adults throughout the year, and in winter most or all of the feathers are white except for the dark tail. The British race (*scoticus*), however, has brown flight feathers and never develops white plumage on the upperparts. Males of all races have scarlet "combs" above the eyes (most conspicuous in spring) and during spring and summer are extensively rusty hazel to chestnut with darker barring above. The tail feathers are dark brown, tipped with white except for the central pair, which resemble the upper tail coverts. In summer females are heavily barred with dark brown and ochre. In autumn the male is usually considerably lighter, and the underparts are heavily barred with dark brown and ochraceous markings, lacking the fine vermiculated pattern found in males of the other ptarmigans at this season. The female in autumn is similar to the male but usually is more grayish above and more extensively white below. In winter both sexes are often nearly or entirely white except for the tail feathers, of which all but the central pair are dark brownish black. In addition, the shafts of the primaries are typically dusky, and the crown feathers of males are blackish at their bases. In first-winter males and females the bases of these feathers are grayish. The British "red grouse" is never white in winter, and in all willow ptarmigans the adult plumage is consistently more rufous than the corresponding plumages of rock ptarmigans.

FIELD MARKS

The dark tail of both sexes at all seasons separates the willow ptarmigan from the white-tailed ptarmigan but not from the rock ptarmigan. In spring and summer the male willow ptarmigan is much more reddish than the rock ptarmigan, and, although the females are very similar, the willow ptarmigan's bill is distinctly larger and higher and is grayish at the base. In fall males are more heavily barred than are male rock ptarmigans, and females likewise have stronger and less grayish markings than do female rock ptarmigans. In winter white-plumaged males lack the black eye markings that occur in male rock ptarmigans, but since this mark may be lacking in females, the heavier bill should be relied upon to distinguish willow ptarmigans.

AGE AND SEX CRITERIA

Females lack the conspicuous bright reddish "eyebrows" of adult males, are more grayish brown and more heavily barred on the breast and flanks than males, and lack the distinctive rusty brown color of males in the summer. In fall females are usually more heavily barred on the breast and flanks than are males. In the British race *scoticus* the breeding females are likewise more boldly patterned than males, with ochraceous on the upperparts, breast, and flanks, and the outer webs of the secondaries and most wing coverts are vermiculated with tawny. In winter, females of at least the American races are like males, but the concealed bases of the crown feathers are more grayish (Ridgway and Friedmann 1946). They can be fairly accurately identified at this time by their brown rather than black tail feathers and central upper tail coverts and by certain wing and tail measurements (Bergerud, Peters, and McGrath 1963). Voronin (1971) reported that, in the USSR, willow ptarmigans in winter plumage can be effectively sexed by using wing measurements (93 percent accurate) or tail measurements (92.6 percent accurate). In the British population, the nonbreeding plumages of females are more ochraceous buff on the upperparts than males, and the foreneck and chest are paler and more heavily barred, while the underparts are lighter and are more distinctly barred (Cramp and Simmons 1980).

Immatures in first-winter plumage tend to have the tenth primary more pointed at the tip than the inner ones, but this is not as reliable as the fact that (1) there is little or no difference in the amount of gloss on the three outer primaries of adults, whereas immatures have less gloss on the outer two primaries than on the eighth, and (2) there is about the same amount of black pigment on primaries eight and nine (sometimes more on eight than on nine) of adults whereas juveniles have more on the ninth than on the eighth (Bergerud, Peters, and McGrath 1963; Semenov-Tian-Schanski 1959).

Juveniles may be identified by the fact that their secondaries and inner eight primaries are grayish brown with pale pinkish buff margins or barring. However, the late-growing outer two primaries are white, often speckled with black, like the first-winter flight feathers that soon replace the secondaries and inner primaries in all races but *scoticus*. Myrberget (1975) provided methods of aging willow ptarmigan chicks on the basis of primary molting patterns and growth curves, which are most reliable up to about 17 days of age.

Downy young of the willow ptarmigan are reported (in the Scottish population) to be darker on both the dark and lighter areas and to have less clear-cut margins between these areas than downy rock ptarmigans (Watson, Parr, and Lumsden 1969). These authors mention other differences that may also serve to separate downy young of these two species, although these may not apply equally well to North American populations. For example, in the Labrador populations, birds under 3 weeks are almost impossible to identify as to species, although young willow ptarmigans are slightly darker and somewhat greenish instead of yellowish on the underparts (Bendire 1892). After 3 weeks they may be distinguished by differences in the bill.

DISTRIBUTION AND HABITAT

The North American breeding range of the willow ptarmigan is primarily arctic tundra,

4. Current North American Distribution of Alaska (Ak), Alexander (Ax), Aleutian (Al), Baffin Island (B), Keewatin (K), Newfoundland (N), and Ungava (U) races of the willow ptarmigan. Stippled area indicates southern wintering limits.

though it extends southward somewhat in alpine mountain ranges and in tundralike openings of boreal forest (Aldrich 1963). The basic habitat consists of low shrubs, particularly willow or birch, in lower or moister portions of tundra. Weeden (1965b) has characterized the general breeding habitat of the willow ptarmigan as follows: Typical terrain is generally level or varies to gentle or moderate slopes but frequently is at the bottom of valleys. Vegetation is relatively luxuriant, with shrubs usually 3 to 8 feet high scattered through areas dominated by grasses, hedges, mosses, dwarf shrubs, and low herbs. The birds usually occur at the upper edge of timberline among widely scattered trees, or they may occur somewhat below timberline in local treeless areas.

In Eurasia the habitat is similar and consists mostly of treeless tundra, moors, heaths, bogs, and occasionally open wetlands or dunelands where heather (*Calluna*) and similar food plants occur. In Great Britain the form *scoticus* is mostly associated with heather moors well away from trees during the breeding season, and with treeless lower-altitude cover during winter, sometimes extending to farmlands. Elsewhere in Eurasia the breeding habitat often includes low willows or dwarf birches mixed with hillocky and berry-rich tundra, but usually not pure grasslands, rocky tundra, or lichen-rich tundra. In winter many of the Eurasian populations move into more heavily wooded cover, including woodlands of willow, birch, and alder, into forest clearings, and even into coppices or groves of birches, aspen-dominated areas, and the like (Cramp and Simmons 1980).

Although the range limits have retreated in a few areas at the southern edge of the European range, in general this species' range has not changed significantly in historic times. However, populations have declined significantly in Britain, Finland, Latvia, Lithuania, and Estonia (Cramp and Simmons 1980).

POPULATION DENSITY

Ptarmigans are among the arctic-dwelling species that exhibit major fluctuations in yearly abundance and are believed by some to exhibit cyclic population changes (Buckley 1954). In any case, major changes in population density do occur, and estimates of density thus may vary greatly by year as well as by locality. Weeden (1963) summarized estimates of population density for various areas in Canada. These estimates ranged from fewer than 1 adult per square mile (2.5 square miles per adult) to as many as 8 adults per square mile, with the sparser densities generally based on large areas that include much unfavorable habitat. He also reported (1965b) that a study area of 0.75 square mile had spring populations ranging from 38 to 150 males during seven years of study, which represents from 3.2 to 12.3 acres per male. Somewhat comparable density figures have been reported from Newfoundland (Mercer and McGrath 1963), who estimated spring 1962 populations on Brunette Island of from 147 to 207 birds per square mile, depending on the technique used. Considerable population work has been done with the Scottish race by Jenkins, Watson, and Miller (1963, 1967), who found some year-to-year differences in densities as well as major differences in densities in different study areas of varying habitat quality. On

27. Capercaillie, male. Photo by Hans Aschenbrenner.

28. Capercaillie, male. Photo by Hans Aschenbrenner.

29. Capercaillie, female and chick. Photo by Hans Aschenbrenner.

30. Capercaillie, female. Photo by Hans Aschenbrenner.

31. Capercaillie, female. Photo by Hans Aschenbrenner.

32. Black grouse, female. Photo by Hans Aschenbrenner.

33. Black grouse, male. Photo by author.

34. Blackgrouse, male. Photo by author.

34. Blackgrouse, male. Photo by author.

35. Black grouse, male. Photo by R. J. Robel.

36. Ruffed grouse, female. Photo by Alan Nelson.

37. Ruffed grouse, male. Photo courtesy Bird Photographs, Inc., Ithaca, N.Y.

38. Ruffed grouse, female. Photo by Ken Fink.

39. Ruffed grouse, male. Photo by author.

40. Hazel grouse, male. Photo by Hans Aschenbrenner.

41. Hazel grouse, male. Photo by Hans Aschenbrenner

42. Hazel grouse, male. Photo by Hans Aschenbrenner.

43. Black-breasted hazel grouse, male and female. Painting by H. Jones, courtesy Zoological Society of London.

44. Hazel grouse, incubating female. Photo by Hans Aschenbrenner.

45. Hazel grouse, female. Photo by Hans Aschenbrenner.

46. Greater prairie chicken, male. Photo by author.

47. Greater prairie chicken, male and female. Photo by author.

48. Lesser prairie chicken, male. Photo by Ken Fink.

49. Sharp-tailed grouse, male. Photo by Alan Nelson.

50. Sharp-tailed grouse, male. Photo by Alan Nelson.

51. Downy young of Eurasian grouse, including (*upper left*) sharp-winged grouse, (*upper right*) black grouse, (*middle left*) hazel grouse, (*middle center*) black-breasted hazel grouse, (*middle right*) Caucasian black grouse, (*lower left*) black-billed capercaillie, and (*lower right*) capercaillie. Painting by Jon Fjeldså.

rich, intermediate, and poor grouse moors the spring density of birds averaged about 95, 66, and 52 birds per square kilometer, respectively (246, 171, and 135 birds per square mile). They also noted that similar spring densities of 54 to 116 birds per square kilometer have been noted in northern Norway, and about 40 to 60 birds per square kilometer in spring were reported from the Timansk tundra of the USSR. In one study area the density remained fairly constant from year to year, whereas in two others it fluctuated in a nonsynchronized manner. In a highland area of Sweden the usual density of breeding birds was estimated as 4 to 7 pairs per square kilometer (10–18 pairs per square mile) by Marcstom and Höglund (1981).

Several studies suggest that population densities in this species sometimes vary cyclically, over periods ranging from 3 to 10 years (Myrberget 1974; Bergerud 1970*b*; Watson and Moss 1979). Myrberget believed these cycles conformed to small rodent population cycles and might result from an interplay between rodents and their predators and corresponding variations in predation effects on eggs or chicks, while Watson and Moss doubted that cycles in predation or disease were necessary causes and instead attributed the cycles to variations in spacing behavior that occur during periods of high population density. These authors (Moss and Watson 1980) have recently suggested that as populations increase in density, genotypes associated with unaggressive behavior proliferate until peak densities are reached, when selection for genotypes associated with aggressive behavior begins and causes a decline in numbers.

HABITAT REQUIREMENTS

Wintering Requirements

Weeden (1965*b*) reported that winter habitat of willow ptarmigans consists of willow thickets along streams, areas of tall shrubs, and scattered trees around timberline and burns, muskegs, and riverbanks below timberline. Bent (1932) noted that in winter willow ptarmigans move to interior valleys, river bottoms, and creek beds, where food is available in the form of tree buds and twigs of willows (primarily), alders and spruces, and such berries and fruits as can be found above the snow. Godfrey (1966) indicates that during winter the birds may be found well south of the tree line, in muskegs, lake and river margins, and forest openings.

Spring Habitat Requirements

Weeden (1965*b*) stated that male habitat preferences for territories include shrubby and "open" vegetation, with the plants lower than eye level for ptarmigans. Elevated sites such as rocks, trees, or hummocks are used by males during display. Resting areas are provided by small clumps of shrubs at the edges of open areas.

At least in Scotland, territories are established during fall, although they may be abandoned temporarily during winter if snow conditions require it. In Alaska some fall display and calling also occurs (Weeden 1965*b*). Continued residence, however, is not

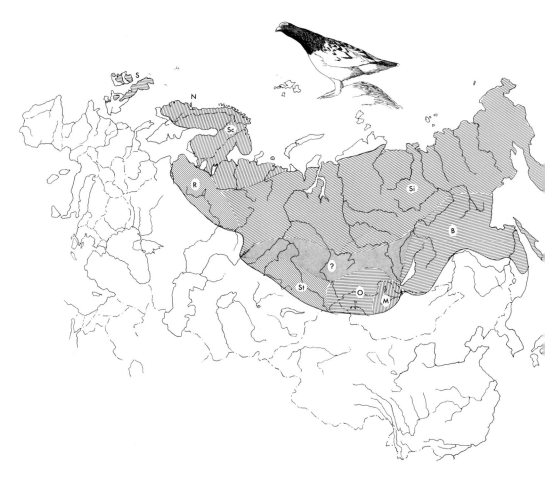

5. Current Eurasian distribution of Baikal (B), Mongolian (M), Norwegian (N), Oriental (O), Russian (R), Scandinavian (Sc), Scottish ("red grouse") (S), and steppe (St) races of the willow ptarmigan. Light stippling indicates area of uncertain racial designation.

typical in Alaska or probably in any part of the North American breeding range, since considerable seasonal movement is usual. Thus, local topography, as it affects snow deposit and rate of snow thaw exposing territorial sites, may have considerable effect on territorial distribution of birds in arctic North America.

Nesting and Brooding Requirements

Requirements for willow ptarmigan nest sites are apparently fairly generalized. Brandt (in Bent 1932) reported that nesting may occur anywhere from coastal beaches to mountainous areas, and nests may be placed beside drift logs, in grass clumps, under bushes, in mossy hummocks, or near similar sources of screening protection. Weeden (1965b) indicates that the nest is usually protected from above and at the side by shrubby vegetation, while one side borders an open area. The nest is within the periphery of the male's territory.

Brooding habitat is similar to nesting habitat, according to Weeden (1965b), with chicks using areas of very low vegetation while older broods use shrub thickets for escape cover. Maher (1959) noted that broods used a variety of habitats with good cover and were common on upland dwarf-shrub and hedge tundra, as well as being sometimes found in riparian shrub and willow shrub at the bases of hills.

FOOD AND FORAGING BEHAVIOR

At least in Alaska, the most important single food source for willow ptarmigans is willow buds and twigs. Weeden (1965b) noted that these constituted almost 80 percent of winter food found in 160 crops from interior Alaska, and Irving et al. (1967) also indicated that winter food consisted almost entirely of the buds and twig tips of willow. Weeden noted that dwarf birch buds and catkins were second in importance, and Irving et al. (1967) similarly found that in wooded areas the birds take some birch catkins and poplar buds. West and Meng (1966) found that 94 percent of the winter diet of willow ptarmigans from northern Alaska consisted of various willow species, and 80 percent was from a single species (*Salix alaxensis*). They also noted that some birch may be used, but that alder, though it is often available and has a higher caloric content than willow, is seldom used.

One exception to the general winter diet of willow for North American willow ptarmigans has been noted, in Peters's study (1958) of the Newfoundland population. He found that the winter diet consists almost entirely of the buds and twigs of *Vaccinium* species, the buds and catkins of birch and alder, and the buds and sweet gale (*Myrica*).

In Eurasian populations the winter diet is relatively varied. In northern Norway the primary winter food is mountain birch (*Betula pubescens*), especially its shoots, but also buds and catkins (Myrberget 1979). The same species is the most important food plant in central Sweden, though tall-growing willow is also important. Elsewhere, birch forests do not reach far into the willow ptarmigan's range, and other species become important. In southern and central Finland, dwarf birch and various willow species are major foods

(Myrberget 1979), and willows as well as birches are also major foods in the USSR (Dementiev and Gladkov 1967). Farther south, as in Newfoundland (Peters 1958) and Britain (Lance 1978), the foods are more variable and usually include ericaceous shrubs or herbs (*Vaccinium, Myrica, Calluna, Erica*). Evidently, the foods taken in different areas vary greatly with availability, food preferences, and perhaps also competition with other ptarmigan species. Willow ptarmigans will sometimes dig down through snow to reach food, but generally they winter in lower and more sheltered areas than do rock ptarmigans, where foods are perhaps usually somewhat more accessible. In most areas the foods taken shift in spring, but in Britain the major food remains heather (*Calluna vulgaris*) throughout the year. Densities of breeding populations as well as breeding success seem to be related to the amount and quality of heather available to the birds (Watson and Moss 1972; Moss, Watson, and Parr 1975; Lance 1978).

With spring, the willow ptarmigan's dependence on willow declines in Alaska, and in addition to the leaves of willow the birds begin to eat a larger variety of leafy materials and the berry seeds of crowberry (*Empetrum*) and *Vaccinium*. Studies on the British population indicate that during spring more than 90 percent of the birds' food consists of heather, with *Erica* spp. constituting most of the rest. Food selection associated with nitrogen and phosphorus content increases from winter to spring and continues throughout spring, until egg laying has been completed. It is thus probable that variations in chick survival are in part related to conditions before their hatching, such as variations in the diet of breeding females (Moss, Watson, and Parr 1975).

With summer, insects and other terrestrial invertebrates sometimes become significant food items, especially in wetter areas where cranefiles and other large, slow-moving insects may be present. However, studies of chicks from the moorland of Scotland suggest that even birds less than 3 weeks old predominantly consume heather shoottips, and after the first 3 weeks following hatching insects may make up less than 5 percent of the total diet. Chicks also consume heather that is richer in nitrogen and phosphorus than that eaten at the same period by adults (Savory 1977).

Summer foods in Alaska consist of various berries, especially blueberries, willow and blueberry leaves, and the tips of horsetail (*Equisetum*), which grows in willow thickets near streams (Weeden 1965b). Peters noted that crowberries, blueberries, and the leaves of *Vaccinium* species, especially *V. angustifolium*, provided major sources of summer foods in Newfoundland. Studies of summer foods in Alaska indicate that both sexes of adults depend largely on willows during this period, eating catkins and buds in June, leaves and developing fruits in July, and leaves and buds in August, along with various grass seeds. In addition to eating willow leaves, the chicks evidently forage on various berries and to a rather limited amount on insects (Williams, Best, and Warford 1980), suggesting that young willow ptarmigans are less insectivorous than are chicks of more temperate-adapted grouse.

In the fall, as the berry supplies are exhausted and leaves fall from *Vaccinium* bushes, the ptarmigans in Newfoundland return to a diet of buds and twigs (Peters 1958). The same situation obtains in Alaska, though it is willow rather than *Vaccinium* buds and twigs that the birds resort to (Weeden 1965*b*). Irving et al. (1967) found a gradual increase in total crop contents of Alaska willow ptarmigans from October to January, followed by reduced contents until April. This population migrates southward in October and November and northward from January until May. Evidently feeding is related to changing patterns of daylight rather than to temperature cycles or to the cycle of migratory activities.

Compared with the rock ptarmigan, it seems that the willow ptarmigan perhaps depends less on low shrubs and other ground-associated vegetation and more on willows and birches, at least during winter. However, neither species depends on any single food plant throughout its range, although the genus *Salix* is probably the most widely utilized single source of food during winter.

MOBILITY AND MOVEMENTS

The willow ptarmigan and its relative the rock ptarmigan are perhaps the most migratory of all North American upland game. Snyder (1957) reports that the willow ptarmigan is migratory to a considerable extent, occasionally wandering as far as the southern parts of the prairie provinces, northern Minnesota, the north shore of Lake Superior, southern Ontario, and southern Quebec. To some degree these southern movements may be related to unusually dense populations in the northern areas (Buckley 1954). Evidently considerable differential movement according to sex occurs in Alaska (Weeden 1964). At Anaktuvuk Pass, for example, most wintering willow ptarmigans are males, while many of the wintering birds in timbered areas to the south are females. Likewise, alpine-fringe areas of the Alaska Range and the Tanana Hills are utilized mostly by males during winter, while females are to be found abundantly in the Tanana valley (Weeden 1965*b*). Weeden (1964) suggested that this differential movement may represent a dispersal mechanism or serve to reduce food competition or perhaps indicate that females may survive better in forested areas under winter conditions.

Irving et al. (1967) have documented the migration of willow ptarmigans through Anaktuvuk Pass in the Brooks Range. Although few ptarmigans nest there, some 50,000 birds pass this point each year. The fall migration reaches a peak in October and is over by December, and the spring migration starts in January and early February, subsides in March, and is renewed in April. The early fall migrants are mostly juvenile males and females, whereas the number of adult males gradually increases to a maximum in March, or two months later than the maximum movement of juvenile males. The authors reported no clear indication of cyclic changes in population numbers annually. A spruce forest area 35 miles south of the pass is one of the areas used for wintering, and breeding occurs on the north slope of the Brooks Range and beyond to the Arctic coast. Some of these breeding areas may not be occupied until late in May.

Among the Eurasian populations, those living in the USSR are certainly the most mobile, though Scandinavian populations also exhibit limited winter movements. Thus, in Norway there are some movements from offshore islands to the mainland or larger islands for the winter, though they seldom exceed 20 kilometers. Compared with the Scottish population, there is more movement in the range of 10 to 30 kilometers, and the longest known distance was a recovery 140 kilometers away in Norway. In Norway most birds have been recovered within 10 kilometers during the first winter of life, but by the second winter more than a third have been recovered at greater distances. Females there tend to move farther than do males, although the extreme distances noted were the same (Cramp and Simmons 1980).

In the USSR there is a substantially greater degree of winter movement, especially in the most northerly populations such as *lagopus*; the birds may move as far as 200 to 250 kilometers into coniferous forest during particularly severe winters. The migration is normally fairly gradual, with departures dependent on food conditions and with birds occasionally moving in groups of 100 to 300 individuals and at heights of up to 200 meters, closely following rivers having stands of willows. The birds sometimes cross rather wide marine bays, and they regularly move between Kolguev Island and the mainland, as well as across Cheshskaya Bay. There is apparently even migration from areas as remote as the New Siberian Islands (Dementiev and Gladkov 1967).

So far, virtually nothing is known of daily movements of willow ptarmigans, and such information will require detailed studies of individually marked birds. Jenkins, Watson, and Miller (1963), studying red grouse, found the birds remarkably sedentary in this nonmigratory population. Of 739 birds banded as chicks, only 5 were recovered more than 5 kilometers away that season, and some of this movement may have been caused by the birds' being driven by hunters. Of 290 birds banded as chicks but recovered as adults, 230 were recovered within 1.5 kilometers of the point of banding. It thus appears that willow ptarmigans move only as far as is necessary to find food and cover during the coldest parts of the year. Weeden (1965*b*) reported that a male and his mate were both found a year after they were banded as adults, defending a brood about 100 yards away from the original point of banding a year previously, which attests to considerable site fidelity in this species. Bergerud (1970*b*) reported that females are more mobile than males, with one banded female moving 61 kilometers in about 3 months.

REPRODUCTIVE BEHAVIOR

Territorial Establishment

Most observations of territorial behavior in this species derive from studies of the red grouse in Scotland by A. Watson, D. Jenkins, and their associates. Likewise, display descriptions are also based on this population, unless otherwise indicated.

Territorial behavior and the success of territorial establishment appears to be a crucial factor in the biology of red grouse populations, judging from the work of Jenkins and Watson. Territories in red grouse are established in early fall, and the numbers of such territorial males that can be accommodated on a habitat apparently limits the density of the breeding population. Nonterritorial males are forced out of the preferred areas into marginal habitats, where they are more heavily exposed to predation, starvation, and disease. However, such losses play little if any role in the success of the population. Since juvenile birds are rarely able to attain territorial status their first fall, early territorial establishment favors reproduction by mature males.

Territorial establishment in the North American willow ptarmigan is presumably in spring, although some fall display and calling by males may occur (Weeden 1965b). However, it is not until late April or May that willow ptarmigans acquire their striking nuptial plumage, which presumably provides important visual signals for proclaiming territory and attracting females. Weeden (1965b) has made the interesting point that whereas the male willow ptarmigan undergoes courtship in this bright brown-and-white plumage, the rock ptarmigan is still in completely white plumage during courtship, which perhaps provides important visual distinction for species recognition between the two.

Territorial size has been studied intensively by Jenkins, Watson, and Miller (1963) for red grouse. They found that in each year some individual territories were larger than others, but in years of high grouse populations the territories in general averaged smaller than in years when grouse were few. Territories selected by previous residents were usually larger than could later be defended against newly colonizing juvenile birds. Sketched maps presented by these authors indicate that territories rarely exceeded a diameter of 300 yards, and most were much smaller. One study area of 56 hectares (138 acres) supported 24 territorial males (2 of which were unmated) in 1961; thus territories averaged 5.7 acres in the area during that year. In 1960, 16 males (2 unmated) occupied the same area, and in 1958 there were more than 40 territorial males (10 unmated) on it. For the study areas as a whole, the breeding density over the years varied from 1 pair to about 5 acres in 1957 and 1958 to about 1 pair per 15 acres in 1960.

Agonistic and Sexual Behavior

In contrast to the species considered previously, it is almost impossible to differentiate completely between male and female behavior patterns in the willow ptarmigan. This is primarily a reflection of their monogamous or nearly monogamous pair bond and a subsequent reduction of sexual selection pressures for dimorphic behavior patterns. Watson and Jenkins (1964) have provided a detailed account of behavior patterns in the red grouse that I will summarize here in the belief that their findings should apply to the North American willow ptarmigan with little or no modification. Although they also discuss displacement activities, distraction behavior, comfort and maintenance activities, and other aspects of behavior, only these patterns directly concerned with reproduction will be mentioned here.

Agonistic behavior patterns of males associated with establishment and defense of territories include sitting on an exposed lookout, such as a hillock or stone, where most of the territory can be seen. Intruders are approached in an attack-intention posture characterized by erect combs, the head and neck stretched forward, the body near the ground, the wings held in the flanks, and the bill open. Before such an approach the bird may fan his tail and droop his wings in a manner resembling the waltzing display. A lesser type of threat consists of standing in one place and uttering *kohway* and *kohwayo* calls. A still weaker threat consists of standing and uttering a *krrow* call, which in turn grades into watchful behavior, flight intention, and finally fleeing by running or flying away.

Several kinds of aggressive encounters may occur. Brief encounters may last only a few seconds and involve birds of either sex, which may or may not occur on a territory. "Jumping" is a communal encounter that also is not limited as to sex and not related to territory. In this, two or more aggressive birds will begin to jump about with wing flapping, causing them to become more fully separated. Prolonged chases may occur when a dominant male follows a subordinate bird for great distances, often beyond his territory, and may in fact kill or wound it. "Facing" occurs when two equally dominant birds face each other with combs erect, heads forward, and wings flicking, with neither one showing signs of retreat. When actual fighting occurs the birds usually do not face each other, but rather face in the same direction and strike each other from the side with their bills, wings, and feet. "Walking-in-line" consists of two birds' walking side by side some 20 inches apart. While so walking they utter *kohway* and *ko-ko-ko* calls that indicate attack intention, and they may also utter the *krrow* threat call. Such a display by two males often occurs at the edges of territories, while hens may perform the same display anywhere in the territory. Occasionally the display occurs outside breeding territories, where up to five or six birds may participate.

Sexual patterns involve pair formation behavior and copulatory behavior. In pair formation the males advertise their territories, and the females are attracted to the more vigorous males. On arriving on a territory, the female may utter a *krrow* call and look for a displaying male to approach. If there is none, she may fly to another territory, until a resident male makes a song flight landing near her and begins to strut toward her. The female then flees but may be driven back to the territory by the male. Sexual activity occurs in Scotland in every month but is most common from February to April, and many pair bonds that had been established earlier are temporary and easily disrupted. When in breeding condition, the male has highly conspicuous red eye combs that can be erected to about 1 centimeter. Although the hen's combs are much smaller and paler, they can also be erected.

The male's approach to another bird of either sex is essentially threatening, and a receptive hen responds by submissive gestures. Thus the sexual differences in display are not so much qualitative as quantitative, in terms of relative dominance and submission. Sex recognition is probably also achieved by the different voice, plumage, and comb development of the hen.

The postures a male performs in the presence of a female but not in the presence of other males may be considered "courtship" displays. Watson and Jenkins (1964) list five such displays: tail fanning, waltzing, rapid stamping, bowing, and head wagging.

Tail fanning is performed by a male when approaching a hen. While cocking his tail, he

20. Male displays of the willow ptarmigan (after various sources), including (1) front and rear view of tail fanning and wing drooping, (2) calling posture of territorial male, (3) song flight posture, and (4) aerial chasing.

may fan it with a rapid flick, at the same time lowering his wings and scraping the primaries on the ground as he moves forward. In this stage the wings are drooped equally and the tail is not tilted. Often the male moves in a slight curve in front of the female, or he may pass in front of her alternately from both sides. Sometimes the under tail coverts are exposed by his turning away from the hen. Such movements grade into "waltzing" during which the male circles the female closely, pivoting around her short, high steps and drooping the wing nearer her, at the same time tilting his tail to expose its upper surface more fully to her view. The body may be tilted toward the hen as well. During "rapid stamping" the males runs toward the hen with his tail slightly fanned, his neck thickened and arched, and his head held low with the bill wide open. In this posture he might pass close beside the other bird and appear to be attacking her, but the differences in wing and neck positions make it possible to distinguish these two types of behavior easily. If the hen does not flee and mounting does not occur at that time, the male will often raise and lower his head, with his body still held low, the tail partly fanned and the nape feathers raised, in a display called "bowing."

The last of the courtship displays is head wagging, which both sexes perform. The bird crouches near its mate, extends its neck forward, and quickly wags its head in lateral fashion, exposing its eye combs and twisting its head slightly with each wag. When a hen approaches a cock, the male may also crouch low, erect his combs, and lower his head, producing a posture strongly suggestive of the precopulatory "nuptial bow" of prairie grouse. Although both sexes perform head wagging, it is not a mutual display, and instead the birds often perform it alternately. When the female performs it, the male may attempt to mount her. However, during actual solicitation, the female crouches without head wagging, opens her wings, and holds her head up. The male immediately mounts, drooping his wings around the hen during copulation. Afterward the male utters several threat calls, displays strongly for a few minutes, and often moves to a lookout post.

Vocal Signals

Watson and Jenkins (1964) describe fifteen different vocal signals of adults that are uttered by both sexes, although the hen's calls may be recognized by their higher pitch. Song flight, or "becking," is uttered as the bird takes off, flies steeply upward for 30 feet or more, sails, and then descends gradually while fanning its tail and beating its wings rapidly. On landing the bird may stand erect, droop its wings, fan its tail, and bob its head. During the ascent phase the call is a loud, barking *aa*, while a *ka-ka-ka-ka* is uttered some eight to twelve times with gradually slower cadence. After landing a gruffer and slower call *kohwa-kohwa-kohwa* (also interpreted as *go-back, kowhayo,* and *tobacco*), is uttered for a varying length of time. Hens and nonterritorial males do not fly as high or call as loudly as territorial males, and no doubt this call is important in territorial proclamation.

In calling on the ground, a similar signal is uttered, often from a song post such as a stone. The bird stretches its neck diagonally upward and utters a vibrating *ko-ko-ko-ko-krrrrr*, up to about twenty syllables, increasingly faster toward the end. Such calls may be used to threaten approaching animals or birds flying overhead and are largely but not entirely territorial advertisement.

During attack, the birds utter a *kowha* sound, like the last part of the flight song but without preliminary notes. It may be given during attack, when trying to mount hens, or immediately after copulation. A similar call, *koway*, is an attack intention, or threat, call and is rapidly repeated as a series of hurried notes. A variant is *kohwayo*, also repeated, but indicating less aggressiveness than the previous call. Still less aggressive notes are *krrow* and *ko-ko-ko*, the latter indicating flight intention. This call is given by a bird about to fly or one being handled by a human and may stimulate other birds to take flight.

When a grouse is charging another bird, a single note, *kok*, may be uttered, especially by the chased bird. The same call may be used as an in-flight alarm note. A similar *kok* note serves as a mammalian predator alarm note, while a *chorrow* note serves for an aerial raptor warning signal. A sexual note, *koah*, the emphasis on the first syllable, is used between members of a pair when crouching and head wagging, when examining nest sites, or when bathing. Hens may also utter it when the nest is approached, but hissing is more often elicited under these conditions. Hissing may also occur when a bird is being handled. A *krow* note is used during distraction display by parents, causing the young to crouch, while a *korrr* or *koo-ee-oo* serves as a call to chick, especially those uttering distress calls. Finally, a harsh, chattering *krrr* note is used as a defense against avian predators that are attacking the bird or its family.

Watson and Jenkins report that the distress *cheep* of chicks is uttered until the young are nearly full grown, but that it gradually changes to a *kyow* note and finally to the adult *krrow* and probably then serves as a contact call. Even newly hatched chicks will utter a chattering call that evidently is aggressive in nature and apparently develops into the adult "ground song." By the age of 10 to 12 weeks, the male begins to acquire a voice that differs from that of females, resembling more the voice of an adult cock.

Nesting and Brooding Behavior

Nest-site selection in the willow ptarmigan is still rather little studied, but Myrberget (1976) reported on the analysis of 254 nests in Norway. There a strong preference for forest edges was evident. Other preferred habitats were those rich in *Empetrum* and junipers. Habitats that were apparently not preferred were heaths and meadows. Cover plants associated with nests were mainly *Empetrum*, mountain birch, and junipers.

Another analysis of nest-site selection was provided by Jenkins, Watson, and Miller (1963) for Scottish red grouse. They studied 163 nests, nearly all of which were in heather cover (*Calluna*). The average height of the heather cover was 27 centimeters, compared with a mean cover height of 17 centimeters. Most nests were partly overhung with vegetation, but 17 percent were completely uncovered and 12 percent were completely covered. Most were on hard, well-drained ground, and 67 percent were on flat ground. Most were shallow scrapes, sparsely lined with various plants, including grasses and heather. Usually the nests were within 500 feet of grit sources, water, and mossy or grassy areas where the chicks could feed.

In Norway there appear to be substantial year-to-year as well as local variations in clutch sizes, averaging 9.8 over a ten-year period but varying between years and also ranging from 7.8 eggs in areas of high density to 10.0 in those having low grouse densities (Myrberget 1972). Considerably smaller average clutches appear to be typical of the British red grouse, which averages about 7.5 eggs but exhibits marked year-to-year variations largely associated with weather differences and resulting differences in green heather availability and onset of breeding (Jenkins, Watson, and Miller 1963, 1967).

Some comparable information is available for North American willow ptarmigans. Kessel and Schaller (1960) reported that 5 nests in Alaska had 6 to 7 eggs, averaging 6.8. Eight clutches from northern Alaska in the Denver Museum average 7.8 eggs. Bergerud (1970b) reported an average clutch of 10.2 eggs in 106 Newfoundland nests. Nests containing up to 17 eggs appear to be the work of at least two females. The incubation period of the North American birds is likewise 21 to 22 days, and the egg-laying interval is somewhat greater than 24 hours (Westerskov 1956). Bergerud (1970b) judged that in Newfoundland renesting probably accounted for between 12 and 18 percent of the young produced.

Observations on captive birds of the Scandinavian race *lagopus* indicate that the incubation period may vary from 21 to 25 days, and although it is certainly normally performed by the female alone, in one of four pairs it was performed entirely by the male (Allen et al. 1977). In another case the male competed with his mate for incubation duties (Allen 1977).

Unlike the other species of ptarmigans, the male typically remains with the female through the incubation period and assists in brood defense. Jenkins, Watson, and Miller (1963) reported that the percentage of broods observed with both parents in attendance ranged from 61 percent to 90 percent in various years and areas. In good breeding years, most broods were attended by both parents until they were at least 2 months old, while in poor breeding years 30 to 40 percent were not attended by parents at any stage. The percentage of parents observed performing distraction displays ranged from 4 to 72 percent. Individual brood sizes ranged to as many as 12, and the average varied greatly in different years. Roberts (1963) reported an average brood size of 6.3 chicks for Alaska willow ptarmigans. This figure is higher than any yearly average reported by Jenkins, Watson, and Miller (1963), whose highest reported brood size was 5.2 for one study area in 1960.

Hatching success in this species is sometimes surprisingly high. Jenkins, Watson, and Miller (1963, 1967) reported hatching success rates of 326 of 395 clutches (82.5 percent), and hatchability of eggs in successful nests is often above 90 percent. Myrberget (1972) reported that over a ten-year period the hatching success in Norway averaged 78.3 percent, with annual means ranging from 55.4 to 91.6 percent. Egg-hatching rates in his study varied from year to year and seem to be strongly affected by the intensity of predation, although this was partially compensated for by second clutches. Renesting may be fairly common in this population; Parker (1981) found that 15 of 57 females attempted to renest 3 to 15 days after losing their initial clutches. Second nests had smaller average clutch sizes (6.1 vs. 10.7 eggs), but hatching success rates were the same. Myrberget (1972) estimated that the average percentage of chicks surviving to late July was 54.4, with yearly variations from 39.4 to 76.2 percent. A similar chick survival rate of about 52 percent was found by Jenkins, Watson, and Miller (1963), and Bergerud (1970b) estimated a 31 percent chick mortality between hatching and autumn over an eleven-year period, with extremes of 12 and 49 percent. Perhaps both the surprisingly high hatching success and seemingly moderate rates of chick mortality can be partially attributed to care by both parents. Most chick mortality apparently occurs during the first week or so after hatching, and it frequently seems to be associated with unfavorable weather during this critical period.

EVOLUTIONARY RELATIONSHIPS

Evolutionary relationships of the genus *Lagopus* as a whole seem very close to both *Dendragapus* and *Tetrao*, as Short (1967) has already suggested. It is perhaps impossible to judge which of these two genera *Lagopus* most closely approaches, and presumably all three differentiated from common stock at about the same time.

Relations within the genus *Lagopus* represent another problem. The white-tailed ptarmigan differs from the rock and willow ptarmigans in several respects, which have been enumerated by Short (1967), and it is clearly the most isolated of the three species. Höhn (1980) suggested such an early offshoot of ancestral white-tailed ptarmigan stock in North America, with which I am in agreement. Höhn judged that the willow and rock ptarmigan ancestral stock also diverged in North America, with the rock ptarmigan moving east to Greenland and both species moving west across the Bering Strait into Eurasia. This kind of speciation model seems unlikely to me, since I can visualize no major barriers that might have allowed for separation of ancestral willow and rock ptarmigan stock in northern North America. It seems more likely to me that one of these types developed in Eurasia and the other in North America after a splitting of common gene pools and that after secondary contact the rather marked ecological differences between them allowed the development of the extensive geographic contact between them that now exists. In contrast, Johansen (1956) suggested that the genus *Lagopus* originated in Asia and reached North America at an early date, during which the ancestral white-tailed ptarmigan separated from pre-*mutus* stock.

In a strictly behavioral sense, I regard the willow ptarmigan as more primitive than the other two ptarmigans, in both of which a breakdown of strong pair bonds and a tendency toward polygamy may be seen. It seems probable to me that mating patterns in the grouse evolved from an originally monogamous situation to a polygamous or promiscuous one, rather than that the monogamous situation of the willow ptarmigan is derived from a nonmonogamous mating type. The retention of monogamy or near monogamy in the ptarmigans seems to me to be an ecological artifact resulting from the greater need for

187

intensive parental care in an arctic situation than in a subarctic or temperate one, in which the duties of incubation and brood rearing can be more effectively undertaken by the female alone. This latter arrangement thus frees the male to fertilize a potentially larger number of females, and these resulting reproductive advantages have led to reduced pair bonds or to promiscuous matings. It is curious, however, that the willow ptarmigan, rather than the rock ptarmigan, has more strongly retained a monogamous and prolonged pair bond, since the rock ptarmigan has an even more northerly breeding distribution and must nest under equally severe breeding conditions. Arnthor Gardnarsson (pers. comm.) has found that in Iceland the males suffer much more predation by gyrfalcons than do females, apparently as a result of the males' more conspicuous plumage during the breeding season. The mating system there is essentially promiscuous, since the females do not closely associate with males or their territories. Such differential sexual predation pressures might account for the rock ptarmigan's less strongly monogamous mating system and the reduced period of contact between the sexes.

Rock Ptarmigan

Lagopus mutus (Montin) 1776

Other Vernacular Names

Arctic grouse, barren-ground bird, rocker (in Newfoundland), snow grouse, white grouse; Alpenschneehuhn (German); lagopède alpin (French); tundravaya kuropatka, gornaja kuropatka (Russian).

Range

Circumpolar. In North America from northern Alaska, northwestern Mackenzie, Melville Island, northern Ellesmere Island, and northern Greenland south to the Aleutian Islands, Kodiak Island, southwestern and central British Columbia, southern Mackenzie, Keewatin, northern Quebec, southern Labrador, and Newfoundland. Also resident in Iceland, Scotland, Spitsbergen, and Franz Josef Land. Breeds in Eurasia in the Scandinavian and Kola peninsulas, from the northern Urals east through Siberia to the Bering Sea, on the Commander Islands and northern and central Kuriles, and in Japan (Honshu). The Siberian range extends from the northern coast south to the Arctic Circle west of the Yenisei, the Angara River, and the mountains of Lake Baikal and south to Mongolia, the mountains of Transbaikalia and Amurland, and Kamchatka. Some populations undergo limited southward migratory movements in winter.

Subspecies

L. m. evermanni Elliot: Attu rock ptarmigan. Resident on Attu Island, Aleutian Islands.

L. m. townsendi Elliot: Kiska rock ptarmigan. Resident on Kiska and Little Kiska islands, Aleutian Islands.

L. m. gabrielsoni Murie: Amchitka rock ptarmigan. Resident on Amchitka, Little Sitkin, and Rat islands, Aleutian Islands.

L. m. sanfordi Bent: Tanaga rock ptarmigan. Resident on Tanaga and Kanaga islands, Aleutian Islands.

L. m. chamberlaini Clark: Adak rock ptarmigan. Resident on Adak Island, Aleutian Islands.

L. m. atkhensis Turner: Atka rock ptarmigan. Resident on Atka Island, Aleutian Islands.

L. m. yunaskensis Gabrielson and Lincoln: Yunaska rock ptarmigan. Resident on Yunaska Island, Aleutian Islands.

L. m. nelsoni Stejneger: Northern rock ptarmigan. Breeds in northern Alaska and northern Yukon south to the eastern Aleutians, the Alaska and Kenai peninsulas, and Kodiak Island and east to the western Yukon. Also breeds in Siberia from the Chukotski Peninsula to the Yamal Peninsula and northern Urals, south to Kamchatka, the Kuriles and the mountains of Amurland and Transbaikalia.

L. m. rupestris (Gmelin): Canada rock ptarmigan. Breeds from northern Mackenzie, Melville Island, northern Ellesmere Island, and southern Greenland south to central British Columbia, southern Mackenzie, southern Keewatin, Southampton Island, northern Quebec, and Labrador.

L. m. dixoni Grinnell: Coastal rock ptarmigan. Resident on the islands and coastal mainland of the Glacier Bay region of Alaska and on the mountains of extreme northwestern British Columbia south to Baranof and Admiralty islands.

L. m. welchi Brewster: Newfoundland rock ptarmigan. Resident in Newfoundland.

L. m. captus Peters: North Greenland rock ptarmigan. Breeds in Greenland from Blosseville Coast on the east northward to the northern tip, and southward on the west coast to Melville Bay. Also breeds on Ellesmere Island, Canada.

L. m. saturatus Salomonsen: West Greenland rock ptarmigan. Resident in Greenland from Upernavik District southward to Egedesminde District. Not found in northern Canada (Browning 1979).

L. m. reinhardti (Brehm): South Greenland rock ptarmigan. Resident in southern Greenland from Godthaab District on the western coast southward and eastward to Blosseville Coast on the east, intergrading with *saturatus* to the west (Browning 1979).

L. m. nadezdae Serebrovsky: Siberian rock ptarmigan. Breeds in the Russian Altai and Sayans east to the Khamar Daban Range south of Lake Baikal, and south to Mongolia.

L. m. ridgwayi Stejneger: Commander Islands rock ptarmigan. Resident on Bering and Madny islands in the Commanders.

L. m. japonicus Clark: Japanese rock ptarmigan. Resident in the Japanese Alps of central Honshu.

L. m. hyperboreus Sundevall: Spitsbergen rock ptarmigan. Resident on Spitsbergen, Franz Josef Land, and Bear Island.

L. m. islandorum (Fager): Iceland rock ptarmigan. Resident on Iceland.

L. m. mutus (Montin): Lapland rock ptarmigan. Breeds in Lapland from the Kola and Scandinavian peninsulas southward to southern Norway, Sweden (Dalarna), and Finland to about 67° N latitude.

L. m. millaisi Hartert: Scottish rock ptarmigan. Resident in northern Scotland and some of the Inner Hebrides.

L. m. helveticus (Thienemann): European rock ptarmigan. Resident in the Alps from Savoie east to Austria, and north to southern Bavaria in Germany.

L. m. pyrenaicus Hartert: Pyrenees rock ptarmigan. Resident in the central and eastern Pyrenees in France and Spain.

MEASUREMENTS

Folded wing: Adult males 172–236 mm; adult females 163–222 mm (males average 10–15 mm longer than females).

Tail: Adult males 97–131 mm; adult females 85–115 (males average 10–15 mm longer than females; measurements of *hyperboreus* are substantially larger than those of other races).

IDENTIFICATION

Adults, 12.8–15.5 inches long. Both sexes carry blackish tails throughout the year, and although the scarlet comb of males is most evident during the spring, it is also apparent to some extent through the summer. In the summer males are extensively but rather finely marked with brownish black and various shades of brown and lack the rich chestnut tone of male willow ptarmigans. In summer females are more coarsely barred and are generally lighter overall but have somewhat finer markings than do female willow ptarmigans. Females have definite barring extending to the throat and breast, rather than having these areas finely barred or vermiculated as males do. In autumn males are generally pale above, with tones of ashy gray predominating (tawny brown predominates in some Aleutian races), and females at this time have relatively more brown and fewer black markings, plus a sprinkling of white winter feathers. Both sexes in winter are mostly white with blackish tails, and males (but not all females) have a black streak connecting the bill with the eye and extending somewhat behind the eye.

FIELD MARKS

The smaller, relatively weaker, and entirely black bill of the rock ptarmigan is sometimes detectable in the field and serves to separate this species from the willow ptarmigan in all seasons. In the winter the presence of a black line through the eyes is also diagnostic, but its absence does not exclude this species. For plumage distinctions useful in separating the willow and rock ptarmigans, see the account of the preceding species. During the breeding season the rock ptarmigan is found in higher, rockier, and drier country than the willow ptarmigan, but they may occur together during winter and intermediate periods. In all seasons the dark tail distinguishes the rock ptarmigan from the white-tailed ptarmigan.

AGE AND SEX CRITERIA

Females lack the reddish ''eyebrows'' of adult males and in summer are more heavily barred, with dark markings both above and below. In autumn the barring is reduced in the female, which is still somewhat more heavily marked than the grayish and finely vermiculated male. In winter the sexes are nearly identical, but females usually lack the black stripe through the eye (loral stripe) that is present in males (Godfrey 1966). A Finland study indicated that about 90 percent of 137 winter specimens could be sexed accurately by using the black loral stripe. Further, all black-tailed birds were males, but nearly 35 percent of the males had some brownish or whitish on the tail, which was typical of all the females studied (Pullianinen 1970*a*).

Immature females are browner and more narrowly barred with blackish brown above and on the breast than are adult females in autumn (Ridgway and Friedmann 1946). The pointed condition of the outer primaries has been reported to be an unreliable indicator (Weeden 1961). Instead, young rock ptarmigans may be distinguished by the fact that in adults the ninth primary (second from outside) has the same amount of pigment as the eighth, or less, whereas immature birds have more pigment on the ninth (Weeden and Watson 1967). This same criterion evidently applies to rock ptarmigans in the USSR, since Semenov-Tian-Schanski (1959) reported that in immatures the outer two primaries are pigmented with black, while the adults the black pigment is lacking.

Juveniles may readily be recognized by the presence of at least one brown primary or secondary feather (the eighth primary is the last to be molted). These feathers are typically mottled with pale buff (Ridgway and Friedmann 1946).

Downy young of the rock ptarmigan are usually paler throughout than those of willow ptarmigan, and the crown is lighter and more chestnut colored than the blackish brown crown of the willow ptarmigan (Watson, Parr, and Lumsden 1969). See willow ptarmigan account.

DISTRIBUTION AND HABITAT

The most arctic-adapted of all the grouse, the rock ptarmigan is more widely distributed in the high arctic than is the willow ptarmigan. It also extends south to Hudson Bay during the breeding season and undertakes considerable southward movement during winter, sometimes occurring as far south as James Bay. Unlike the willow ptarmigan, the rock ptarmigan breeds as far north as Ellesmere Island and on adjacent Greenland to its northern limits at approximately 83° N latitude. Also unlike the willow ptarmigan, this species can survive in the rocky desertlike habitat of the high arctic, which may be a limiting factor in the northern distribution of the willow ptarmigan. Weeden (1965*b*) reports that typical breeding terrain of the rock ptarmigan consists of moderately sloping ground in hilly country, such as the middle slopes of mountains. Typically the vegetation is fairly complete, but it may be sparse on the highest and driest slopes. Shrubs are usually from 1

6. Current North American distribution of Adak (Ad), Amchitka (Am), Atka (At), Attu (Au), Canada (C), coastal (Co), Iceland (I), Kista (K), northern (N), Newfoundland (Ne), North Greenland (No), South Greenland (S), Tanaga (T), West Greenland (W), and Yunaska (Y) races of the rock ptarmigan. Stippling indicates southern wintering limits.

to 4 feet tall and are concentrated in ravines or other protected sites, while most plants are usually less than 1 foot tall. Many creeping or decumbent woody plants are typical, as well as rosette forms, while sedges and lichens are usually abundant. Breeding terrain rarely extends below the upper limits of timberline and usually occurs from 100 to 1,000 feet above timberline in hilly country.

There have probably been few changes in the distribution of the rock ptarmigan in historical times, since it is the species least likely to be affected by human activities.

POPULATION DENSITY

Weeden (1963) has summarized population density figures for rock ptarmigans based on various studies in the Northwest Territories. These estimates range from as many as 8 adults per square mile to 4,000 adults on 12,500 square miles. Based on a five-year intensive study on a 15-square-mile study area in Alaska, Weeden (1965*a*, *b*) reported yearly spring densities of males varying from 5.9 to 11.3 per square mile. Slightly lower estimates of female populations were obtained for the same period.

In a study of Scottish ptarmigans, Watson (1965) estimated spring populations to be as high as 1 pair per 2 to 3 hectares (approximately 5 to 7.5 acres) in peak years on the best habitats. However, unlike the fairly uniform heather (*Calluna*) habitats favored by red grouse, the arctic-alpine breeding vegetation is typically more varied, and an area of 100 or more acres rarely contains no unfavorable habitat. Thus, extrapolating local density figures to large areas is unprofitable; this also helps explain the wide differences in densities reported on small, favorable areas and those estimates based on large regional surveys. Watson (1965) estimated that, in peak years, spring numbers on his study area of 1,220 acres were as high as 15 to 18 birds per 100 hectares (247 acres), and as low as 5 in one year.

In an area of the central Alps, the breeding density of rock ptarmigans was estimated by Bossert (1980) at an average of 4.7 males or 3.2 pairs per square kilometer (a male per 52.5 acres or a pair per 77 acres), with little yearly variation since 1973. However, there are marked annual variations in many areas, and the species appears to be distinctly cyclic in some areas such as Iceland, where a ten-year population cycle is apparently present (Watson and Moss 1979). A ten-year cycle may also be present in Alaska (Weeden and Theberge 1972), while in Scotland there is some evidence of a six-year cycle, according to Watson and Moss (1979).

HABITAT REQUIREMENTS

Wintering Requirements

In Alaska, rock ptarmigans winter in such locations as shrubby slopes at timberline, in large forest openings where shrubs, especially birch, project above snow level, and,

7. Current Eurasian distribution of Commander Islands (C), European (E), Lapland (L), northern (N), Pyrenees (P), Siberian (S), Spitsbergen (Sp), and Scottish (Sc) races of the rock ptarmigan. Stippling indicates areas of uncertain distribution.

rarely, in riparian willow thickets (Weeden 1965*b*). Watson (1965) noted that in Scotland the birds moved down from their arctic-alpine breeding grounds into a moorland zone of heather that was used by red grouse during the breeding season. Ptarmigans can scratch through a few inches of soft snow to reach plants, but Watson did not find them burrowing under the snow to forage. Local variations in topography caused areas to be blown fairly free of snow periodically, exposing food plants, and the birds will move from one such area to another in search of food. Watson noted little if any competition for food between ptarmigans and red grouse, since the two species remained almost completely separated during winter. As I mentioned in the willow ptarmigan account, considerable separation of the sexes occurs in North American willow and rock ptarmigans during winter, with males remaining in more alpinelike habitats while the females tend to move into relatively protected situations.

Spring Habitat Requirements

Territorial requirements for the rock ptarmigan consist of a large proportion of relatively open vegetation than is the case for willow ptarmigans (Weeden 1965*b*). Some territories contain no shrubs at all, and males utilize rocks, knolls, or similar elevations for territorial display and for resting. Watson (1965) reported that ptarmigans were most common where large boulders or outcrops occurred on stunted heath or a mixture of stunted heath and grassy vegetation. The birds rarely took territories on pure grassland, tall heaths, bogs, or stone fields without healthy vegetation. Favorite areas for territorial establishment were usually on varied heaths or a mixture of varied heaths and grasses. The highest territorial densities were on areas of nearly continuous heath broken up by large boulders, slightly lower densities were found on scattered patches of heath, and much lower densities occurred on areas of continuous heath with only a few boulders present. Territorial densities were lowest on bare, gravelly places with only scattered vegetation and boulders.

Nesting and Brooding Requirements

Nest sites for the rock ptarmigans may have less overhead concealment than those of willow ptarmigans, but some overhead protection is usually present (Watson 1965). Parmelee, Stephens, and Schmidt (1967) indicated that the nesting habitat is usually dry and rocky and sometimes is barren and high but may consist of wet tundra sites with heavy vegetation where willow ptarmigans also breed.

Brooding habitat is similar to nesting habitat, but broods tend to gather in swales on ridges and upper slopes (Weeden 1965*b*). They avoid dense shrubs and after beginning to fly at 10 or 11 days of age escape by flying out of sight over knoll ridges.

FOOD AND FORAGING BEHAVIOR

The best source of information on rock ptarmigan food habits in North America is the research by Weeden (1965*b*), based on 482 crop samples from interior Alaska. Winter foods there consist primarily of dwarf birch buds (*Betula*) and catkins, followed by willow buds and twigs (*Salix*). Dried leaves of shrubs extending above the snow are also taken in limited quantities.

In most areas where the species has been studied in winter, it takes its foods at that season mainly from the ground. Particular foods taken seem to vary greatly and are strongly influenced by snow conditions. Generally, low-growing species such as *Vaccinium* spp., evergreen herbs such as *Andromeda*, berry-producing species, and the buds and twigs of birches and willows are important foods (Myrberget 1979). Winter foods are typically those that are easily accessible and have a high sugar content, although those with a high protein content are distinctly preferred when they are available (Bossert 1980). The preference for foods that are high in such nutrients as nitrogen and phosphorus continues into the breeding season, particularly among breeding females (Cramp and Simmons 1980).

Spring foods, based on relatively few samples, appear to include variety of plant materials, including the new growth of shrubs, horsetail tips (*Equisetum*), and a small amount of birch and willow materials. Summer foods include an even greater array of plant foods, which consist largely of leaves and flowers in early summer and berries and seeds later on. Blueberries (*Vaccinium*), crowberries (*Empetrum*), and mountain avens (*Geum*) are important food sources during this time. During fall, blueberries and heads of sedges (*Carex*) are important, and dwarf birch begins to assume the great importance that will continue throughout winter.

Reporting on birds taken on Baffin Island, Sutton and Parmelee (1956) noted that in the crops of 8 adults taken in May about 60 percent of the total food materials consisted of buds and twigs of willow, 32 percent was the leaves and twigs of dryas (*Dryas*), and the remainder consisted of *Saxifraga, Draba,* and willow galls. A newly hatched chick had eaten leaves of crowberry (*Empetrum*).

Moss (1968) has made an interesting nutritional comparison of foods taken by rock ptarmigans of the Icelandic and Scottish populations. In Iceland the birds have a diet predominantly of twigs of willow, leaves of dryas, the leaves and bulbils of *Polygonum*, which are relatively high in nitrogen and phosphorus, and berries of *Empetrum*, which are high in soluble carbohydrates. By comparison, the Scottish ptarmigans subsist on a relatively nutrition-poor diet of heather (*Calluna*), *Vaccinium*, and *Empetrum*. Correlated with this is the fact that in Iceland the ptarmigans have an average clutch size of about 11 eggs, whereas in Scotland the clutch is usually 6 to 7 eggs, averaging 6.6. The average clutch size in Alaska, based on studies made by Weeden (1965*a*), is essentially the same as in Scotland. Significant annual differences in clutch sizes do occur in Alaska and apparently also in Scotland, but they have not yet been adequately correlated with population density or food quality. Lack (1966) has suggested such a possible correlation between clutch size and heather conditions. Watson (1965) believed that annual differences in clutch sizes were unimportant compared with variations in chick survival. At least in the red grouse, chick survival may be related to the physical condition of the hens as determined by food supplies.

A possibly significant point related to food supplies and reproductive success is that although the rock ptarmigan is the most northerly breeding of the ptarmigans, it is considerably smaller than the willow ptarmigan. Likewise, the alpine-breeding white-tailed ptarmigan is much smaller than either the rock or willow ptarmigan, in contrast to what might be expected with arctic-breeding birds (Bergmann's principle). The possibility exists, therefore, that smaller body size in the rock and white-tailed ptarmigans is an adaptation to reduced food supplies and has evolved relatively independently of selective pressures related to environmental temperatures. Yet Irving (1960) reported that willow ptarmigans collected in arctic localities of Alaska averaged 90 grams heavier then those from subarctic points some 600 miles south. Further, winter birds tended to be heavier than summer birds, and males, which averaged 10 to 40 grams heavier than females, wintered in more hostile environments.

Whereas Irving (1960) found that the willow ptarmigans at Anaktuvuk Pass are migratory, the rock ptarmigans there are not, and in winter they feed on high, rounded slopes where low vegetation is exposed. Also, although willow ptarmigans often retreat with their crops filled with from 50 to 100 grams of food to burrows some 1½ to 2 feet under the snow, this behavior is apparently not typical of rock ptarmigans. Manniche (cited in Bent 1932) does indicate that in Greenland the birds may spend the night in holes about 20 centimeters deep on the lee side of rocks or in narrow snow-filled ravines in the rocks. MacDonald (1970) noted that the birds would dig roosting forms deep enough that only their heads remained above the snow or would use the depressions caused by humans walking across the snow.

MOBILITY AND MOVEMENTS

The relatively large heart size (Johnson and Lockner 1968) of the rock ptarmigan suggests that it may be capable of considerable movements, but there is little detailed information on actual daily or seasonal movements in the species. Snyder (1957) stated that the bird is migratory to an appreciable degree in arctic Canada, and Weeden (1964, 1965b) reported that some low-altitude wintering grounds of the species are a minimum of 10, and probably 15 to 20, miles from the nearest alpine breeding areas. Weeden believed that, at least in the lower parts of the wintering range, rock ptarmigans move in an unpredictable fashion. By March and April, however, movements are quite limited and consist of visits to various feeding areas separated by up to half a mile or more, the stays at these areas lasting varying lengths of time. Irving (1960) reported that at Old Crow, Alaska, wintering birds might convene from a nesting area some 30 miles in diameter, but no actual evidence for a regular migratory pattern was indicated. Bent (1932) indicated that although the majority of rock ptarmigans withdraw from the northern limits of ther summer range, they do not usually retreat beyond the southern limits of their breeding range. Nelson (cited in

Bent 1932) reports a regular fall evening migratory movement across Norton Sound, via Stuart Island, and a comparable spring flight in April.

Movements of rock ptarmigans in Eurasia appear to be highly variable according to area. In much of Europe including Russian Lapland movements during late fall and winter are largely altitudinal and depend on local weather conditions. When there is deep snow, Scottish birds may descend to the heather moorland zone, where they come into contact with red grouse, and when weather is severe they may even reach the edges of forests and cultivated areas. In northern Norway, nine autumn movements of banded birds averaged only 8.4 kilometers and were in random directions, while two north-south winter movements were of 22 and 104 kilometers (Myrberget, cited in Cramp and Simmons 1980). Longer movements may occur in such high-arctic populations as those of *hyperboreus* on Spitsbergen, where interisland movements sometimes occur. Similarly, in Greenland the races *captus* and *saturatus* are distinctly migratory. These birds develop immense flocks during migration and sometimes winter surprising distances away. One bird, banded on Disco Island in July, was recovered more than 1,000 kilometers to the south the following February (*Ibis* 114:582). Similarly, nomadic flock movements seem to be typical of Icelandic birds during the winter, though sexual segregation in winter is not yet reported for western Palaearctic birds (Cramp and Simmons 1980).

Weeden (1965b) noted that in Alaska the rock ptarmigans disappear from their wintering areas at low altitudes in March and April and that in 1962 the first migrants arrived at their Eagle Creek breeding ground study areas on March 29. This movement continued through April, and during April males begin establishing territories, in advance of the arrival of most hens. In the study area, northeast of Fairbanks, egg laying begins in the second to fourth week of May. Farther north at Old Crow and Anaktuvuk Pass, the males become territorial in late April and May. By comparison, the first flocks of rock ptarmigans that Parmelee, Stephens, and Schmidt (1967) saw on Victoria Island arrived in mid-May and were all males. The first territorial flights were noted on May 19, and the first female was seen on May 23. Fresh eggs were noted from June 3 until late June, or nearly a month later than in central Alaska. Interestingly, the weights of spring males collected on Victoria Island averaged about 100 grams more than Irving reported for Anaktuvuk Pass and Old Crow, and females averaged about 90 grams heavier.

REPRODUCTIVE BEHAVIOR

Territorial Establishment

The period of breakup of winter flocks and establishment of territories probably varies greatly by locality and year. In Scotland, Watson (1965) noted that this behavioral transition occurs with the coming of spring thaws and sunny weather, which may be as early as the first part of January or as late as the end of April. In North America, where the birds usually move out of their breeding areas during the winter, there is probably a fairly short lag between the arrival of the males on the breeding ground and the establishment of territories. The observations of Parmelee, Stephens, and Schmidt (1967) indicate that this lag may be as short as a few days. Both yearling and adult male ptarmigans establish territories; Weeden (1965a) found that the percentage of first-year ptarmigans in male breeding populations varied from 41 to 67 percent. Yearling females made up from 17 to 75 percent of the breeding populations, and there was no evidence of any nonbreeding by females.

Agonistic and Sexual Behavior

MacDonald's recent observations (1970) on Bathurst Island indicated that individual males there may defend surprisingly large areas of about 1 square mile, which include several lookout prominences adjacent to moist hummocky tundra with heavy vegetation. From these points the male watches for other ptarmigans, attacking males and courting females. During the early stages of territoriality the male spends much of his time advertising his location with song flight displays. As his aggressiveness increases, the size and brilliance of his eyecombs also increase. Territorial males, on seeing a rival male, engage in aerial chases with tails spread, combs erected, and bodies rocking from side to side while in flight. Aerial chases of females were not seen by MacDonald but have been reported by Weeden.

The basic territorial advertisement display of the rock ptarmigan is the song flight. MacDonald noted that the height of this display flight varies from as little as 4 feet early in the season to an estimated 250 feet observed in a highly aggressive male. The display may be performed spontaneously or may be elicited by a disturbance of some kind within hearing or visual range of the male. The bird typically leaps into the air, uttering a loud, belching call, and swiftly flies forward and upward with alternate wing flapping and sailing. At the end of the climbing flight, the male sets his wings, fans his tail, and begins an upward soaring glide until he finally reaches stalling speed. At this point he swells his neck and begins to utter a series of staccato, belching notes. As the bird begins his descent on bowed wings he utters a second series of belching notes and slowly parachutes downward toward the ground. Just before landing he tilts his spread tail vertically downward, and as he alights he quickly cocks it back upward to a near-vertical position. The wings are held to the side of the body and drooped toward the ground as the male stands with an erect neck or runs forward a short distance while uttering a staccato call. Then the neck is deflated, the primaries are lowered so they drag on the ground, and the tail is fully spread while being tilted at an angle of 45°. Next the bird begins a short forward run, simultaneously extending his neck and making a single slow bowing movement with his head. When a female is newly present on his territory, the male may run in an arc toward her, tilting his tail toward her and extending one wing away from her. The head is also tilted toward the female, exposing the enlarged eyecombs. After a female has become established on a male's territory, this ground display is omitted. Females evidently gradually associate themselves with a specific male and his territory, initially following

21. Male displays of the rock ptarmigan (after MacDonald 1970), including (1) parachuting descent from a flight song display, (2) posture taken immediately after landing, and (3–5) stages in ground display following a flight song.

him in flight and later being followed by him. MacDonald noted that at least one male mated with three females in one season, all of which nested in his territory.

When two territorial males meet, violent fights may ensue. Threats may be uttered as the birds sleek their plumage, inflate their necks, and close their tails so they are nearly hidden. Their crowns may be raised or lowered, the combs erect or concealed. During attacks the birds attempt to grasp each other with their bills while striking with their wings. Often feathers from the neck may be pulled out, and sometimes the eyecombs are torn.

Pair formation in rock ptarmigans is apparently gradual, judging from MacDonald's observations. He noted that while the resident male drives other males off his territory, the female becomes more submissive and dependent on him, increasingly relying on the male to warn her of danger. When near the female he continuously utters a contact call consisting of ticking notes, which change to a ratchetlike alarm call when there is possible danger. When a female is thus alerted, she flushes and is immediately followed by the male, which may perform a song flight before landing. As the male returns to the female after the song flight he may perform the head bowing and tail tilting described earlier. He typically circles the female at a distance of up to 2 feet, with his head held low, his wings dragging, and his tail tilted toward her. Apparently he attempts in this manner to direct the female into a tundra depression, seemingly trying to induce her to crouch in it. In four observed instances of copulation, the female crouched in such a depression, partially extending her wings and exposing her white wrists. The male then stepped on her back and pecked at her nape but did not grasp her neck feathers. Rather, he remained with his body rather upright during copulation, finally bending forward and walking off her back over her shoulder. Then, with his head lowered and held forward, his tail spread and held vertically toward the female, and his wings dragging, he walked in a circular path around the female, with his combs greatly enlarged and his bill open. The female remained crouched for a time, then stood up, shook her plumage, and preened. In two cases the female ran from the male before he completed his postcopulatory display, and in one case the male circled around the female twice while she remained crouched.

MacDonald obtained some data indicating that males were more highly attracted to mounted specimens of females that had piebald brown-and-white plumage than to whiter females, which is of special interest since females molt into their brown nuptial plumage much earlier than males, which remain white and highly conspicuous throughout the pair-forming period.

Vocal Signals

MacDonald (1970) was the first person to try to describe the vocalizations of adult rock ptarmigans, and subsequently Watson (1972, and in Cramp and Simmons 1980) provided a more complete analysis. Most or all of the large and complex vocabulary of this species is common to both sexes, but there is a considerable sexual dimorphism in the vocalizations that is probably the result of the male's special inflatable tracheal air sac. This sac is evidently inflated during vocalizations and presumably is responsible for the male's

202

low-pitched vocalizations (MacDonald 1970). The value of low-frequency sounds to the rock ptarmigan seems to be correlated with the apparently large territories that males typically hold, and also with their long-distance visual signals in the form of the striking black-and-white plumage pattern of breeding males.

The flight song of males is a loud belching call that is uttered at the start of the descent phase of the song flight, which is primarily a loud *AA*, followed by a series of loud *ka* notes that often leads to a cackling *ka-ka-ka-ka* . . . as the bird approaches the ground and alights. This is sometimes prolonged after landing, but a softer, slower, and hollow-sounding *kwa-kwa-kwa* typically ends the song. The call is uttered primarily by territorial males. A flushing song, similar to the flight song, is uttered when the bird is flushed by humans or by another male and usually sounds like *AAr-aa-ka-ka* and may be followed by prolonged cackling. The ground song or perch song is a loud snoring call that usually leads to a repeated cackling or is followed by a phrase similar to the flight song and is usually uttered during threat situations. The attack call is a loud, repeated *kwa-kwa-kwa*, much like the last part of the flight song, and the threat or attack-intention call is a fairly high-pitched *ko-Wa-o*, *ko-Wa-ka*, or *ko-Wa-ee-ak*. While threatening other males in a lateral stance, the male often utters a coarse, low-pitched, rattling *krrr* call, and this is sometimes also directed toward humans or toward females. The flight-intention call is a rapid, hollow-sounding repeated *kuk* or *kwa* note, similar to but sharper than the attack call. The chase call, used mostly by females being chased by males but also at times by the pursuing male, is a sharp, popping, repeated *ko-Wa* note. When approached by a fox or dog, adults utter a single, low *kwuk* or *kwa* warning call that causes other adults to utter the same call; when uttered by males in summer this call stimulates females and young to crouch. A similar hawk-alarm call is also uttered. Both sexes utter a "sexual call" during sexual activity, while digging nest scrapes, and while dusting, which is a soft, rapid and chattering *KOO-koo-koo*, or *KOAH-koah-koah*. It has been also heard uttered by captive females while laying. A defensive hissing call is produced by sitting females when disturbed and by both sexes during distraction display or while being handled. The female's typical distraction call is a high-pitched *kik* note. When under attack by an aerial predator, the birds also utter a defense call. Finally, females utter an assembly call to gather together chicks that have been hiding or those uttering distress calls.

Nesting and Brooding Behavior

Female ptarmigans locate their nests within the territorial boundaries of the male. In Scotland at least, the numbers of females associated with territorial males is rarely more than 50 percent (Watson 1965); thus few if any males are normally likely to acquire more than one female. Weeden (1965*b*) reports that in Alaska two females may sometimes mate with a single cock, and presumably both hens nest within his territory. To what extent the male defends the female and her nest is still not very clear for the rock ptarmigan. Höhn (1957) described how, when two female rock ptarmigans were shot, the male quickly approached and displayed to the corpses, but this kind of behavior clearly does not belong in the category of female defense. Weeden (1965b) noted that about 1 brood in 20 will have a male in attendance, but he never observed any actual brood defense by males. However, MacDonald (1970) reported several cases of brood defense by males, including both attack and distraction behavior.

Rock ptarmigan females build simple, shallow nests, the depressions often being little more than might be caused by the weight and movements of the brooding hen (Weeden 1965b). Clutch sizes vary considerably by locality and by year. Weeden (1965a and unpublished Alaska Fish and Game Department *Game Bird Reports*, vols. 7–10) noted clutch sizes varying annually between 1960 and 1969 from 6.4 to 9 eggs, and the average size of 195 clutches was 7.2 eggs. In the more arcticlike environment of Victoria Island, Parmelee, Stephens, and Schmidt (1967) found three nests, two containing 11 and one containing 13 eggs, suggestive of somewhat larger clutch sizes at higher latitudes. Judging from Weeden's data (1965a), about two-thirds of the nests hatch during an average year. Renesting is apparently not common enough to affect overall productivity. Weeden (1965a) provided data indicating an average brood size in August of 5.3 for 208 broods, with yearly averages ranging from 4.8 to 6.1 between 1960 and 1964. By comparison, Watson found that the average size of full-grown broods between 1945 and 1963 was from 1.2 to 6.2 young. Watson found that, on the average, 38 percent of the females went broodless each year, but in different years it varied from none to over 80 percent. Weeden (unpublished Alaska Fish and Game Department *Game Bird Reports*, vol. 8, 1967) reported that, between 1963 and 1966, 60 percent of 130 year-old females were seen with young, while 77 percent of 185 older females had young; thus, incubating or brooding efficiency evidently increases with age of the female.

The female is highly attentive to her young and when disturbed by humans utters a throaty *krrr* during distraction behavior (Sutton and Parmelee 1956). When calling chicks toward her, she utters a clucking *kit* or *krit* call. Weeden (1965b) indicates that by imitating the distress peeping of a chick, he could elicit a low, crooning note that carried up to 100 yards and helped locate brood hens.

Weeden (1965b) noted that one brood seen in 1960 moved about 4,200 feet in 5 days, while another was found only about 50 feet from the point where it had been seen 10 days before. Of two broods that were seen again after 28 days, one had moved about 50 feet and the other 7,800 feet. In general the broods stayed within an area of about ½ square mile but did not appear to be attracted by the male's former territory. By late July most broods had moved to areas higher than the nesting sites, congregating on moist and gentle slopes where sedges, grasses, forbs, and low shrubs predominated. Weeden also found several indications of transfer of individual chicks between broods. Hens that have lost their clutches or broods join the flocks of males that gather on high, rocky ridges or in streamside willow thickets. As the broods mature, they tend to combine, and these flocks in turn attract groups of males and nonproductive hens. In time, flocks of 50 to 300 individuals may build up. However, at the same time, there is some calling and displaying among the males and an apparent resurgence of territoriality. The possible signifcance of this fall behavior is still unknown.

EVOLUTIONARY RELATIONSHIPS

Some general statements on the evolutionary history of the ptarmigans have been made under the willow ptarmigan account. In addition it might be noted that the rock ptarmigan not only is the most northerly and most widely distributed of all the ptarmigans but also might perhaps be considered most representative of an ancestral ptarmigan type adapted for high-arctic breeding. From such a type the evolution of an alpine offshoot, represented by the white-tailed ptarmigan, and a subarctic type, represented by the willow ptarmigan, might easily be imagined.

White-tailed Ptarmigan

Lagopus leucurus (Richardson) 1831

Other Vernacular Names

Snow grouse, snow partridge; lagopède à queue blanche (French); Amerikanische Alpenschneehuhn (German).

Range

From central Alaska, northern Yukon, and southwestern Mackenzie south to the Kenai Peninsula, Vancouver Island, the Cascade Mountains of Washington, and along the Rocky Mountains from British Columbia and Alberta south to northern New Mexico (A.O.U. *Check-list* 1957). Recently introduced in California, Oregon, Utah, and New Mexico.

Subspecies (ex A.O.U. *Check-list*)

L. l. leucurus (Richardson): Northern white-tailed ptarmigan. Resident above timberline from northern Yukon, western Mackenzie, British Columbia, and west-central Alberta south to the northern border of the United States.

L. l. peninsularis Chapman: Kenai white-tailed ptarmigan. Resident above timberline from south-central Alaska to Cook Inlet and the Kenai Peninsula, extending east and southeast to Glacier Bay and White Pass.

L. l. saxatilis Cowan: Vancouver white-tailed ptarmigan. Resident above timberline on Vancouver Island, British Columbia.

L. l. rainierensis Taylor: Mount Rainier white-tailed ptarmigan. Resident above timberline in Washington from Mount Baker south to Mount Adams and Mount Saint Helens.

L. l. altipetens Osgood: Southern white-tailed ptarmigan. Resident above timberline in the Rocky Mountains from Montana south through Wyoming and Colorado to northern New Mexico. Released in the Wallowa Mountains, Oregon, in 1967–69, the Sierra Nevada, California, in 1971–72, and the Uinta Mountains, Utah, in 1976 (*Southwestern Naturalist* 23:66). Recently also released (1981) in the Pecos Wilderness area of New Mexico (Clait Braun, pers. comm.).

MEASUREMENTS

Folded wing: Adult males 164–94 mm; adult females 155–92 mm (males average 5 mm longer than females).

Tail: Adult males 85–109 mm; adult females 83–98 mm (males average 8 mm longer than females).

IDENTIFICATION

Adults, 12–13.5 inches long. In any nonjuvenal plumage the white tail will serve to separate this species from the other two ptarmigans. Adult males in summer plumage are vermiculated and barred or mottled with black, buffy, and white dorsally, with a buffy or pale fulvous tone predominating on the lower back, rump, and upper tail coverts, and the underparts are mostly white. Unlike the other ptarmigans, the wings as well as the tail (except for the central pair of feathers) are completely white at this season. Females are similar in plumage but have a heavily spotted and more yellowish color dorsally. In the fall both sexes are mostly pale cinnamon rufous above, with fine spotting and vermiculations of brownish black and a lighter head and neck. A few breast feathers are usually marked with white, and the abdomen, under tail coverts, tail, and wings are white. In the winter both sexes are pure white except for a black bill, eyes, and claws.

FIELD MARKS

A small alpine ptarmigan with white wings and tail in summer, or an entirely white plumage in winter, is of this species. It is usually extremely difficult to see against a lichen-covered rocky background and is therefore overlooked unless forced to fly.

AGE AND SEX CRITERIA

Females exhibit eyecombs (unlike the two other ptarmigan species) virtually identical to those of adult males, but in summer hens are more coarsely and regularly barred with black and rich ochraceous buff markings on their brownish back and side feathers, while feathers of males in these areas are finely vermiculated with brown and black. In addition, although males retain their white lower breast, abdomen, and under tail coverts through the summer, females have yellowish buffy brown feathers with some black barring in these areas (Braun and Rogers 1967a). In the autumn differences between the sexes diminish, but for a time females retain a few of their coarsely barred nuptial plumage feathers, especially on the nape, sides, inner wing, and upper tail coverts. In winter birds of both sexes are identical in plumage but may differ slightly in wing length, length of the outer five primaries, and outer rectrix length (Braun and Rogers 1967a). In spring, males can be recognized by their distinctive black-tipped head and neck feathers, which give a "hooded" effect that is lacking in females as they gradually acquire their brown, black, and yellow nuptial plumage (Braun 1969).

Immatures may be recognized by the pigmentation of their two outer primaries (Taber, in Mosby 1963). If black pigment occurs on either the ninth or the tenth primary the bird may confidently be called an immature. Likewise, pigmentation on the outer primary covert is an indication of an immature bird, whereas lack of pigmentation in these areas is typical of adults (Braun and Rogers 1967a).

Juveniles have tail feathers that are yellowish brown centrally or white with mottled brown edges (Ridgway and Friedmann 1946). Until they are all molted, the secondaries and inner eight primaries are also brownish in juveniles (see willow ptarmigan account). Giesen and Braun (1979b) have provided a method for aging young birds (to about 90 days) by using lengths of postjuvenal primaries. By using a single postjuvenal primary, it was possible to determine ages of birds between 17 and 90 days old to within a few days of actual ages.

Downy young of the white-tailed ptarmigan are the least rufous dorsally of all the ptarmigans and have only a suggestion of the usual chestnut crown with its black margin. The two black dorsal stripes are also indistinct, and instead the back has an indefinite blending of buff, gray, sepia, and black shades. The feathered toes will separate downies of this species from any non-*Lagopus* forms.

DISTRIBUTION AND HABITAT

The current distribution of the white-tailed ptarmigan in North America conforms closely to that of alpine tundra, although it does not extend southward along the Cascade and Sierra ranges into Oregon or California, nor does it apparently include the Brooks Range of northern Alaska, both of which would seem to provide suitable habitat opportunities for the species. In the Rocky Mountains of the western states the range of the species is highly disjunctive because of the limited elevations above timberline, and it must be presumed that these southern populations became isolated during Pleistocene times. These southernmost populations are probably the ones most vulnerable to extirpation. Ligon (1961) noted that although the New Mexican range of this species once included all the alpine ridges of the Sangre de Cristo range from Lake Peak to the Colorado line, the birds are now found only on a few peaks near the Colorado line. Braun (1970) reported finding them on Costilla Peak in 1970 and has also verified their occurrence on Baldy Peak near Santa Fe.

8. Current distribution of Kenai (K), northern (N), Mount Rainier (R), southern (S), and Vancouver (V) races of the white-tailed ptarmigan. Populations in Utah, Oregon, and California are the result of introductions of uncertain success.

Braun (1969) concluded that, although the birds may once have occurred in Oregon, Idaho, and Utah, their recent natural occurrence in these states is unproved. Attempts have been made to establish them in the Wallowa Mountains of Oregon, in the Sierra Nevada of California, and in the Uinta Mountains of Utah, as well as in the Pike's Peak area of Colorado (Braun, Nish, and Giesen 1978). These attempts have apparently all been successful; a 1981 release in New Mexico is still of uncertain success.

Except for Alaska, Colorado is the state with perhaps the greatest amount of white-tailed ptarmigan range in the United States. Rogers and Braun (1968) estimate that more than 4,000 square miles in the state are occupied by this ptarmigan.

Weeden (1965b) reported that typical terrain of this species consists of steep slopes and ridges, often around cirques and stony benches, where ledges, cliffs, and outcrops are common. The vegetation is generally sparse, with shrubs nearly absent and dwarfed when present. The birds in Alaska are usually from 500 to 2,000 feet above timberline. In Montana, Choate (1963) found that ptarmigans are not present in timber or in shrubby vegetation more than 18 inches high. Rather, they prefer areas of rocks and moist ledges with alpine vegetation that is low-growing but well developed. Rocks from 6 to 24 inches in diameter provide optimum habitat, since they provide protection from bad weather and cover from visual predators. Ptarmigans are rarely found in boggy areas or areas where the vegetation is taller than the birds themselves. They usually frequent gently sloping areas where moisture is abundant and vegetation is present. Preferred cover plants include willow, heath (*Phyllodoce* and *Cassiope*), and mosses.

Braun (1969, 1970) concluded that in Colorado the distribution and abundance of alpine willow are the key factors determining ptarmigan distribution. Willow represented the bulk of the ptarmigan's food from late September until May, and its occurrence in snow-free areas in late May is an essential component of breeding territories.

POPULATION DENSITY

Choate (1963) reported the overall density of breeding birds on a 2-square-mile plot at 17.5 birds per square mile, but if unsuitable habitats are excluded from consideration, the density could be calculated as 50 breeding birds per square mile. On study areas totaling 8.41 square miles, Rogers and Braun (1968) reported 52 and 56 breeding pairs plus 11 to 25 unmated birds in 1966 and 1967, or 15.2 to 15.5 birds per square mile. In 1968 there were 55 pairs and 21 unmated males on areas totaling 6.93 square miles, or 19.2 birds per square mile, and in 1969 there were 60 pairs and 28 unmated males on 8.41 square miles, or 17.8 birds per square mile (*Colorado Game Research Review*, 1968 and 1969). Summarizing six years of censuses, Braun and May (1972) noted that ptarmigan populations had remained relatively stable on five different Colorado study areas, with breeding densities ranging from 5.2 to 30.8 birds per square mile and averaging about 17 per square mile for all areas. Apparently they exhibit short (six- to eight-year) cycles of low amplitude, varying only 30 to 40 percent between highs and lows (Clait Braun, pers. comm.).

HABITAT REQUIREMENTS

Wintering Requirements

Braun (1969, 1971*b*) reported that wintering areas for ptarmigans in Colorado must contain alpine willows (*Salix nivalis* and *S. anglorum*), and alpine areas lacking these species cannot support ptarmigans for long periods. Braun and Pattie (1969) reported that the Beartooth Plateau of Wyoming almost completely lacks willow in this timberline zone, and willow stands that do occur are snow covered during winter. The birds evidently do not occur there or in certain northern New Mexico peaks where willow is also absent (Braun 1970).

In general, the availability of willow directly controls the distribution of ptarmigans in winter, at least in Colorado. There the major areas preferred for winter use are drainage basins at or above treeline and stream courses below tree line. Birds apparently have a high fidelity to wintering areas, just as they show strong attachment to breeding sites (Braun, Hoffman, and Rogers 1976).

Spring Habitat Requirements

Braun (1969, 1971*b*) reports that willow is an essential habitat characteristic of successful male territories. In Colorado breeding territories are adjacent to the spruce-willow alpine timberline (krummholz) zone and also include small windblown areas. In the Beartooth area of Wyoming this combination of habitat characteristics in the alpine zone is lacking; thus the area is apparently unsuitable as a breeding ground (Braun and Pattie 1969). In Colorado territories are established in suitable habitats where the snow is gone by early May (Braun 1969). Willow bushes over ½ meter tall are extremely important for food until early June, when the tundra vegetation turns green and territories become larger and less confined to areas containing willow (Braun 1971*b*).

Nesting and Brooding Requirements

Nest-site characteristics for the white-tailed ptarmigan are evidently rather broad, judging from the diversity of nest sites that have been found (Schmidt 1969). Of 25 nest sites studied by Braun (1971*b*), 8 were in various krummholz situations, and the rest were in four different vegetation types some distance from krummholz. Aspect seemed to be of little importance, but most nests had rock nearby. Generally, nests were in areas that were somewhat protected from wind and that became snow free by early June.

Brooding areas for females and suitable summering areas for postterritorial males as well as unsuccessful hens occur where the vegetation is short and where rocks 6 inches in diameter or larger cover more than 50 percent of the ground (Braun 1969). The vegetation of suitable meadow areas adjacent to rock fields consists principally of sedges *(Carex)* and forbs such as *Geum* and *Polygonum*.

Fall Habitat Requirements

During late summer and early fall the birds occupy summer sites, but they begin to form flocks and wander over all vegetation types between summering and wintering areas. At this time they favor areas where snow is starting to accumulate, and their patchwork plumage during fall seems to match closely the snow-and-tundra background of such areas at this time. Furthermore, the areas where snow first accumulates are those that tend to provide the last remaining green plants, and they are also the last to become snow free in spring (Braun 1971*b*).

FOOD AND FORAGING BEHAVIOR

Weeden (1967) has reported on the analysis of 167 crops of this species collected from Colorado to Alaska. Winter foods of Alaskan populations differ from those in Colorado in that alder (*Alnus*) catkins are an important part of the winter diet, with willow (*Salix*) and birch (*Betula*) of secondary importance. In contrast, Colorado ptarmigans subsist largely on the buds and woody twigs of various alpine willows (Quick 1947). Weeden attributed this difference to the increased availability of alder in northern areas, and to possible competition from other species of ptarmigans in Alaska.

May and Braun (1972) reported that among 58 winter food samples from Colorado, willow occurred with a 100 percent frequency, but alder also occurred in samples from areas where that species was locally abundant. Coniferous food sources (*Picea, Pinus, Abies*), although readily available, are rarely taken in winter (May 1970). With spring the birds eat a diversity of green leaves and flowers, though willow remains the most important food. The leaves and flowers of *Potentilla, Ranunculus, Saxifraga,* and *Dryas*, all of which are high in protein, are other important spring foods. During summer the birds also eat a diverse array of leaves and seeds, and the bulbils of *Polygonum viviparum* are an important summer food for adults. During their first 2 weeks juveniles feed largely on invertebrates, then they too begin to feed extensively on these bulbils. Gradually willow gains importance over *Polygonum* for both juveniles and adults, and eventually the birds go back to a diet consisting almost entirely of *Salix* buds and twigs (May and Braun 1972; May 1970).

MOBILITY AND MOVEMENTS

Relatively little is known of white-tailed ptarmigan movements, but certainly there normally is little lateral movement. During winter the birds typically descend to the edge of the tree line, where food is more readily available. In Colorado, ptarmigans gather in flocks of 5 to 30 birds in high alpine basins where willows are abundant (Quick 1947). Single birds also sometimes occur in alpine firs (*Abies lasiocarpa*), limber pines (*Pinus flexilis*), or on steep rock slopes during winter, but when flushed they usually drop down into the snow basins below. Weeden (1965*b*) indicated that in Alaska most birds remain above the timberline, feeding in areas such as steep cliffs, ridgetops, and benches that are blown fairly free of snow. In parts of southwest Colorado the birds go to low valleys every winter regardless of snow cover (Braun and Rogers 1967*b*). During early winter in Colorado, flocks of up to 50 ptarmigans can be found in areas containing available willow, but later the sexes tend to segregate, with males moving nearer timberline and females remaining in the large willow expanses at lower elevations. Birds may move as much as a mile in a day during winter and up to 15 miles over longer periods (Braun and Rogers 1967*b*).

Hoffman and Braun (1975) reported on 45 movements of birds to wintering areas and 99 return movements to breeding areas in Colorado. They noted that females were consistently more mobile than males, with subadult females moving the greatest distances and adult males the shortest. Average male movements were under 2.5 kilometers for adults and 4.0 kilometers for subadults, while these respective age groups of females move 6.8 and 8.5 kilometers. The longest recorded movement was 22.7 kilometers for a subadult female.

In spring, Colorado ptarmigans move back up to the breeding areas, which in the case of males may be a distance of less than a mile. Movements of both sexes are very restricted during the breeding and nesting periods, with birds rarely moving more than 500 yards (Braun and Rogers 1967*b*). When broods appear, males and broodless females move uphill into higher rocky summering areas that may be up to 2 miles from nesting areas, where the birds once again become fairly sedentary. Hens may also move their broods as much as 1/3 mile to such summer brood-rearing areas (Braun and Rogers 1967*b*). Subadult males and unsuccessful hens move considerably farther than adult males or brooding females, and fall movements of females may exceed 10 miles (Braun 1969).

Daily movements probably differ considerably according to sex, age, time of year, and weather conditions. Minimum daily movements may occur among brooding females caring for young chicks. Schmidt (1969) noted that one brood moved about 800 yards in 10 hours, and another moved 300 to 400 yards in 3 hours. Similarly, males on breeding territories move very little. Schmidt found in 1967 that males had an average territory size of 19 acres, with maximum use occurring in 5.3 acres, and in 1968, with a better sample, territories averaged 36 acres, with maximum use in a 9.5-acre area. These territorial areas were used over 2½ months, during the entire pair-bond period.

REPRODUCTIVE BEHAVIOR

Prenesting Behavior

Virtually all that is known of the reproductive behavior of the white-tailed ptarmigan comes from the work of Schmidt (1969), which is still unpublished. The following summary is based on Schmidt's observations.

Territorial Establishment

With the return of the males from their timberline wintering areas to the alpine breeding grounds, territories were gradually established, ranging in size from 16 to 47 acres. Within these fairly large defended areas, which overlapped slightly, males were usually to be found in areas of maximum use, of from 3.2 to 15.7 acres. Males typically returned to their territories of past years, and females usually returned to the same territory and the same male each spring. Territorial activity was not strong until the females arrived on the breeding areas, and males would often feed together until that time.

Males were typically monogamous, and Schmidt found that, although males were sometimes found with two females, this was less common than seeing unpaired males. Territories were usually held by males at least 22 months old, with subadults successful in obtaining territories only if they were vacated by older birds. Territorial defense and proclamation became spirited in late April or early May when the females arrived, and the pair-forming period occurred at the same time. The most intensive territorial activity was typically in very early morning or after feeding in the evening, but during foggy periods or snow squalls activity was intense, apparently as a result of restricted visibility.

Male Territorial and Pair-Forming Behavior

Male displays and calls may be discussed according to whether they serve the dual prupose of warding off other males from the territory and attracting females, or whether they are performed only in sexual situations. The basically agonistic territorial signals may be considered first.

Schmidt classified the territorial behavior of males into three general types—the "scream flight," "ground challenging," and intimidation displays—noting, however, that they form a continuum of functions and have certain merging characteristics. The male scream flight, which corresponds to the song flight of willow ptarmigans, consists of the birds' taking off and uttering a raucous call containing four syllables, *ku-ku-KIIII-KIIERR*, lasting about 1 second and being repeated at intervals of about 1 to 3 seconds. Choate (1960) had noted that this flight was sometimes characterized by a steep rise followed by a shallow glide, which Schmidt did not see. This display clearly attracted females and warned rival males of the territorial location. However, the display was sometimes seen in midsummer after territories had been abandoned, and females sometimes uttered a homologous call while the male was calling or when defending chicks.

Ground challenges were uttered from convenient calling posts, and the associated call varied considerably in emphasis, such as *duk-duk-DAAK-duk-duk* or *DAAK-DAAK-duk-DAAK-duk-duk-duk*. Some "long ground screams" closely resemble the flight scream in their last four notes. Intimidation displays performed on the ground included two major postures. These were a flat posture assumed during running and an upright threat posture held during slow walking or while standing still. During these displays the eyecomb was exposed by raising the crown feathers, and low clucking sounds were typically uttered. During territorial border disputes males would usually face on another at distances of from

214

22. Adult displays of the white-tailed ptarmigan (after Schmidt 1969), including (1) male ground challenge posture, (2) male upright intimidation posture, (3) male strutting while circling female, (4) male "pursuit strutting," (5) male postcopulatory posture, and (6) "attack" posture of brooding female.

5 to 30 feet in the upright postures, sometimes making short flights while calling. Aerial chases occasionally occurred.

With the arrival of females on a territory, the responses of resident males changed. Males would chase the individual females that entered their territories and perform several specific postures and calls. The "courtship chase" and associated strutting were much like an aggressive attack toward another male, but the head was held more upright, the tail and under tail coverts were more strongly lifted, the breast feathers were fluffed, and the wings were slightly drooped. When the female attempted to escape from the approaching male, chases typically ensued.

Males sometimes varied their strutting approach to females with a "slow approach" and a rhythmic "head bowing" that resembled the ground pecking "displacement" display of male spruce grouse, but the bill was lowered only partway toward the ground. Frequently the male performed a "waltzing" display as he approached the female and attempted to circle in front of her. While so doing, he tilted his tail toward the female and dragged both wings, with the wing nearer the female held lower than the other. This waltzing display lasted from 1 to 5 seconds and was usually repeated several times in a 20- to 40-second period. No calling was heard during this display.

Evidently pair formation was achieved by the repeated performance of these displays, after which the female followed the male closely, the two birds feeding and resting at the same times. While the female fed, Schmidt heard the male utter "assurance clucks" from fifty to eighty times a minute. When the female rested near the base of a rock, the male typically stood on top of that rock or an adjacent one.

Copulation and the associated behavior patterns were observed only a few times and occurred just before the period of egg laying and incubation. On one occasion Braun (cited in Schmidt) observed an apparent instance of precopulatory invitational "tidbitting," during which the male pecked the ground and uttered a series of low-pitched clucking sounds that stimulated the female to rush over and join in the pecking. As the pair began pecking head to head, the male raised his head, exposed his eyecombs, fluffed his feathers, and drooped his wings. He then began bowing his head over the female while uttering "churring sounds." Then he walked around the female and grabbed her nape, causing her to drop to the ground with her neck extended forward. When mounting and during copulation the male lowered his wings and crouched down on the female. When released, she ran forward in several short dashes, stopping between dashes to shake. The postcopulatory display of the male resembled normal strutting, but the wings were more strongly drooped, and the bird walked in slow steps. In each of four cases, the male moved from 10 to 50 feet before resuming normal feeding. In one case several short dashes were made by the male as well.

One other display Schmidt noted was "tail wagging," which apparently occurred as a displacement activity in times of stress. Schmidt found that it occurred in adults of both sexes and in young only 6 weeks old. Females typically performed tail wagging when approached by a courting male, but only when approached from the side or behind. Displacement feeding movements were also noted in stress situations.

Vocal Signals

In addition to the several calls mentioned earlier, Schmidt noted several other vocal signals. Females emitted hissing sounds when defending the nest, and when performing distraction displays females typically uttered a harsh *craaow* note that apparently served as an alarm call to the chicks. Females also uttered a loud *brrrt*, apparently with a similar function. When the young were older, females uttered "alert calls," running to the cheeping distress calls of young and uttering high clucks in an upright alert posture. Females also uttered soft contact calls in the presence of their broods, and while pecking they made cackling noises that served to attract the young. Schmidt noted that such functional tidbitting behavior had earlier been reported for both willow ptarmigans and sage grouse. It is of interest that so far only in the white-tailed ptarmigan has tidbitting been reported as an adult display pattern, where it possibly serves as a precopulatory attraction signal.

Nesting and Brooding Behavior

Relatively few nesting studies have been made on this species. Choate (1963) reported on 11 nests in Montana that had from 3 to 9 eggs, averaging 5.2. Giesen, Braun, and May (1980) reported an average clutch size of 5.9 eggs for 48 clutches, and a range of 2 to 8 eggs in 56 total nests. Choate (1963) found one known instance of renesting in Montana, and Giesen and Braun (1979*c*) found two definite cases during twelve years of intensive research. They believed that renesting accounted for about 11.5 percent of the broods seen during twelve years in Rocky Mountain National Park, and that it occurred during at least eight of these twelve years. The average size of second clutches was 3.6 eggs, versus 5.9 for initial clutches.

The egg-laying interval of this species is slightly under 1½ days, and the incubation period is 22 to 23 days (Braun 1969). During incubation, females in one study stayed on the nest more than 95 percent of the time between sunrise and sunset and fed primarily at dawn and dusk. Incubating females sometimes eat white feathers and retrieve eggs from as far as 18 centimeters away; both behaviors may tend to reduce nest predation rates, which averaged 43 percent of 60 nests in one study (Giesen and Braun 1979*c*).

In a summary of Colorado studies, Giesen, Braun and May (1980) reported on data from 62 nests and 673 chicks observed during twelve years of study in Colorado. The time of egg deposition was determined for 12 females and was found to range from 8:30 a.m. to 4:30 p.m. MDT, with no evidence of any egg deposition after dark. Females deposited their eggs later during the day on successive days and skipped every third or fourth day. Incubation did not begin until the final egg was deposited. In 5 of 6 observed cases it lasted 23 days, and pipping of eggs occurred between 24 and 48 hours before hatching.

Geisen, Braun, and May (1980) reported that adult birds tend to produce larger average clutches (6.2 vs. 5.5 eggs) than do yearling birds, and since the incidence of yearling birds in the population varies annually, this may influence the productivity of the species.

Slope characteristics were measured for 60 nests and found to range from 0 to 70 percent, with a median slope of 20 percent. Slopes of less than 40 percent were typical of more than 93 percent of the nests. There was no indication that particular slope directions were favored, although more nests were found on south-facing slopes. Only a few nests were below the tree line, and most were within 250 meters of the tree line, in the krummholz zone. About 40 percent of the nests were in rock or boulder fields, and 12 of 20 nests found in turf areas were near rocks or boulders. Most of the nests in krummholz were at the edge of a shrub clump or near an opening; such locations served to conceal and protect the female while still providing an opening for escape. Most of the nests were in natural depressions, which the female may have enlarged slightly. Lining materials were gathered from within 40 centimeters of the nests. All nests were within the mate's territory, which may serve to limit the female's choice of elevation and aspect, but in general snow-free areas of moderate slope and with protection from prevailing winds appear to be preferred for nest sites.

Choate (1963) found an incubation success of 70 percent for nests studied in Montana and a hatching success of 85.5 percent of eggs observed. Braun (1969) reported a nearly identical hatching success of 81.1 percent in Colorado.

The male apparently normally remains with the female until the time of hatching, judging from observations of Schmidt and Braun in Colorado, although Choate (1963) indicated that the pair bond may last only 2 or 3 weeks. Females regularly perform strong nest and brood defense displays, and Schmidt (1969) noted that males may also defend the nest site. Early in the incubation period, a female disturbed from the nest typically skitters over the ground for 10 to 50 feet, with her wings dragging and her head low in a distraction display. As hatching approaches she is more likely to remain at the nest, hissing and spreading her wings. Schmidt never found a male defending a brood, but female brood defense may take several forms. She may attack the intruder, with eyecombs expanded and white carpals exposed, running with the wings extended and head raised and uttering hissing sounds. When the chicks were still very young the female often performed distraction behavior and led the intruder away from the brood. When the chicks were older the female usually uttered "alert calls" or placed herself between the observer and the brood, running back and forth and hissing. When they were from 10 to 21 days old the chicks could fly from 20 to 150 feet, after which they would run and utter cheeping calls. Lost chicks also uttered loud calls, which gradually changed to hoarse *cheer-up* sounds in older birds. When captured, birds up to 12 months old sometimes uttered similar sounds.

Concentration of females with broods occurred on certain favored areas that combined rocky habitat and an abundance of low, rapidly growing herbaceous vegetation. Brood mixing commonly occurred on such areas. Hens remained with well-grown young through the autumn as the birds gradually moved closer to wintering areas (Braun 1969).

EVOLUTIONARY RELATIONSHIPS

General comments on ptarmigan relationships have been made earlier (see willow ptarmigan account). Recent authorities (Höhn 1980; Braun 1969) appear to agree that the white-tailed ptarmigan must have been derived from a relatively early offshoot of ptarmigan stock that became isolated in western North America. Braun agreed with Johansen (1956), who thought the white-tailed ptarmigan originated from ancestral stock of *Lagopus mutus* that arrived very early in North America. Judging from plumage characteristics of downy young as well as adults, I favor the view that such a separation of pre-*leucurus* stock occurred before a subsequent splitting of gene pools that gave rise to the modern rock and willow ptarmigans; thus I believe these two species are more closely related to one another than either is to the white-tailed ptarmigan. Differences in bill size among the three species where they occur together in Alaska and western Canada may be advantageous in reducing foraging competition; thus, indirectly, selection for differences in body size among the three species may have occurred. Weeden (1967) has already suggested that winter foods taken by white-tailed ptarmigans in Alaska may be influenced by competition from the two other species of Alaskan ptarmigans.

Capercaillie

Tetrao urogallus Linnaeus 1758

Other Vernacular Names

Capercailzie; Auerhuhn (German); grand tétras (French); glukhar (male), glukharka or koppala (female) (Russian).

Range

Northern Europe and Siberia east to the basins of the Vilyuy and upper Lena and mountains south of Lake Baikal, south to the Alps, the Balkans, the southern Urals, the steppes of western Siberia, and Altai and northwestern Mongolia, with isolated populations in the Pyrenees and Cantabrian Mountains. Reintroduced in Scotland, where it formerly bred (Vaurie 1965).

Subspecies (after Vaurie 1965)

T. u. aquitanicus Ingram: Pyrenees capercaillie. Resident in the Pyrenees.

T. u. cantibricus Castroviejo: Cantabrian capercaillie. Resident in the Cantabrian Mountains, where very rare.

T. u. urogallus Linnaeus: Northern capercaillie. Resident in Europe (excepting the areas occupied by *aquitanicus* and *taczanowskii*) east through northern Siberia to the basin of the lower Kureika River, and from the northern limits of coniferous forest in Russia, Finland, and Scandinavia southward through Germany and central Europe to eastern France, Switzerland, northern Italy, northern Albania, Yugoslavia, and locally to Bulgaria and Romania, and southward in Russia and Siberia to about 60–63° N latitude. Also resident in Scotland, where reintroduced from Swedish stock. Now declining throughout its European range.

T. u. taczanowskii (Stejneger): Siberian capercaillie. Resident in Russia and Siberia east of *urogallus*, extending south to about 52° N latitude in the west and in Siberia to the gallery forests along the Irtysh and the Ob, the Altai and Sayan ranges, the Tunkan Alps, and the mountains south of Lake Baikal, and east to the Patom Hills northeast of Lake Baikal.

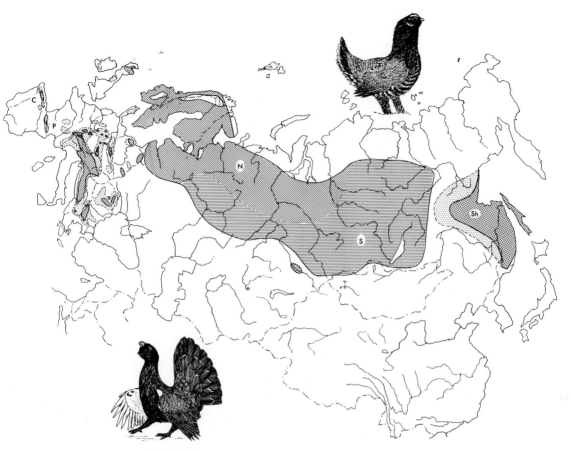

9. Current distribution of the sharp-winged grouse (Sh), and of the Cantabrian (C), northern (N), Pyrenees (P), and Siberian (S) races of the capercaillie. Probable historic distributions of these species are indicated by stippling.

MEASUREMENTS

Folded wing; Adult males 373–425 mm; adult females 290–320 mm. Females average 80–110 mm less than adult males in wing length, but first-winter males range 340–85 mm (Dementiev and Gladkov 1967; Witherby et al. 1944; Glutz 1973). According to Moss, Weir, and Jones (1979), the bill depth (maxilla plus mandible at edge of feathering) is 20–22 mm in first-year, 23–25 mm in second-year, and 25–30 in older birds.

Tail: Adult males 280–355 mm; adult females 155–99 mm. Adults of *urogallus* average 312 and 184 mm in males and females respectively (Cramp and Simmons 1980).

IDENTIFICATION

Length, 24 inches (females) to 34 inches (males). The large size and very dark coloration of this species help to separate it from all other grouse except the black-billed capercaillie. The adult male is generally dark slaty gray with fine vermiculations; the wing coverts are dark brown; the throat and sides of the head are black; the breast is dark glossy blue green; and the tail is black with some irregular and variable white markings. The upper tail coverts have white tips, and the belly and under tail coverts are also marked with white. There is an area of bare red skin above the eye; the bill is whitish; and the legs are feathered to the base of the toes. Females are barred and mottled with buff, black, and grayish white, as are juvenile males. Males in their first winter resemble adult males but are smaller and duller (Witherby et al. 1944).

FIELD MARKS

Like other grouse, this species flies with alternating wingbeats and prolonged glides and often is quite noisy when initially taking flight. The larger size and absence of a forked tail in both sexes may separate this species from the black grouse where both species occur. Young capercaillies closely resemble black grouse in flight but take off and gain altitude much more slowly than do individuals of that species. Where this species occurs with the black-billed capercaillie, the latter's extensive white markings on the upper wing surface and larger white spots on the tips of the upper tail coverts will help to identify males. The more rufous and unbarred condition of the breast region of female capercaillies will help to separate them from females of the black-billed species, which are heavily barred on the breast. The "songs" of the two species are also apparently quite different, with the male black-billed capercaillie uttering only "crackling" sounds (Dementiev and Gladkov 1967).

AGE AND SEX CRITERIA

Females may be easily distinguished from adult males by their much smaller size (maximum wing measurements 305 mm) and absence of glossy blackish feathers.

222

First-winter males resemble adult males but are smaller (maximum wing measurements 385 mm) and have narrower tail feathers. The longer coverts are more vermiculated than in adults, have less-white tips, and the tail feathers have ill-defined whitish mottling. The outer two primaries are retained through the first year (Helminen 1963), and the beaks of young males are dusky (Lumsden, 1961*b*). Young males also have weaker beaks than adults (Moss, Weir, and Jones 1979).

First-winter females have narrower and less richly chestnut tail feathers than adults, and the buffy portions of the head and neck are paler and more yellowish. As in males, the outer two primaries are retained, and their tips are speckled with reddish brown extending at least 2 centimeters down the inner web, while there is much less or none on adults (Moss, Wier, and Jones 1979).

Juveniles have narrow pale shaft streaks on some of the mantle feathers, which also are tipped with pale wedge-shaped markings. Juvenile females resemble males but are not so gray and blackish on the head and neck and are more buffy and barred dorsally (Witherby et al. 1944). Even newly hatched male chicks have more dusky bills than do females, and by 6 days of age the emerging secondaries are more uniformly colored in males. By 14 to 21 days the male's gray ear coverts and tail feathers are also visible and distinctive (Högland 1956).

Downy young lack a brown crown cap, and instead have a single large dark v-shaped forehead spot, as well as numerous other blackish spots and speckles (Fjeldså 1977). Otherwise the upperparts are generally pale yellowish buff, with ill-defined markings of black and rufous, and the underparts are pale buffy yellow.

DISTRIBUTION AND HABITAT

Almost throughout its entire range the capercaillie is associated with coniferous forests, a major exception being the population of the Cantabrian Mountains of Spain, where the species occurs in broadleaved woodlands and where holly (*Ilex aquifolium*) provides the evergreen component (Castroviejo 1975). Otherwise it ranges from tall, dense forests of spruce (*Picea*) and fir (*Abies*) through lighter and often more open forests of pines (*Pinus*) and larch (*Larix*) to mixed forests (Cramp and Simmons 1980). In the USSR the species ranges from pure coniferous forests (excepting those of pure larch) to pure broad-leaved forests, but in the latter area the densities are relatively low. In some areas of oaklands the species leaves such habitats for the winter, moving into pinewoods or sections of forest containing small elements of pines, and remains until spring. In other areas, however, the birds winter in oak woodlands (Dementiev and Gladkov 1967).

Collectively the distributional limits of this species lie between the July isotherms of approximately 53° F in the north and 70° F in the south (Voous 1960), and its limits are further nearly circumscribed by those of the Scots pine (*Pinus sylvestris*) (Seiskari 1962). Its preferred habitats are extensive, shady, and heavy forests of pure coniferous forest or mixed coniferous-deciduous forest, especially those with a dense brushy understory that are near bogs or small forest glades (Voous 1960). Mature primeval and rather moist coniferous forests are probably the most important habitat types (Dzieciolowski and Matuszewski 1982).

Ideal year-round habitats are those where extensive areas of well-grown trees alternate with glades or other open terrain having ericaceous berries or other berries such as *Vaccinium*. In Scotland the preferred habitat consists of mature open natural pinewoods on hillsides with an undergrowth of heather and bilberries. Plantations having trees at least 25 feet tall or about 20 to 30 years old are less frequently utilized, but such plantings are usually too dense to allow much ground vegetation to develop. In natural pine forests there are many older trees, or "grannies." These trees typically have rounded or irregular rather than conical profiles and, like trees that have been damaged by flood, fire, or disease or are otherwise stunted, seem to be preferred for winter feeding. Needles of such preferred trees have more protein than do others (Pullianinen 1970*b*; Moss, Weir, and Jones 1979).

In Scandinavia the capercaillie relies almost entirely on pine for its winter food (Seiskari 1962), while the black grouse feeds equally on birch. However, summer foods of the two species do not differ greatly, and yet the habitat preferences of broods of the two species are significantly different. In general, capercaillie broods are less frequent than black grouse broods in areas of clear-cutting and other relatively open places, and they prefer more mature forest stands. About half of the 34 capercaillie broods observed by Børset and Krafft (1973) were in middle-aged to mature stands, in which the approximate age of the spruce trees was probably more than 70 years. These stands usually have sparsely developed brush strata, and in general capercaillie broods seemed to be less associated with particular grass and herb communities than were the broods of black grouse. In this particular study area spruce-peat bogs were a relatively rare habitat type, but other studies have indicated that these are also a preferred summer habitat type for capercaillies. Likewise, dwarf, open heathland above timberline is locally important as brood habitat (Cramp and Simmons 1980).

In Norway, clear-cutting of large areas (40–50 hectares) of forest has had severe effects on capercaillie populations, and when display grounds are cut the birds soon cease to use them. After clear-cutting, the blueberry understory is also invariably replaced by grass cover, which further renders the area unattractive as a brood habitat. For these and other reasons, there has been a progressive decline in Norwegian populations of capercaillies, primarily in the southern and east-central areas (Wegge 1979). Likewise during the past thirty years in Sweden, the populations of woodland grouse have declined, and the capercaillie's habitat has been greatly reduced by clear-cutting of older forests. Its best remaining habitat there is in large areas of mature variable-aged coniferous forest that often contain some deciduous trees such as aspen. In such habitats there are frequently bogs, marshes, and damp mossy areas as well as blueberries and bare rocky areas or windfalls (Marcstrom 1979). In Finland the capercaillie population has also declined, especially in southern areas (Rajala 1979).

The population status of the capercaillie in other European countries is summarized later in this book (Appendix 3) and in general is declining almost everywhere that accurate records are being kept, probably primarily because of forest destruction, but perhaps in part also because of excessive shooting, disturbance, and other habitat changes, and possibly even because of climatic fluctuations (Cramp and Simmons 1980).

POPULATION DENSITY

Densities of capercaillies in the USSR evidently vary markedly in various habitats, judging from figures for males observed on display grounds as presented by Dementiev and Gladkov (1967). Lowest densities are found in the southern populations (*"major"*) occupying European broadleaf forest habitats. An area of about 106,000 hectares of forested habitats (alluvial plains, forests of oaks and conifers, and swamplands of the Belovezhskaya forest) supported only 100 male capercaillies, or 0.95 per 1,000 hectares of forest. A mixed coniferous-broadleaf forest in the Carpathians containing 15,000 forested hectares supported 20 male capercaillies, or 1.3 per 1,000 hectares. However, in a game preserve in Gatchinskaya-Okhota consisting of transitional conifer-broadleaf forests, 536 males were counted on about 98,000 hectares, or 5.5 males per 1,000 hectares. A similar density was reported in a south Ural pine-birch-larch forest of 9,000 hectares, which supported 56 male capercaillies, or 6.2 per 1,000 hectares. The densest populations were reported for a sanctuary in wooded steppe habitat, mixed broadleaf and pine forests, where 47 males were seen in 5,200 hectares, or 9 per 1,000 hectares. In Norway some 8 million hectares of forest habitat exist, nearly all of which supports capercaillies (Wegge 1979). In 1960 the total Norwegian population was estimated at 300,000 to 400,000 birds, equal to an average density of 37 to 50 birds per 1,000 hectares of habitat, or 50 to 67 acres per bird. This is presumably a maximum (fall) density, so by spring it might be in the vicinity of 25 birds, or perhaps no more than about 10 adult males per 1,000 hectares in spring.

In a Scotland study, Moss, Weir, and Jones (1979) reported that the average winter density of capercaillies in three natural forests was 24.3 birds per square kilometer, while three planted forests averaged 8.0 birds per square kilometer. These figures indicate an unusually high local density range of from about 75 to 230 birds per 1,000 hectares (10 to 30 acres per bird), or well above those estimated for Norway and the USSR. Jones (1982), in reporting a population decline in Scotland, compared earlier densities of as many as 30 birds per square kilometer with recent ones of under 10. The average population of the capercaillie in Finland has been about 600,000 birds (Rajala 1979). A somewhat lower estimate of 240,000 pairs was made by Merikallio (1958) for the early 1950s, and his range map indicates that in some parts of Finland the average density reaches 1.2 birds per square kilometer. Rajala (1974) estimated densities during late summers of the early 1960s at about 12.2 birds per square kilometer (15.52 in central Finland), or about 24 to 37 per square mile. In Sweden the total mid-1970s population was perhaps 100,000 pairs (Cramp and Simmons 1980), and some 23 million hectares of forest still exist in the country. This suggests an average countrywide population density of a pair per 230 hectares of forest, or about 284 acres per bird.

Population densities of central and southern European countries are probably all appreciably lower than these. Gindre (1979) estimated a 1975 population in France of about 6,050 birds, spread over an area of 11,200 square kilometers, or 1.85 square kilometers (480 acres) per bird.

It is likely that a positive if crude correlation exists between the average population density and the average number of males present per display ground in spring. Dementiev and Gladkov (1967) present data indicating the following relation between the average number of males per 1,000 hectares of forest and the average number of males per display ground: 1.3:2.0; 5.3:3.6; 5.5:11.9; 6.2:9.3. A more detailed analysis of the possible use of display-ground numbers as an index to population densities would be useful. Such studies would have to take into account the daily or seasonal variations in numbers of males visiting each ground, which in a study by Lumsden (1961b) ranged from 1 to 7 birds over a two-week period. Most of this variation was caused by the variable presence of up to 4 nonterritorial males, all of which were apparently immatures, while the number of territorial birds consistently remained 3.

Müller (1974) found that "display areas" in Hesse, Germany, varied from 5 to 29 hectares (average 16.8 hectares) and were from 170 to 3,200 meters apart. Almost invariably they were on hilltops or other elevated sites. In spring the territory may be up to 12 hectares in area, with a smaller portion, or "preferred ground" up to 3 hectares. The total home range varied with age and sex. Adult males had home ranges averaging 53.3 hectares, and subadults had home ranges of 33.3 hectares. Those of females averaged 44.7 hectares. The home range of brooding females was only a fourth of their normal home range and was in its center.

HABITAT REQUIREMENTS

Probably the major habitat requirements for the capercaillie are suitable wintering areas having appropriate foods, and adequate brood-raising areas. Display occurs within the same habitats that are used for wintering, and summer and fall habitats are essentially identical to those needed for rearing broods. In general, capercaillies prefer natural mixed woodlands where various growth stages are present. The presence of *Pinus sylvestris* and *Vaccinium myrtillus* was found by Müller (1974) to be of special significance. Other important components of the general habitat include sites for dust baths, drinking places, ant hills, and gravel suitable for use as grit.

Wintering Requirements

One of the most important aspects of wintering habitat for the capercaillie is the presence of food trees in a suitable stage of development, primarily pines (*P. sylvestris*) in Scandinavia (Seiskari 1962). The same basic winter food sources are used in Scotland (Zwickel 1966a), and there the felling of older "granny" trees was followed by a decline in population densities (Moss, Weir, and Jones 1979). Not only are populations there higher in natural forests than in planted ones, but breeding is more successful in such habitats. In at least one study area, the abundance of capercaillies in the autumn was determined mainly by the density of breeding birds during the previous spring and was scarcely related to the percentage of young birds in the harvest (Moss, Weir, and Jones 1979).

According to Seiskari (1962), the winter habitats of male capercaillie are concentrated on dry heaths and pine peat bogs, characterized by a predominance of pines, the height and density of which are not critical. Females usually frequent pine woods on dry heaths, typically in stands that have relatively high density and often are between 7 and 18 meters high. In north Finland, where snow roosting is common, open areas in the environment are needed, while in more southerly areas the birds roost in spruce branches. Winter habitat requirements of males seem to center on the availability of pine trees for food, whereas females seem to be more influenced by structural features of the environment, including tree height and density. Throughout Eurasia, the distribution of the capercaillie seems to be confined within the limits of the distribution of Scots pine (*Pinus silvestris*), with a few exceptions in the south Urals and east of the Urals, but in such areas the birds often make winter migrations into the coniferous forests during the winter.

By late winter the females begin to increase their feeding, and this is also stimulated by cold weather and high pressure. Simultaneously there is gradual disintegration of male flocks, probably as a result of hormonal changes and decreased intraspecific intolerance (Koskimies 1957).

Spring and Summer Requirements

Display during spring evidently occurs in moderately heavily wooded habitats, but a small clearing is apparently desirable and often serves as a mating site (*Woodland Grouse* 1978, p. 129). However, as noted earlier, when display grounds are cut over the birds soon cease to use them (Wegge 1979; Hjorth 1982).

Nesting and brood-raising habitats of the capercaillie are still inadequately studied, but in Norway the brooding habitats are typically those associated with a mature forest having a sparsely developed bush stratum. Capercaillie broods also seem to be attracted to swampy areas. Factors affecting brood habitat preferences presumably include both food and cover characteristics, but according to Børset and Krafft (1973) cover may be more important than specific food requirements. In Finland, spruce peat bogs and heaths are the favorite summer habitats of capercaillies, and the birds seem to depend on spruce. The optimum height of the trees is apparently 13 to 15 meters, with a density of about 50 percent. The lower vegetation is typically abundant, with bilberry in particular an important summer food plant (Seiskari 1962).

After the display season has finished, there is a period when males aggregate during summer. Very probably the summer and early autumn groups of males are derived from display aggregations. In contrast to black grouse, fall and winter flocks of capercaillies tend to be unisexual. As the broods disintegrate and begin to move into winter habitats, the females become more mobile and vagrant, while the males tend to remain closer to the home area.

227

FOOD AND FORAGING BEHAVIOR

Food Requirements

Just as the black grouse is intimately associated with the birch, so the capercaillie has a strong tie to pines during much of the year. In Finland the staple food of both sexes is pine needles; during the winter months this food constitutes nearly 100 percent of the diet, both quantitatively and qualitatively. In central Finland the capercaillie begins to eat winter food at about the end of November and does not shift to summer food until the end of April. During periods of food shifts, and to a very limited extent during the winter, the needles and berries of junipers (*Juniperus communis*) supplement the pine diet (Seiskari 1962). Similarly, in the USSR a winter diet of pine needles is consumed from September until the end of March, and even as late as May pine may be present in most crops. In some parts of the USSR the cembra pine (*Pinus sembra*) is used in addition to Scots pine, and junipers serve as important interseasonal foods between winter and summer periods (Semenov-Tian-Shanski 1959; Seiskari 1962).

Farther south in central Europe, the winter food consists of a mixture of pine and interseasonal foods, especially among females. In such areas females have greater access to the food resources of the forest floor than is the case farther north. Males, however, continue to show a high tendency to concentrate on pine needles during this period. However, males consume interseasonal foods as late as November in Finland and apparently as early in spring as the absence of snow cover allows. In Finland females evidently shift over to winter feeding earlier than do males and likewise finish feeding on pine earlier in the spring. However, both sexes feed on pine for about the same total amount of time, 5 months, and the birds also eat pine during interseasonal periods as well (Seiskari 1962).

Although pine needles constitute about 80 percent of the birds' diet in Finland, pine shoots make up about 18 percent, and a small proportion consists of very small pine cones. Specific conifer trees are preferred during winter, and the birds are apparently able to select those trees having needles highest in nitrogen content (Pullianinen 1970*b*; Moss, Weir, and Jones 1979).

During the summer months and in interseasonal periods the birds eat a much wider array of foods. These include the leaves, stems, and berries of bilberries, leaves and stems of bog whortleberries, crowberries, cowwheat (*Melampyrum pratense*), cloudberry (*Rubus chamaemorus*), sedges, marsh andromeda (*Andromeda polifolia*), horsetails (*Equisetum*), mosses, and leaves of woodrush (*Luzula pilosa*). In the autumn the leaves of aspens as well as grain may be locally important (Cramp and Simmons 1980).

Among adult birds, insects or other invertebrates are probably eaten only accidentally. However, among a group of hand-raised birds, more than half of the food taken by birds up to 20 days old consisted of animal materials, including such items as spiders, springtails, beetles, and ants (*Suomen Riista* 13:143–55).

Foraging Behavior

During the winter in northern Finland, capercaillies have to feed throughout the entire period of daylight to obtain enough sustenance to survive. They begin to feed at dawn and cease soon after sunset. As the days grow longer in late winter, the feeding rhythm becomes bimodal, with peaks in the morning and evening. The crop of a female capercaillie can accommodate about 135 grams of food and that of an adult male about 160–240 grams, which is inadequate to provide enough energy for a 24-hour period. Thus the birds tend to keep their crops as full as possible, eating more as soon as the stored food has been transported to the gizzard. When tree feeding, males favor foraging very high, mostly in a zone about 2 meters from the top downward, while females prefer to feed somewhat lower, mostly in a zone about 3 to 4 meters below the treetop (Seiskari 1962).

MOBILITY AND MOVEMENTS

Recent radiotelemetry studies of capercaillies in Norway (Larson, Wegge, and Storass 1982) indicate that during the late spring (April to the end of May) yearling males tend to move about considerably, sometimes visiting several different display grounds. One male visited at least three, and perhaps four, in less than a month. These birds are apparently familiar with the leks and regularly travel from one to another, even though they may be several kilometers apart. The longest distance that was determined to be traveled in a single day was 7 kilometers. These mobile birds typically appear on the leks during the morning, after the adults have already begun their displays. By late May the birds did not regularly visit the leks, and therafter they moved about almost continually. They seldom remained in any one place for more than 14 days and had home ranges in the vicinity of 600 to 800 hectares (6–8 square kilometers). By contrast, the adult males were distinctly sedentary during the spring display period, each having a small display territory and an exclusive daytime territory that ranged from 20 to 100 hectares and extended outward from its display area. During the postdisplay period these older males enlarged their territories to areas of from 70 to 300 hectares, also near their display areas. When young birds wandered onto the summer territories of breeding males the younger ones moved on voluntarily or perhaps were excluded from the area. Evidently male capercaillies are solitary during the summer, and the wandering behavior of young birds is a reflection of social inferiority. On the other hand, the older birds are very familiar with their own habitats and are able to survive in a relatively small area.

Like the black grouse, this species evidently periodically erupts in northerly areas, including the USSR and Scandinavia. In the USSR, more southerly populations apparently do not show these eruptions. Instead the birds perform regular seasonal movements between breeding areas in deciduous woodlands to wintering localities in conifers, although in the south Urals some birds evidently remain and winter in oak woodlands. Where the birds do migrate, they arrive on their breeding grounds much later than the beginning of courtship for birds in highland pine woods. The two sexes arrive on their breeding grounds simultaneously and begin courtship immediately. In such areas, as in nonmigratory populations, the displays occur in the same locations each year, proving that the birds behave like true migrants rather than wanderers (Dementiev and Gladkov 1967).

During their first autumn and winter, females tend to move considerable distances. As the broods disintegrate, the females gradually move farther from the hatching place, while males tend to remain closer to their home areas. Young males probably remain more or less together, forming a nuclear element for further flock formation later in the fall (Koskimies 1957). Some females may move as far as 25 kilometers during this time, and a summary of recoveries from birds banded in Finland and Sweden indicates that 5 of 30 females recovered were at least 20 kilometers from the point of banding, while only 2 of 39 males recovered were farther away than 9 kilometers. The longest known movements are from Sweden, where at least 6 birds were known to have moved 1,000 kilometers or more to the south during the autumn (Cramp and Simmons 1980).

A few extreme long-distance movements have been documented for Swedish birds in autumn. One female was recovered more than 1,000 kilometers south of the point of banding, and a male moved some 500 kilometers, including crossing a marine sound 20 kilometers wide (Cramp and Simmons 1980).

REPRODUCTIVE BIOLOGY

Mating System and Territoriality

Although this species has been studied for a very long time, opinions differ on whether it should be considered promiscuous or polygynous, with loose, temporary pair bonds between males and haremlike groups of females. Hjorth (1970) quite clearly regards the species as promiscuous but considers the females "mated" for a few days while visiting a male or group of males. Because three females, all recognizable individually, visited only a single male during 6 consecutive days, Pukinsky and Roo (1966) regarded this as a haremlike system, which would be unique in the grouse family. Certainly it is well known that females exhibit specific preferences for males when given choice situations on display grounds (Moss 1980). They thus might indeed visit or associate with a particular male until fertilization occurred, but this might rather questionably be regarded as a real pair bond. Junco Rivera (1975) observed that in Spain the males maintain permanent territories and live with their harems within these territories, which is probably not typical for the species in general. Recently harem formation has also been observed in Scotland (Jones 1981).

A second equally important point is whether or not the display groupings of male capercaillies should be considered leks. Again, Hjorth (1970) believes that neither *arena* nor *lek* is an appropriate term, since the capercaillie differs from the typical lek-forming grouse in several ways. First, it selects wooded areas rather than open areas for display, which has influenced its evolution of ecologically effective signals and also the relative development of territoriality. Second, in Hjorth's opinion, among true lek-forming grouse (excepting the sage grouse) the males establish a rather fixed system of territories before the onset of display, but again the capercaillie is an exception. I am not certain this second point is valid, and in any case it is not highly relevant to the definition of arena behavior. I believe the three major features of lek behavior, namely the localized competition of males for territories and mating opportunities, the restriction of the territory to essentially the function of a mating station, and a strong dominance-to-fitness relationship among the males, in which territorial success is positively related to mating success, all are readily fulfilled by the social system of the capercaillie.

Territorial Establishment

Like the true lek-forming species, male capercaillies gather at traditional locations for territorial establishment and display. Müller (1979) reported that a lek was occupied every spring and fall for fifteen years and that the territories of at least some males did not change at all for as long as four years. Those of some high-ranking males were so well defined and recognized by their rivals that they were respected even during the owners' absences. Changes in territorial rank occur only gradually but, having occurred, may last for several seasons, suggesting that there is indeed a dominance/fitness gradient in territorial size or location. Only in the territory of the highest-ranking male can females gather for an undisturbed copulation, since rival males respect his territorial borders. This resulted in the "alpha" male's performing more than 90 percent of the total copulations in Müller's study, with the remainder performed by a single "beta" male and none by lower-ranking cocks. In Müller's major study site the "alpha" male also maintained the largest territory, having a maximum area of 12 hectares in spring and 10.5 hectares during fall.

The studies of Hjorth (1982) in Sweden convinced him that capercaillie territories are rather indefinite in size and are "radially arranged outward without limitation." Thus they are shaped somewhat like pieces of pie, with the apex representing the display ground where the birds display and defend their areas. Thus the territory in Hjorth's view is simply a small part of the bird's home range, situated at its apex. In coming and going from the display area, the birds never crossed the arena but simply moved in and out from the periphery. The actual defended area was only about 0.8 hectare, in his view.

Similar small territories were estimated by Lumsden (1961b) in Scotland, consisting of approximately 300, 800, and 1,000 square yards in three cases. These central display areas were separated by rather wide boundary zones that in Lumsden's opinion were left undefended and thus in this view did not seem to represent typical lek situations. However, elsewhere in Scotland much larger territories of a hectare or more have been observed (A. M. Jones, in Cramp and Simons 1980), and similar-sized (0.5 hectare) well-defined territories were reported by Hainard and Meyland (1935) in the Jura Mountains of Switzerland. Müller (1979) noted that 3 males held territories in an area of 24 hectares in the western Rhön Mountains, although a single "alpha" male controlled nearly half of this entire area. In 33 display grounds near Leningrad, an average of 6 to 7 males occupied an area of about 50 hectares. One display ground occupied by 20 to 25 males covered 1 square kilometer, representing an average of about 4 to 5 hectares per male (Pukinsky and Roo 1966). Castroviejo (1975) observed that in Spain 4 males occupied overlapping territories averaging 0.5 hectares.

As noted earlier, the number of males attending a particular display area probably varies directly with population density. Frequently only 2 or 3 territorial males and an equal number of immature nonterritorial males are present in a given area, but in rare instances a dozen or more males may be present. The largest numbers seem to occur in Siberia, where 10 to 20 males are commonly present, and at times up to 50 (Johansen 1961). However, such large groups tend to break up into smaller subunits, each of which probably represents a mating center.

Where males winter near their display areas, they probably only gradually begin to concentrate their daily activities in the display site as spring approaches, spending the rest of the time foraging or sleeping. Frequently the display site may be a kilometer or more away from the foraging grounds, and each male flies to a favorite tree in the evening, where he spends the night preparatory to his display activity. In late evening the bird displays briefly from his chosen site, then roosts there. He begins his morning display an hour or two before sunrise from the "headquarters" tree, though moonlight may stimulate a somewhat earlier start (Hjorth 1970).

The seasonal onset of display doubtlessly varies greatly in different parts of the capercaillie's range. It may begin sporadically as early as the start of the year, but it probably increases very gradually as sex hormone changes associated with increasing photoperiods increase aggressiveness among the males and decrease their social tendencies (Koskimies 1957). Müller (1979) reported territorial occupation of a lek in West Germany as early as January during three of fifteen years, and in every year the lek was occupied by the latter part of March. Lek activity usually terminated there by the last week of May, probably ending with the onset of molt in adult males. During the summer molting period only yearlings were observed displaying in the study area. Fall activity sometimes began as early as August (six of fifteen years) but usually was initiated between the end of September and the middle of October. November normally marked the end of the fall display period, with December observations occurring in four of the fifteen years, depending on the first snowfall. All told, the adult males spent about a third of their annual time at the lek, which indicates the great significance of this activity.

In Müller's study area, the "high season," or period of maximum male activity and the time of visits by females, was from late March to early May in most years. Hjorth (1970) stated that the peak of the display season in south-central Sweden was in late April in 1965 and 1967, and Lumsden (1961b) observed copulations in Scotland in late April. In the southern Urals of the USSR, females visit the display grounds from about the middle of April until the latter part of May, while in the northern Urals they begin to visit the grounds from mid-May to late May (Dementiev and Gladkov 1967). Probably farther east and north in Siberia the high season is even later, if Andreev's (1979) observations on the black-billed capercaillie can be also applied to this species.

Characteristics that make an area suitable for display sites are still uncertain, though there are records of individual areas being used for as long as one hundred years. There are also cases where as many as seven displaying males have been shot out of the same tree over a period of ten years (Cramp and Simmons 1980; Lumsden 1961*b*). Thus some structural features of the habitat are evidently favored. Small knolls or slopes are often used by males, suggesting that visibility is very important. Scherzinger (1976) found that there is a strong preference for upper slopes and peaks having flat areas where ground vegetation is low, especially where there are eastern exposures and dense undergrowth nearby. Hjorth (1982) reported that arenas usually have a certain amount of "vegetational or topographic curtain," some aged pines or stands of such pines, some tall shrubs, and some open space. Castroviejo (1975) reported that display sites have many characteristics in common. They are in primitive forests that are not dense and that include spacious and grassy clearings. The sites are also oriented toward sunrise, and there are usually piles of stones or rock formations. Likewise, in all of the twenty-three sites studied *Vaccinium myrtillus* was growing.

Tree density probably also has a definite effect on territorial size and desirability; according to Müller (1979), the denser the trees the smaller the territories, and mating usually occurs in an open site. In his study area the ground vegetation was apparently relatively dense but low and was not tall enough to obscure males standing on the ground. Display usually begins with the bird singing in the tree in which he spent the night, but he soon flies down to the ground directly below or to an area up to several hundred yards away, where the main display occurs.

It is not clear how old a male is when he first becomes territorial. Males less than a year old apparently sing imperfect songs early in the spring period but begin to perform typical displays toward the end of the period. They probably begin to visit the display grounds later than the older cocks and probably do not become territorial during their first spring, even though their testes may by then be large enough to ensure fertility (Lumsden 1961*b*). They continue to display during the summer while the adult males are molting, and probably by the second fall of life they attempt to establish territories. The territorial activity of the males during fall is believed to be important in strengthening the male ranking order and in transferring traditional information on the lek to younger males (Müller 1979).

Territorial Advertisement

Territorial advertisement takes three recognizably distinct forms: tree display, flight display, and ground display, which tend to occur in that approximate sequence, though there are many variations.

During very early morning hours, when light levels are extremely low, a male capercaillie will usually begin display by calling from the tree in which he spent the previous night. This display is apparently exclusively vocal, with none of the elaborate posturing typical of ground display. According to Lumsden (1961*b*), capercaillies began to sing nearly an hour before sunrise in late April, though these early songs were usually incomplete. During such singing the tail apparently is usually kept low, in a normal resting position, and the neck is held upright and appears thin, as in the typical "thin-necked upright" ground display (Hjorth 1970). Lumsden stated that the birds in trees often performed the "clicking" phase of song display from the tree for some minutes before flying down to the ground to perform the complete display, but other observers have heard the complete song display in early morning and late evening hours.

Flight display, or "wing beating display" as it is called by Hjorth (1970), takes two general forms in capercaillies. The first is the noisy flight down from the perching tree to the ground, where display continues. This flight is characterized by a heavy drumming and rattling noise, often resembling the sounds made by a falling tree. Later, during ground display, the bird may also perform short springing flights into the air, called "drumming jumps" by Hjorth, or "drumming flights" in the case of the somewhat longer versions. This flight is begun from a thin-necked upright posture, after the "cork-note" stage of the song. At that point the bird suddenly springs into the air, appearing as if it "had seized the hook of a fishing-line and is being pulled upwards (and somewhat forwards), all the time intensively beating its wings" (Hjorth 1970). The noises made during this display are a double drumming sound, and they carry the farthest of all the noises made by displaying capercaillies. They are probably all mechanical sounds; the bill appears to be closed during the entire display. Lumsden (1961*b*), who calls the display a flutter-jump, noted that the birds rarely rise more than 3 feet from the ground at such times but might cover 7 to 10 yards in horizontal distance. Like other observers, he noticed that the display is often performed when other male or female capercaillies are moving about on the display ground, such as arriving or leaving. Yearling males that were nonterritorial very seldom performed this display, according to Lumsden.

Ground display is the typical territorial display of male capercaillies and may be performed in complete isolation or in the company of others. The posture assumed during this time is the "thin-necked upright," in which the tail is fully spread and held vertically upward, the neck is stretched nearly vertical, and the head is tilted upward with the "beard" protruding. The wings are slightly lowered, exposing the white "shoulder patch," the alula is sometimes extended, and the white-tipped under tail coverts are visible from behind. The red combs above the eyes are small and narrow and not noticeably engorged (Hjorth 1970).

The male's song consists of four phases, of which the first may be omitted, and the occurrence of the third varies geographically. Many observers have described the song, but the descriptions of Hjorth (1970) and Lumsden (1961*b*) are the basis of the account given here.

The song, or "canto," begins with a series of double-noted clicks, *ki-kop*, about 0.13 second apart, similar to the sound made by a mechanical "cricket" toy. The bill is opened on the first click and closed on the second, the head jerked upward, and the esophageal area sucked in and released. In some cases nictitating membrane moves across the eye occasionally, but not in regular synchrony with the clicking notes. These double clicks gradually increase in speed until the sounds nearly merge into a continuous sound train, or "roll," lasting nearly a second. In the western European population this phase is terminated by a sudden loud sound called a "cork note," since it resembles the noise made when a cork is withdrawn from a bottle. This note can be heard for several hundred yards. At this time the head and bill are suddenly jerked backward to a point beyond the vertical, and the beak is shut immediately after the cork note. Siberian capercaillies evidently completely omit the cork note from their display (Hjorth 1970).

The final phase of the song is performed with the beak held vertical. The head jerks upward and backward several times, and the beak opens and shuts as the bird utters from three to five high-pitched scraping notes that sound like a scythe being whetted. This "whetting phase" lasts about 3 seconds, and during that time the bird is apparently relatively oblivious to his surroundings, probably because of reduced hearing ability. At times the nictitating membrane may cover the eye during this phase, but the bird nevertheless is fully aware of his visual surroundings.

Recently Moss and Lockie (1979) found that the song has an infrasonic component that can probably be heard by other capercaillies for more than 200 meters. Moss also (1980) observed that female capercaillies seem to prefer to mate with males that are superior fighters, which would explain why capercaillie males are the largest of all grouse and why the species exhibits a high level of sexual dimorphism in body weight in spite of small lek size and presumably reduced intermale competition.

Territorial Defense and Fighting

Territorial encounters between rival males result in several possible behaviors, according to Hjorth (1970). A ritualized "display flight" may occur, with the birds in bill-to-bill confrontation, there may be additional display features and associated belchlike vocalizations, and an actual fight may finally ensue.

As rival males approach, their neck feathers are greatly ruffled, and the fanned tail may be tilted toward the opponent. In this "oblique" posture the bird utters a forceful belching call, the "belching cantus," which lasts about 0.8 second and consists of a cluster of four guttural sounds, with the second the loudest. As the males confront one another they make repeated bowing movements more or less in synchrony, with their beaks almost touching as they reach the lowest point of the bow with the head held only a few inches above the ground. In this crouching posture they may peck at each other with jerky head movements but do not actually touch one another.

Such ritualized confrontations may grade into real fighting as the pecking becomes more intense and blows are exchanged. At this time their tails may be raised or lowered rapidly, and their beaks often strike one another. Unlike some of the smaller grouse, these birds do not jump while beak fighting but rather stand close to one another. If a male

23. Male displays of the capercaillie (after Hjorth 1970 and Glutz 1973), including (A) calling in thin-necked upright posture, (B) calling in thin-necked upright posture with tail spreading, (C) precopulatory posture, and (D) aggressive behavior between males.

successfully grabs the other's beard or neck skin, he often follows up with an attack of wing beating, producing clashing sounds resembling gunshots (Hjorth 1970). The use of feet during fighting has not been reported. Fights may last up to 2 minutes and often end inconclusively. When there is a definite winner, the vanquished male retreats, often pursued by the victor, which may rarely leap on the loser's back and continue to beat him with his wings. Fatalities rarely occur during such combats, but subsequent deaths may be fairly common, and three were reported during three years of study (Jones 1981).

Sexual Displays and Copulation

Females probably do not normally roost in the display area but instead typically fly in sometime after display activity is underway. Lumsden (1961b) noted that the females usually arrived about 15 minutes after the first song was uttered by a male, and they usually landed high up in trees. From these perches they gradually work down to the ground and make their way to the territory of the dominant male. At times several females will be near the "alpha" male simultaneously (Moss 1980). These "packs" of females sometimes cackle from trees, stimulating the males below to increased display activity (Hjorth 1970). I have observed up to 9 females in a single cluster.

When females arrive on the display ground, the males perform directed displays toward them. This posture resembles the thin-necked upright, but the primaries are spread so that their tips drag on the ground and the tail is usually tilted toward the females. Single females or groups of females are thus encircled by the males, and the noise level is increased by the congested sounds of singing and wing scraping. Unlike their response to other rivals, males rarely raise their neck feathers when displaying toward hens (Hjorth 1970).

A female that is ready for mating responds to a circling male by stopping in his path, crouching so that her breast just touches the ground. Her tail is raised slightly above the horizontal and her wings are often drooped slightly. The male then may perform several full song displays beside her and circle her in progressively smaller loops. He finally mounts her from behind, grasping her nape with his bill and stepping onto her back. Treading may last for as long as 30 seconds, although about 6 seconds is more typical. The male then slides off and resumes singing. The female rises, runs ahead a few feet, then pauses to shake and preen before flying into a tree (A. M. Jones, in Cramp and Simmons 1980). Probably one copulation is sufficient to fertilize an entire clutch, but at times a male will mate with a female repeatedly, or females may visit and mate with a male over a period of several days (Glutz 1973). I observed 8 copulations by a single male in a 90-minute period.

As I noted earlier, several observers (Lumsden 1961b; Müller 1979; Moss 1980) have noted that nearly all copulations are performed by a single dominant or "alpha" male. Perhaps the importance of physical strength in male capercaillies in establishing and maintaining prime territories is one of the sources of selection for large size among males of this species, and females certainly show specific attraction to particular dominant males

(Moss 1980; Hagen 1980). The capercaillie is thus an extreme among the lek-forming grouse, where there seems to be a general relation between the weight ratio of males to females and the degree of sexual selection associated with the mating system. Evidently both aspects of sexual selection—the heterosexual attraction of females to males on the basis of relative size, strength, or behavior and the intrasexual competition among males in territorial establishment and maintenance—play important roles in the evolution of sexual dimorphism as it relates to reproductive fitness.

It has been observed that in captive situations with only a few birds, such as 4 males and 10 females, preference for a single male still exists, even though that particular male may have lost most of his tail feathers. Fighting among the females for "possession" of a dominant male has been observed as well. In Sweden there seems to be no tendency for the occurrence of a specific dominant male on leks studied by Hjorth, but older and more experienced males have a better mating strategy than do younger and less experienced birds (*Woodland Grouse* 1978, pp. 129–30).

Nesting and Brooding Behavior

After successful copulation, females probably immediately leave the display grounds and begin nesting. Their nest sites are usually not very far from such display grounds and at times may be somewhat clumped. As many as 5 nests have been found within an area of 1.36 square kilometers, and another 5 within an area of 1.10 square kilometers (A. M. Jones, in Cramp and Simmons 1980). There is likewise no indication of individual distances maintained between nests, which have been found as close as 10 meters apart (Glutz 1973).

The nests are shallow scrapes, in thick cover, often at the base of a tree. Rarely a female will lay in the abandoned nest of another species, as high as 5 meters aboveground. From 5 to 12 eggs are laid, with larger clutches of up to 16 almost certainly multiple clutches. The smallest average clutch sizes are those found in northern USSR, averaging 6.2 eggs among 265 clutches (Dementiev and Gladkov 1967). In the Urals the largest clutches (averaging 7.6 eggs) are typical of the southern region (Malafeev 1970). In Europe the average clutch seems to range from about 7.1 eggs in Bavaria to 8.4 eggs in Croatia (Glutz 1973). The laying interval between eggs is 1.2 to 2.2 days, so that a complete clutch is likely to require approximately 10 days. Incubation begins with the last or penultimate egg and requires 24 to 26 days (Cramp and Simmons 1980). During incubation the female typically leaves the nest twice a day to feed. She is absent an average of about 35 minutes, usually during early morning and late afternoon (Pullianinen 1971). Most clutch losses, which ranged from 6 to 16 percent in a Finnish study, are probably the result of predation or human influences (Siivonen 1957). Linden (1981) estimated a nest mortality rate of 34 percent, with 94 percent of the eggs hatching in successful nests.

Hatching of the young is synchronous, and they soon leave the nest to feed themselves. They are capable of short flights when 2 to 3 weeks old and are full grown at 2 to 3 months. During the first few weeks after hatching their survival is greatly affected by precipitation and temperature (Brüll et al. 1977). In Finland, 59.7 percent of the eggs laid resulted in young surviving to late August (Rajala 1974). Linden (1981) estimated the first-winter mortality rate at about 76 percent.

EVOLUTIONARY RELATIONSHIPS

As noted in the account of the black-billed capercaillie, these two species obviously constitute a superspecies and are incompletely reproductively isolated. It is worth noting that the Siberian population of the capercaillie has a song that tends to approach that of the black-billed capercaillie in that it omits the distinctive "cork note" of the western European populations of capercaillie, which is one of the most distinctive differences between the species. Whether this is a secondary loss in the Siberian population, perhaps resulting from contact and hybridization with the black-billed capercaillie, or whether the west European population evolved this song component independently cannot be stated with certainty. However, it is the vocal element that is loudest and carries the farthest, suggesting that it might be important in rather densely forested areas. A second interesting cline in the capercaillie is the trend toward a white-bellied condition in males in the southeastern parts of its range. This would of course be in sharp contrast to the black-bellied condition of the black-billed capercaillie and presumably might help maintain species distinctiveness where the ranges overlap.

Black-billed Capercaillie

Tetrao parvirostris (Bonaparte) 1856

Other Vernacular Names

Rock capercaillie, Siberian capercaillie; Felsenauerhuhn (German); tétras à bec noir (French); kemennyi glukhar (Russian).

Range

Eastern Siberia, from the basins of the lower Yenisei, Kokechuma, and upper Kureika rivers eastward, north to the limits of the taiga and east to the region of Markovo on the upper Anadyr and the Penzhina basin, southward to northern Mongolia, Transbaikalia, Manchuria, Amurland, Ussuriland, Sakhalin, and Kamchatka (Vaurie 1965).

Subspecies (after Vaurie 1965)

T. p. parvirostris (Bonaparte): Siberian black-billed capercaillie. Resident in areas indicated above, except for Kamchatka.

T. p. kamtschaticus Kittlitz: Kamchatka black-billed capercaillie. Resident in Kamchatka.

MEASUREMENTS

Folded wing: Males (*parvirostris*) 382–402 mm, average of 4, 396 mm. Males of *kamtschaticus* apparently average slightly smaller, the mean of 15 being 348 mm (Dementiev and Gladkov 1967); females 298–324 mm, average of 4, 314 mm.

Tail: Males 248–345 mm, average of 5, 302.6 mm; females 210–235 mm, average of 4, 218.5 mm.

IDENTIFICATION

Length, 20 inches (females) to 30 inches (males). Males are similar to those of *urogallus*, but white markings are limited to the upperparts rather than the underparts. The upperparts are blackish (*parvirostris*), or light gray with a sulfurous tinge on the interscapulars and upper coverts (*kamtschaticus*). The secondaries have broad (*kamtschaticus*) or narrow (*parvirostris*) white tips, and the larger wing coverts have white markings ranging in size from broad and triangular to small droplike markings, as do some of the scapulars and the longest upper tail coverts. The head and neck are blackish gray, and there is a small area of red skin above and behind the eye. The bill is blackish, and the legs are feathered to the bases of the toes. Females resemble those of *urogallus* but are more heavily barred on the upper breast with black, yellowish, and white and have large white markings on their upper wing coverts and secondaries (Dementiev and Gladkov 1967).

FIELD MARKS

This species is found in much the same habitats as is *urogallus*, and confusion between them is possible in the area of sympatry. The large white markings on the back, wings, and upper tail coverts of the adult male *parvirostris* should aid identification, and the heavily barred breast of the females is probably the best field mark for that sex. The calls of the displaying males lack melodic and metallic elements and consist only of clicking sounds uttered in rapid series.

AGE AND SEX CRITERIA

Adult males are readily distinguished from females by their dark gray and white coloration.

First-winter males have not been specifically described; presumably they resemble adult males but are smaller, as in the Eurasian species. Like that species, they probably retain the two outermost primaries through the first year, and this trait should aid in recognizing young birds.

Juveniles are said to resemble adults (females) after losing their down; their underparts are barred with black, whitish, and light yellow (Dementiev and Gladkov 1967). Presumably as in the Eurasian species the mantle feathers also have pale shaft streaks.

Downy young are similar to those of the Eurasian species of capercaillie and have yellow underparts, a brownish yellow spine, and an orange brown head (Dementiev and Gladkov 1967). Fjeldså illustrated (1977) the upperpart pattern as similar to that in the capercaillie but with a rufous crown patch, surrounded laterally and posteriorly with a more definite dark band, especially to the rear. His color illustration (plate 51) suggests that the back color is also more rufous and is less distinctly streaked with black.

DISTRIBUTION AND HABITAT

The distribution of this species of capercaillie is generally complementary to that of the more widespread Eurasian species, but the two overlap over a considerable area, generally between the upper portions of the Yenesei and Lena rivers. According to Andreev (1979), the black-billed capercaillie's distribution is associated with that of the larch (*Larix gmelini*, previously *L. dahurica*) taiga forest, but with a secondary occurrence in the birch (*Betula ermani* and *B. japonica*) forests of Kamchatka. Dementiev and Gladkov (1967) reported that the species' general distribution includes the forest zone and wooded tundra in most of its range but extends to upland woody steppes in southern Transbaikalia. It occurs most often in larch forests on gently sloping terrain of broad watersheds, avoiding the steeper slopes of small river valleys.

More recent studies in Kamchatka indicate that the birds occur throughout the year from the wet meadows of the Kamchatka River to the tree line, both in groves of birches and in pine forests. Evidently females tend to winter at the tree line, while males predominate on the wet meadows near the river at that time of year. During summer both sexes occur in all habitat types (Lobachev, in Kirikov and Shubinkova 1968). Another Kamchatka study reported that the birds occur exclusively in tall forests on the seaward slopes of volcanic valleys, at elevations 10 to 650 meters above sea level. There the birds occur in birch woods having an undergrowth of rowans (*Sorbus*), preferring well-drained areas and avoiding those with too much moisture or excessive melting snow. Thus the soil conditions affecting drainage tend to influence local distribution patterns (Markov, in Kirikov and Shubinkova 1968).

On Sakhalin Island the species inhabits coniferous forests of firs (*Abies sachalinensis*?), and birches (*Betula japonica*). It also occurs on slopes where fires have been followed by a regrowth of birches, alders, rowans, and bushes (Vshivtse and Vononor, in Kirikov and Shubinkova 1968).

In China the birds are confined to upland forests of the Great Khingan Mountain region and never leave the limits of the thick forests of *Larix gmelini*. They breed in forests of this type having high undergrowth of *Rhododendron dahuricum* and thick moss cover (Cheng 1979b).

In the vicinity of Lake Baikal, the birds are associated with taiga of larches and larches mixed with pines, with associated blueberries and red bilberries. The birds favor mountain slopes and ravines having springs and swamps with low-growing woods (Isailov and Tarolov, in Kirikov and Shubinkova 1968). In the lower Amur basin they likewise occur in forests of larches, pines mixed with larches, firs, and mixed deciduous-coniferous forests, rarely extending into sphagnum swamps having scattered larches (Nechayev, in Kirikov and Shubinkova 1968).

POPULATION DENSITY

Few good sources of information exist on the subject of population density. The species is

10. Current distribution of the Siberian (S) and Kamchatka (K) races of the black-billed capercaillie.

apparently quite local in distribution, and in Kamchatka the availability of hot springs tends to produce unusually high local concentrations over much of the year, particularly during winter (Markov, in Kirikov and Shubinkova 1968). Two crude estimates of abundance on Kamchatka are 26 birds within an area of 15 square kilometers in January and 12 "mating" birds within an area of 10 square kilometers in May (Lobachev, in Kirikov and Shubinkova 1968). However, these densities are probably considerably higher than is typical; in most areas of favorable habitat the average density is only about 1 bird per 100 square kilometers, and under the best conditions it does not exceed 5 per 100 square kilometers, according to Markov (in Kirikov and Shubinkova 1968). Although there is no specific information on densities, the Chinese population has been steadily decreasing and is in need of protection (Cheng 1979*b*).

HABITAT REQUIREMENTS

Practically no specific information exists on habitat requirements, and one must assume that this species required much the same conditions as the more common capercaillie. A winter source of larch buds, birches, or similar foods is obviously needed, and the birds apparently are well adapted to fairly heavy snow cover. According to Cheng (1963), the birds roost in trees during nights when the snow is not thick, but when the cover is heavy and the nights are very cold they will spend the night in the snow. Andreev (1979) reported that snow provides the best substrate for displaying males and that in general the species seems better adapted to a fairly open environment than the other species of capercaillie. The visual and acoustic elements of its display favor fairly long-distance transmission of signals, which would tend to promote population maintenance in spite of low densities.

FOOD AND FORAGING BEHAVIOR

According to Andreev (1979), these birds spend the worst part of the winter (November to March) in larch forests of the Kolma basin, feeding on wild rose (*Rosa*) berries and large twigs. In southwestern Transbaikalia they also feed largely on larch buds. The crop of a female taken on the lower Amur at that time of year contained birch buds and larch shoots. Before winter, and even during early winter, the birds feed on berries, frequently digging under the snow to obtain them. Thus they favor forests having such berry crops. Probably berries become important as soon as they ripen; the crops of two birds obtained in the Chita are in mid-July contained blueberries as well as ant eggs (Dementiev and Gladkov 1967).

Observations on Kamchatka are more complete, and there the birds evidently concentrate on the catkins, buds, and tips of the twigs of stonebirches during winter. However, the buds and shoots of willows and alders are used very little. During spring the birds evidently return to eating berries; those killed on display grounds in May had eaten birch buds, juniper berries, winter cowberries (*Vaccinium vitis-idaea*) and crowberries (*Empetrum nigrum*), the stems of horsetails (*Equisetum*), and the leaves of blueberries and other

244

herbaceous vegetation. Later in summer three birds were collected that had eaten the berries and seeds of crowberries, cowberries, and other plants as well as catepillars and midges (Dementiev and Gladkov 1967).

Cheng (1963) reported that their winter diet includes the needles and buds of larch, Korean red pine (*Pinus koraiensis*), fir, and birch, and also the seeds of Korean pine. Berries are eaten during the summer, and in the fall the birds often fly to river areas, borders of burned-over areas, and cultivated fields in search of food.

MOBILITY AND MOVEMENTS

Virtually no information is available on movement. Andreev (1979) noted that before the period of display activity, males could at times be seen considerable distances (5–6 miles) away from the lek, suggesting appreciable mobility. In Kamchatka there are obviously considerable altitudinal movements associated with the season, with females wintering at higher elevations than males (Lobachev, in Kirikov and Shubinkova 1968).

REPRODUCTIVE BEHAVIOR

Territorial Establishment

The period of reproductive behavior in this species is very extended, and mating begins at different times even within relatively small geographic regions. In Kamchatka the mating period lasts from early April until the beginning of June, and its initiation is associated with the first thawed patches in display areas. The locations of these leks are quite constant, changing little from year to year, and in Kamchatka the usual number of males present is from 3 to 5, or occasionally as many as 9 (Markov, in Kirikov and Shubinkova 1968).

North of Kamchatka, on the Kolyma drainage (66° N latitude), sexual behavior begins about mid-April, when males move from their wintering areas in flood larch forests to mountain larch forests. From late April to early June is the period of high sexual activity. In the first phase, lasting until about mid-May, adult males regularly appear at the lek, but females and subadults appear only irregularly. At this time true rather than ritual fighting sometimes occurs. In one lek studied, the number of territorial males ranged from 6 to 10 over a two-year period. At this time the weather is usually fine and the winds are calm, forming ideal conditions for visual and acoustic communication. Snow is present and rather hard, providing an optimum substrate for displaying males (Andreev 1979).

In China, display begins in late March, with a peak in mid-April, and is mostly over by the first of May, with a few males still calling in mid-May. Observations there so far do not indicate that the birds form lek assemblies (Cheng 1979*b*).

Individual males establish territories that are rather small and include a very small area (about 10×10 meters) where displaying is most intensive, within a larger area from 20 to

50 meters in diameter from which they expel other males. Judging from a sketch map provided by Andreev (1977), 8 males were territorial within an area of approximately 37,000 square meters, about 4,700 square meters per male. The average minimum intermale distance was approximately 50 meters for these 8 birds.

Territorial Advertisement

During the period of most intensive sexual activity in mid-May, when females regularly visit the lek, males begin to arrive on the display grounds about 6 P.M., usually walking in. There is a period of evening display that is sporadic until sunset but vigorous thereafter until midnight, dropping slightly, then becoming stronger after about 1 A.M. and remaining almost continuous until about 8 or 9 A.M. Thus, display may occur for as much as 12 hours each day (Andreev 1979).

The usual display posture generally resembles the thin-necked upright of the more common species of capercaillie, but there are several differences. The bill is kept permanently open, exposing the bright pink palate, and the "crop" area is distinctly swollen. Additionally, the white markings on the scapulars, wing coverts, and upper tail coverts are highly conspicuous, forming a distinctive pattern that is lacking in the other species. The black-billed capercaillie lacks white markings on the under tail coverts and the tail feathers themselves, but both species have white "shoulder patches."

There are also important acoustic differences between the species. The song of the black-billed capercaillie is much louder and consists of rhythmic clicking components of two types, simple clicks and vibrating clicks. The former resembles the click note of the capercaillie, while the latter includes three-jointed or vibrating clicks. The entire canto includes two parts, beginning with four to seven individual clicking phrases followed by a terminal roll that is an extended vibrating click. Thus the entire canto sounds as a repeated *tack-track-tack-tack*, terminated by a more prolonged *tr-r-rack*. The duration of each canto ranges from 4 to 12 seconds, and is usually 5 to 7 seconds. When a female is present, 15 or 16 cantos may be uttered per minute. During calm weather this vocalization can be heard for more than 500 meters. Evidently the individual phrases of this song correspond to the "double clicks" of the common Eurasian capercaillie, and the roll component may be analogous to the rolling portions of that species' canto. There are not apparent geographic variations in the vocalization (Andreev 1979).

Hjorth (1970) reported that, based on a tape recording he heard, the vocalization of this species may be well imitated by pattering a matchbox with one's nails, and he noted that the notes are arranged in groups of five or six rather than triplets.

In contrast to the common capercaillie, the "wing beating" or flutter-jump display of this species is relatively quiet and produces a visual rather than an acoustic effect. The bird flies from a minimum of 4–6 meters to a maximum of 15–20 meters, 1 to 3 meters above the ground. The intensity of the display is greater after dark, which does suggest some acoustic significance.

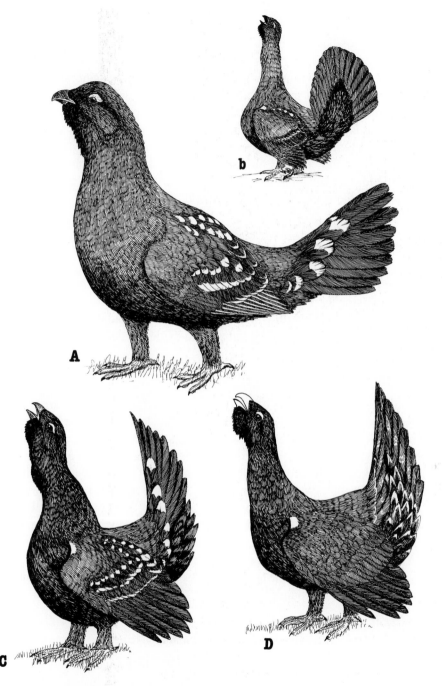

24. Male displays of the black-billed capercaillie (after Andreev 1979 and photographs), including (A) low-intensity wide-necked upright posture, (B and C) thin-necked upright posture with calling and tail spreading. The corresponding display of the capercaillie is also shown (D).

Aggressive Behavior

Ritual confrontations between males are frequent, especially in the absence of females. These usually begin at the boundaries of adjacent territories. Initially males maintain an upright attitude and may continue vocalizing. However, during aggressive excitement the song becomes less rhythmic and sometimes resembles the "belching cantus" of the more common capercaillie. Additionally, as the confrontation continues the birds may assume a wide-necked attitude, lift their wings, and attempt to strike each other with wingstrokes. Probably some pecking also occurs, since males are sometimes seen with bald spots on the neck or with parts of the "beard" missing. Occasionally territorial males will leave their territories to feed, and subadults may temporarily invade them. On returning the owner responds with a wide-necked threat attitude and attacks by running or flying (Andreev 1979).

Copulation

As in the other lek-forming species, a master cock is present. This bird may be readily recognized by his high degree of activity, with most females gathering within his territory. Before copulation the female crouches, and for a time the male continues to display in his thin-necked upright posture with wing dragging. During copulation the male spreads his wings on either side of the crouching female and evidently does not attempt to grasp her nape (Andreev 1979).

Nesting and Brooding Behavior

In Kamchatka, egg laying occurs during the second half of May or in early June. At this time of year snow still completely covers the ground at elevations above 250 meters, so the nesting occurs at fairly low elevations. Nests are placed on snow-free ground, sometimes under the crown of a fallen tree. The clutch size there is usually 6 or 7 eggs, with extremes of 5 and 8 (Markov, in Kirikov and Shubinkova 1968; Dementiev and Gladkov 1967).

According to Cheng (1963), females remain in pine forests to build their nests, while males disappear after the courtship period. The nest is often built below the drooping branches of a fir, under fallen logs or branches, or under the top of a fallen tree. It is a small depression lined with needles, mosses, grasses, and some feathers. Egg laying in China occurs in the latter part of May, and clutches there are from 6 to 10 eggs. The incubation period is about 24 days, and by the time the chicks are 10 days old they can fly up into trees 2 or 3 meters above the ground.

EVOLUTIONARY RELATIONSHIPS

There can be not doubt that the two species of capercaillies are extremely closely related and constitute a superspecies. Short (1967) came to this conclusion after his study of morphological characteristics, although he noted that in one area of overlap 12 percent of the displaying males were hybrids (Kirpichev 1958). Thus the species are not completely isolated reproductively, in spite of the considerable acoustic and visual differences in male displays. Andreev (1979) has suggested that some of these differences have ecological significance and thus they may not have been specifically selected for as a mechanism of maintaining species isolation.

Black
Grouse

Tetrao tetrix (Linnaeus) 1758

Other Vernacular Names

Blackcock (male), greyhen (female); Birkhuhn (German); Tétras lyre (French); korhoen (Dutch); orre (Swedish), Tetrev-kosach (male), teterka (female) (Russian).

Range

Great Britain and Eurasia from Scandinavia east at least to the Kolyma, the Lena, and Ussuriland, south to the Ardennes, Alps, Balkan Peninsula to Montenegro and Bulgaria, southern Russia, northern Kirghiz steppes and Turkestan Mountains to the Tian Shan, northern Mongolia, Manchuria, and northwestern Korea (Vaurie 1965).

Subspecies (after Vaurie 1965)

   T. t. britannicus (Witherby and Lönnberg): British black grouse. Resident in Great Britain from Devon, Somerset, Wales, the Midlands, and northwestern Lincolnshire to Scotland and some of the Inner Hebrides.

   T. t. tetrix (Linnaeus): European black grouse. Resident in Europe from the Scandinavian Peninsula and Finland southward to southern France and northern Italy, and through central Europe east to the Carpathians and the Balkan Peninsula to Bulgaria and Romania, and eastward across Russia and Siberia to the Kolyma River in the northeast and the Lena in the southeast, south to the western Siberian steppes and to about the Angara and southern Yukutia, probably intergrading with *viridanus* to the west and *ussuriensis* to the east.

   T. t. viridanus Lorenz: Russian black grouse. Resident in southeastern Russia from the basins of the Don and Sura, eastward to the Siberian steppes, where it intergrades with *ussuriensis*, northward to the lower Kama, and southward to about 50° N latitude on the Ural, Ilek, Turgai, and upper Yenisei rivers.

T. t. ussuriensis Kohts: Siberian black grouse. Resident in southeastern Siberia, south of *tetrix* and east of *viridanus*, from Lake Baikal eastward to Amurland, Manchuria, Ussuriland, and northwestern Korea, and south to northern Mongolia.

T. t. mongolicus Lönnberg: Mongolian black grouse. Resident in the Russian Altai, northwestern Mongolia, south to the Tian Shan, west to the Kirghiz Range, and east to at least 82° 30′ E longitude in Chinese Turkestan.

MEASUREMENTS

Folded wing: Adult males 251–76 mm; females 214–34 mm. First-year males 245–65 mm; females 226–33 mm (Witherby et al. 1944; Glutz 1973).

Tail: Adult males 70–100 mm at center, outer feathers 65–100 mm longer (45–75 mm in first-year males); females 70–93 mm (Witherby et al. 1944). Cramp and Simmons (1980) report longer (96–102 mm) tails for females of *tetrix*, with age differences insignificant, and 98–108 mm for adult males.

IDENTIFICATION

Length, 16 inches (females) to 21 inches (males). Breeding adult males of this species are unique in possessing a lyre-shaped tail and a uniformly glossy black plumage except for a white wing bar, white under tail coverts, a white patch on the wrist joint, and a bright red wattle above the eye. Females are barred and freckled with brown and black, and unlike the female capercaillie have a white wing bar, are less heavily barred, and are distinctly smaller. They also have a slightly forked tail, which is noticeable only in flight. Adult males acquire a somewhat femalelike plumage in late summer, with mottled brown and black feathers on the head and neck, some whitish on the throat, and some mottling on the mantle. Juvenile males resemble females but are smaller. The first-winter plumage of the male is like that of the adult but is browner and duller, and the outer tail feathers are less curved (Witherby et al. 1944).

FIELD MARKS

In flight these birds alternate quick wingbeats and prolonged glides, and in their dark coloration may resemble capercaillies, but they are distinctly more agile and have forked tails. They are more prone to be found in the edges of moors, pastures, and relatively open habitats than are capercaillies. During the display period the crowing and dovelike "rookooing" calls of the males are distinctive, but the calls of the females are more like those of the capercaillie (Witherby et al. 1944).

Females may be easily distinguished from adult males by the absence of glossy black feathers and less elongated outer tail feathers.

First-winter males resemble adult males but have less uniform iridescence, with rufous vermiculations on the scapulars, the inner wing coverts, and the outer webs of the secondaries. The vermiculations often extend to the mantle and upper tail coverts. The outer tail feathers are narrowly tipped with whitish; there is often white also on the tips or shafts of the chin, throat, and breast feathers, and the tips of the outer two primaries are minutely freckled with buff to rufous buff. *First-winter females* closely resemble adults, but the outer two primaries are usually more sharply pointed and more vermiculated at the tips (Witherby et al. 1944).

Juvenile males are barred with black and buff on the upperparts, the tail feathers are narrow and short, barred with rufous buff, and whitish shaft streaks are present on the scapulars and wing coverts. *Juvenile females* resemble males but are not so rufous on the upperparts and wing coverts, and the breast is less rufous and more widely barred (Witherby et al. 1944).

Downy young have a brown crown cap bordered by black, which separates them from the capercaillie and hazel grouse; they more closely resemble ptarmigan young, but there is a narrower median zone of black on the back of the forehead. The upperparts are generally more finely mottled, and the toes are naked (Fjeldså 1977).

DISTRIBUTION AND HABITAT

In general the distribution of this series lies between the July isotherms of approximately 51° F in the north and 71–75° F in the south, thus broadly overlapping the capercaillie and locally replacing it in less densely wooded areas and along the edges of forests. It occurs in swampy heathlands, moorlands, and bogs, preferably those along the edges of forests or having a light growth of birch and pine, and in generally rough and open areas close to or surrounding coniferous and birch forests. Although it mainly occurs in lowlands, it locally reaches nearly 6,000 feet above sea level in the French Alps (Voous 1960). To a large degree the availability of birch, the preferred winter food, probably determines the species' overall distribution. Seiskari (1962) has shown that the total range of the black grouse falls within the limits of the white birch (*Betula pubescens*) and silver birch (*B. verrucosa*).

In Finland the high dependence of the black grouse on birch, which occurs in a variety of forest types, means that it is not bound to specific forest types, and thus its winter habitats are less easily characterized than those of the pine-dependent capercaillie (Koskimies 1957). However, birch is almost invariably present in winter habitats, and alder is usually also available. On the other hand, spruce seems to be avoided, since its branches offer obstacles to falling snow, and this perhaps conflicts with the black grouse's need for deep snow in which to roost (Seiskari 1962).

In addition to suitable winter habitat, the species also needs adequate nesting and brood-rearing habitats. Nesting habitats are not yet well defined, but brood-rearing habitats have been studied in Norway (Børset and Krafft 1973). It appears that, in common with its more general adaptations, for brood-rearing the black grouse prefers habitats in an earlier stage of forest succession than does the capercaillie. Almost half the black grouse broods seen were in forest in the early stages of regeneration after cutting (average height of trees no more than 1.3 meters). Additionally, almost three-fourths of the broods were seen in habitats characterized as a "*Dryopteris* subassociation" (containing *D. linnaeana, D. phegopteris,* and *Anemone nemorosa*). The authors suggested that a preference for a special plant, a combination of plants, or some special microclimatic properties of this association might account for the species' apparent concentration in this habitat type.

Throughout the species' range it typically occurs in habitats transitional between woodland or forest and some nonforested habitat type such as heath, fen, bog, steppe, or marginal cultivation. The presence of some trees is apparently almost essential, but they should be neither dense nor very high and should adjoin larger open areas. The collective requirements for special food plants, open or sparsely vegetated areas for display, suitable roosting sites, and arboreal perches all tend to concentrate birds on mixed terrains. The transitory nature of such habitats often requires population shifting with time. However, major ecological changes over much of the species range, as well as excessive hunting, human disturbance, and perhaps other factors have brought about major range contractions in much of Europe (Cramp and Simmons 1980; Glutz 1973). A short summary of the population status or trends in various European countries is provided in Appendix 3.

POPULATION DENSITY

Probably the highest population densities of the black grouse in Europe are now found in Scandinavia, where rather large areas of forest still exist. In Sweden, which is about 60 percent forested, the black grouse has locally benefited from clear-cutting practices that have tended to break up mature forests, but on the other hand afforestation of heathery areas and a mosaic of small farmlands have had the opposite effect, as has the removal of birch woods and the creation of coniferous monocultures (Marcstrom 1979). There are now possibly 300,000 pairs of black grouse in Sweden, and some 23 million hectares of forests, or about 1 bird per 38 hectares (95 acres) of forest. In Norway there are about 8 million hectares of forest habitat, and in 1960 there may have been as many as 500,000 to 600,000 birds present in that country (Wegge 1979). This suggests a very high country-wide density of about 1 bird per 13–16 hectares (33–40 acres), though in recent years the population has apparently declined greatly. In Finland the population in the late 1960s was perhaps 900,000 individuals (Rajala 1979). Merikallio (1958) estimated a smaller total population (310,000 pairs) in the 1950s and provided a density map indicating an average density of up to 1.8 birds per square kilometer. Seiskari (1962) stated that the density ranges from about 1 or 2 pairs per square kilometer (65–130 acres per bird) to as much as 3 pairs (43 acres per bird) in eastern Finland. Collectively these figures suggest a density range of from 33 to 95 acres (13–38 hectares) per bird over relatively large areas, and Rajala (1974) reported average late-summer densities of about 16 birds per square kilometer, or about 16 acres per bird.

Observations in the USSR indicate population densities of from 36 to 146 birds per 1,000 hectares (17–90 acres per bird), although yearly fluctuations there seem to be typical, and in one case population dropped from maximum density to 4 percent of that level in only three years (Dementiev and Gladkov 1967).

Densities on smaller study areas provide an alternative and probably better method of estimating actual population densities. Studies in four areas in the French Alps (one hunted, three not hunted) provided average densities of from 1.8 to 4.0 males per 100 hectares (or 1 male per 62–137 acres) in spring, with the hunted area the lowest in estimated density (Ellison 1979). These areas ranged from 682 to 1,045 hectares, and the spring sex ratio in the hunted area was 1 male per 2.3 females. If such sex ratios are typical of the region as a whole, the total spring population density would be about 6 to 13 birds for 100 hectares (or 1 bird per 19–41 acres). A study on a 500-hectare unhunted area in the Swiss Alps indicated a density of about 7 males per 100 hectares (or 1 male per 35 acres) (Pauli 1974).

HABITAT REQUIREMENTS

Habitat requirements for this species can be divided into two general types; those associated with winter and those needed for breeding. In Finland, at least, the black grouse occurs on a rather wide variety of forest and bog types during winter but is especially abundant in "dry" heaths and "fresh" heaths, with the former favored in southern areas and the latter in more northerly areas. Fresh heaths are those formed on a mineral soil that is rather poorly permeable to water and fairly rich in nutrients, while dry heaths are those formed on a more permeable mineral soil. Dominant tree species are most often birches, and birches are almost invariably part of the wintering habitat, with pine and alders (*Alnus incana* and *A. glutinosa*) typically also conspicuously abundant. Relatively sparse stands of trees are favored over dense stands, and the most favored tree heights are from 7 to 18 meters. The winter habitat of this species is thus largely dependent upon structural characteristics of the forest, with deciduous or mixed forests preferred over coniferous ones and young and sparse stands preferred over older and denser ones (Seiksari 1962).

During the breeding season, mixed forests dominated by deciduous trees are preferred by black grouse in Finland. The dominant tree species in the usual summer habitat is either birch or spruce (*Picea excelsa*), with trees relatively sparse and having an average height of about 11 meters. However, nesting sometimes occurs in a wholly treeless environment (Seiskari 1962).

Similarly, in Norway the habitat preferred for raising broods is one of relatively low and open woodlands, especially those with well-developed ground cover. Black grouse broods tend to be found in areas with higher and denser ground vegetation than those used by capercaillie broods (Børset and Krafft 1973). Likewise, in Denmark and Holland, where the species occurs on almost treeless heaths, it is typically associated with taller heather and bushes or small trees.

In Denmark the black grouse is now virtually extirpated, but the remaining population occurs in two rather different biotopes. The more common one consists of rather flat and dry heathland dominated by heather (*Calluna vulgaris*) and other low ericaceous bushes such as crowberry and cranberry, but with only a few trees. The second type is bogs, mostly raised bogs, whose vegetation is more varied but that usually have extensive growths of willows and birch (*Betula pubescens*). This latter biotope was probably the preferred one, but in recent years activities causing bog destruction have made such habitats extremely scarce. Additionally, successional changes can alter a bog's attractiveness to black grouse, and in particular the invasion of moutain pine (*Pinus mugo*) can produce a dense shrub cover and render the area unsuitable (Degn 1979).

In southern Europe the optimum biotope of the black grouse seems to be one having a high diversity of plants, including different levels of heather, short grass, open land, and taller vegetation (*Rhododendron, Pinus, Vaccinium*). In the French Alps the birds occupy moderately dense forests of spruce-fir or larches, provided an understory of shrubs or herbaceous plants is present and open ground is available for display sites.

FOOD AND FORAGING BEHAVIOR

Food Requirements

Throughout its range and during the entire year the black grouse is essentially vegetarian, and feeding generally can be divided into a spring, summer, and fall period of ground feeding, and a winter period of shrub and tree foraging. During late fall and early spring ("interseasonal" periods) a mixture of these foods and feeding behaviors occurs.

In the USSR winter food consists largely of the catkins of birches, with birch buds and shoots of secondary importance and the vegetative parts of Scots pine also important (Semenov-Tian-Schanski 1959).

In Scandinavia birch is also the primary winter food. Seiskari (1962) found that all but 3 of 47 crops analyzed contained birch materials, with catkins apparently preferred over buds. Birch catkins are the favorite winter food of the black grouse, and fluctuations in the amount of such catkins may cause density fluctuations in the black grouse. Similarly, in Norway preferred winter foods in one study were found to be the needles and male flowers of *Pinus mugo* and the male catkins of birch (*Betula pubescens*), followed by the shoots of birch and the needles, cones, and male flowers of Scots pine (*P. sylvestris*) (Haker and Myrberget 1969). In another study birch catkins and juniper berries were predominant winter foods, especially from January through March, and *Vaccinium* (especially berries) was of special importance from April through December (Kaasa 1959).

Dependence upon birch is not so clear farther south in the species' range, and in both Scotland and the Netherlands a greater use of dwarf shrubs (especially *Calluna*) and grasses is apparently typical, no doubt as a reflection of less snow cover in more temperate areas and less need for arboreal feeding. In the Swiss Alps, ericaceous plants provide the preferred food even in winter, with bilberry stems preferred over the young leaves and buds of rhododendron and these in turn supplemented by larch and pine needles when the dwarf shrubs are covered by snow (Zettel 1974). In another study area of the Swiss Alps, the stems of bilberry were also the preferred food, but the buds of rowan (*Sorbus*) were the initial substitute, with pine needles a third choice and spruce needles of hardly any significance. In that area larch shoots are rarely available, and likewise alder catkins are almost unattainable (Keller, Pauli, and Glutz 1979). Preferred winter foods in the Swiss Alps are usually high in sugars and disgestible crude protein, though winter foods tend to be protein-rich, which is understandable in terms of relative energy needs and requirements for egg production and molting during these different seasons (Pauli 1979).

The importance of birch in more northerly areas begins to decline in late winter. From January through March birch is maximally used, but by April it constitutes slightly over half of the food and by May its contribution is insignificant (Seiskari 1962). Interseasonal and summer foods in Norway consist of the berries and stems of bilberry, the berries and leaves of bog whortleberry, the berries of cowberry and crowberry, the shoots of heather, the bulbils of alpine bistort (*Polygonum viviparum*), cowwheat (*Melampyrum*), and the nutlets of sedges (Kaasa 1959). In Finland their principal foods in summer are berries (*Vaccinium* spp., *Juniperus communis*) and the shoots of low shrubs (e.g., *Vaccinium myrtillus*). The period of shift from summer feeding to winter tree-feeding there probably is related to the amount of food on the ground, the leaf conditions of the trees, and weather and snow conditions (Koskimies 1957).

Although foods of young are not yet extensively studied, birds smaller tham 100 grams tend to eat mostly invertebrates such as insects and spiders, while older chicks predominantly forage on plants such as berries, drupes, nutlets, and the like (Kaasa 1959).

During the short winter days in northern Finland, the total hours of daylight available for feeding are limited, and black grouse concentrate most of their foraging near noon, in contrast to the capercaillie, which forages during the entire daylight period. Both species begin to feed at dawn, a little before sunrise, and stop soon after sunset. Apparently the crop of a black grouse is able to accommodate all the bird's daily food requirements, so that it need be filled only once each day. Feeding is correlated with the changes in the weather, which stimulate general activity in the birds but tend to reduce foraging. On the other hand, feeding is stimulated by cold weather and high pressure, whereas warm weather and low pressure reduce it (Seiskari 1962).

MOBILITY AND MOVEMENTS

This species is largely resident over the entire range, though in some northerly areas it is apparently eruptive. These periodic movements may be caused by local food shortages and sometimes involve several hundreds of birds. Several such migrations have been observed in the USSR, frequently during early autumn. Often females greatly outnumber males in such movements, sometimes three- to seven-fold. Home ranges of winter flocks in the USSR are usually from 3 to 6 square kilometers, although during periods of heavy snow the birds are prone to wander nomadically (Dementiev and Gladkov 1967). In addition to these periodic or eruptive movements, actual migration may occur in some parts of the USSR. For example, in the Ussuri District small populations regularly perform seasonal migrations, which at times extend up to 300 kilometers from nesting areas (Vorobiev 1954). However, Semenov-Tian-Shanski (1959) noted that in his study area 28 of 39 banded birds were recovered within 2.5 kilometers of the place of banding, and the maximum displacement was 27 kilometers.

In Switzerland the winter home range of a group of 7 to 10 males from one area covered some 90 to 120 hectares. Females and immature males were more mobile, but even these usually remained within an area of 5 square kilometers (Pauli 1974). Winter flocks of black grouse are typically bisexual, in contrast to those of the capercaillie. As capercaillie broods disintegrate, the females tend to become much more mobile, while the adult males tend to remain closer to their home areas. However, in black grouse, entire broods join older birds in early fall to form winter flocks, which probably "condense" around groups of older males. In both species flock size is related to temperature, with larger flocks typical at colder temperatures. However, black grouse are much more prone to form large flocks than are capercaillies, and in a Finnish study as many as 75 to 100 birds were sometimes seen in black grouse winter flocks, but never were as many as 20 present in capercaillie flocks (Koskimies 1957).

In a Scottish study, 1 first-winter male moved 17 kilometers, and 3 moved up to 5 kilometers from the point of marking. Analysis of banding recoveries from Norway indicated that all 29 recoveries of marked males were obtained within 5.5 kilometers from the point of banding, and 83 percent were within 2.5 kilometers, with an average movement of 1.6 kilometers. However, of 18 female recoveries the average movement was 4.4 kilometers, and the maximum distance was 20 kilometers (Cramp and Simmons 1980). Five black grouse marked in Finland during spring were recovered during a later autumn no more than 1.5 kilometers away (Koskimies 1957).

REPRODUCTIVE BIOLOGY

Seasonal Display Activities and Lek Persistence

The seasonal occurrence of territorial activity on display grounds certainly is subject to geographic variation. In southern Sweden the birds are totally silent only during two months of the year—in July, when they are in the middle of their molting period, and during midwinter in January. However, the length of the winter break in activity is weather dependent, and under ideal conditions adult males may be present on the display grounds throughout the winter. After the summer pause for molting, adult males begin to make short and sporadic visits to the lek in August, with fall display reaching a peak from late October to early December. Loose snow accumulation inhibits activity, and when it is deeper than 6 centimeters the arena will remain unused, although the birds may perform from trees. However, crusty snow favors display activity, which gradually increases during spring until it peaks about the first of May. High activity continues for about 2 more weeks, but by late June it has declined to essentially zero (Hjorth 1970).

In other countries the birds follow a similar pattern. In the USSR spring display usually beings in late March, reaches a peak in early May, and terminates at the end of that month (Dementiev and Gladkov 1967). In Finland males may visit display grounds as early as late January but do not reach "full speed" until early April, with hens usually visiting the grounds during the first half of May. Males continue to use the grounds until late May, but by June their visits are only occasional. Clearly, daily weather conditions have a considerable influence on the behavior of males, but in females the readiness for mating seems to develop rather independently of weather and is more probably timed by increasing photoperiod, which remains constant from year to year. The role of fall display perhaps is to allow surviving adult males to reclaim territories held during spring and to provide opportunities for young males to visit the leks. By spring such individuals will have established territories, but these are probably always only "third-class" males, of the lowest hierarchical rank (Koivisto 1965).

Because of their nearly continuous usage, leks tend to persist from year to year with little or no change in location. Koivisto (1965) mentions one area that was first used in the late 1940s when the area was subjected to clear-cutting. It gradually grew up to saplings, and by 1963 it was still being visited by only a single male. One major requirement of black grouse leks is that they be in fairly open to entirely open landscapes. In Koivisto's study areas, 15 of 40 arenas were on the ice of lakes, 10 were in forest saplings, 9 were on moors, and 6 were on fields. After ice breakup, the birds displaying on the lakes moved to other places, which were usually openings by the lakeshore or in the close vicinity of the lake. Where display grounds occur on partially vegetated areas, the center of the arena is in the place having the least vegetational cover. This preference for openness is indicated by the fact that the distances from the center of the area to the nearest forest edge ranged from averages of 40 meters on moors to 110 meters on fields and forest openings and 110 meters on ice. In some areas where succession is very slow, particular arenas may be used for many decades. Hjorth (1970) mentions that in some Swedish bogs the history of use by black grouse extends back at least a century. Once an arena is established, the birds show great persistence in keeping it active, even in the face of vehicular or aircraft traffic nearby.

Lek Aggregations and Territorial Sizes

As reviewed by Hjorth, the number of male grouse that consitute a lek varies greatly, from a very few birds in areas of low density or marginal suitability for leks to as many as 200 individuals, as has been reported in the USSR. On some 300 arenas inventoried in Finland, the average number of males in attendance was 9.4, with a trend toward reduced numbers toward the northern part of the country (Koivisto and Pirkola 1961). Koivisto (1965) found that larger groups of males do not attract relatively more females than smaller ones, nor is copulation apparently more efficient at the larger leks. Apparently a major advantage of social display in the species is increased protection from possible predators, as well as the potential for assortive mating based on the hierarchical dominance territorial system. Although a few copulations may occur outside the limits of the lek, in Koivisto's view these few cases are not sufficient reason for believing that the highly structured lek system is not closely related to reproductive fitness in males.

Within the lek, the size of individual territories varies greatly. Koivisto (1965) reported that individual territories in Finland varied from 2 to 200 square meters, with no significant differences in those of first-class and second-class males. Central territories tend to be small and precisely defined compared with the peripheral ones, which are often rather fluid and indefinite. In large Swedish leks the central territories tend to be rather constant in area, ranging from 170 to 260 square meters (Hjorth 1970). However, topography and vegetation seem to affect territory size, with territories larger where there is a maximum amount of visual contact and smaller where visual contact is low and movements are limited because of vegetation. Even weather may affect territories, the birds remaining closer together during severe cold or on rainy days. Nevertheless, each male normally does defend a discrete territory that, once claimed, may be kept for the rest of his life. Although boundaries may shift somewhat from year to year, the general position tends to remain much the same.

Koivisto (1965) classified territory holders into three hierarchical classes, based on behavior and territorial characteristics. First-class males are coequals; they tend to occupy central territories and show ''respect'' for their common boundaries. They may, however, intrude into the territories of second-class males with impunity. Second-class males occupy peripheral positions but are clearly weaker than first-class birds. Third-class males do not hold permanent territories; they are juveniles that move about in marginal areas and sometimes even move through the middle of the area while feeding. They display occasionally but apparently never directly court females. Koivisto reported that, of 57 copulations observed, all but one were performed by first-class males, which thus correspond to the ''alpha'' males of the capercaillie or the ''master cocks'' of the sage grouse. Judging from marked birds, Koivisto concluded that young males probably attain second-class status just before reaching 2 years of age, and by the time a bird is three years old he will be able to reach first-class status. Once a first-class male, he will probably remain so indefinitely, unless, for example, he is moved to another display ground. The relationship between age and social status was proved by Johnston (1969), who found that no first-year males held territories but that all older ones did. The average period of territorial ownership was 2.5 years per male, with about a third of the territorial males being replaced each year. Males with larger than average territories (12 averaged 76 square meters) were the most successful in attracting females.

In a important recent study, Vos (1983) reported that males normally become territorial in their second year, but sometimes not until three or four years old. Older males are more successful in copulating than are younger ones, and the highest percentage of central territories were held by five-year-olds, followed by four- and three-year-olds. Few males older than five years held central territories, and two males lost their territories after their sixth year of life. Annual male survival rates averaged nearly 70 percent in this protected population.

Hjorth (1970) doubted that such a male hierarchy existed in the Swedish leks he observed and believed that Kruijt and Hogan (1967) were correct in questioning the variable social rank of territorial males indicated by Koivisto, suggesting instead that all males have equal territorial status. Kruijt and Hogan did confirm the observation that a few (four) central males did more than 85 percent of the actual mating they observed in two years, but they attributed this to attraction value of the center of the lek and to the better courting strategy of the males there. Hjorth even suggested that most copulations might occur outside the arena. He provided no positive evidence on this point except that he personally observed few matings on the arenas that he observed for five years.

Johnston (1969) believed that individual variations in male mating success might be affected by their "attendance index" (relative daily lek attendence), behavioral variations (general activity and courtship intensity), differences in the size, position, or appearance of the males' territories, or differences in the males' appearance. The most successful males not only displayed more often than others but also were involved in disputes more often. Most female visits went to males in the "middle-age" group (those in their third or fourth year), and the generally larger territories held by these dominant males gave them more freedom from harassment by other males when courting females.

More recently, Kruijt, de Vos, and Bossema (1972) have examined this problem and have concluded that, apart from the nonterritorial males, the arena structure consists of "central" and "marginal" males. The central males have more desirable and more competitive territories, but no consistent differences in aggressiveness inside or outside the arena could be proved for central versus marginal males. The females' preference for mating with males at the center of the arena was believed to perhaps result from a combination of their preference for the relative clustering of males near the center and their attraction to males that were actively disputing territorial boundaries, although genetically determined differences in the aggressiveness of central and marginal males were ruled out.

Territorial Advertisement

Male advertisement in this species takes two general forms; aerial display and ground display. Aerial displays include "flutter-flights" and "flutter-jumps" in Hjorth's terminology. Before the flight the male lowers his breast slightly and may flap his wings or utter one or more hissing calls. He then rapidly ascends while uttering harsh hissing notes in staccato rhythm. After reaching the peak of the flight, some 2 to 6 meters high, the bird rapidly descends with fluttering wingbeats. During the entire display the white wing bar and under wing linings sharply constrast with the otherwise mostly blackish plumage, and the tail is only slightly spread. During such flights the bird remains within his own territory and thus is unlikely to fly forward more than 15 to at most 30 meters. Besides the loud vocalization, the fluttering noise of the wings is conspicuous. Flutter-jumping is a shorter and lower form of the flutter-flight, in which the bird may rise only a few inches above the ground and perhaps move forward only a few feet. However, even in these short flights the bird often turns slightly, so that he usually faces a slightly different direction upon landing. In the simplest form of the display, the bird does not leave the ground at all but simply performs wing beating in place, accompanied by the hisslike calling. This thus becomes a ground display, which Hjorth has called the "thin-necked upright cum hissing and wing flapping."

Fluttering-jumping stimulates other males to perform the same display and also is often provoked by the sudden appearance of a new bird on the arena, either male or female. Of 200 occasions when flutter-jumping was observed by Koivisto (1965), the largest number of displays (58) were performed without apparent external stimulus, but the next largest number (37) occurred when a hen or hens arrived at or departed from the arena.

Ground displays of the black grouse have been described by many writers (e.g., Selous 1909–10; Lack 1939; Höhn 1953; Kriujt and Hogen 1967), but Hjorth's (1970) description is most complete and recent and is the basis for the following discussion.

As noted earlier, aerial displays grade into ground displays, and the thin-necked upright posture with wing flapping and hissing is sometimes also performed within wing flapping. This hissing call may also be performed without noticeable neck extension or tail cocking. Stationary hissing is often uttered when two males walk toward one another, and hissing is likewise often performed in response to hissing by other birds, or even tape recordings of the call.

Soon after arriving on his territory, the male assumes a "forward" posture, in which the neck is held diagonally upward, with the head fairly low and the bill horizontal, and the tail is cocked and spread, maximally exposing the white under tail coverts. The primaries may also be slightly spread, but the wings are held close to the body, more or less in a resting position. From this preliminary posture the bird utters his "rookooing" call, a two-part song with a total length of 2.2–2.7 seconds, separated by a 0.15-second break. In the first part the male closes his bill, inflates his esophagus, and utters a series of bubbling notes of increasing frequency and amplitude, lasting about a second. The second part of the song is more shrill, and while uttering it the male's entire body quivers, especially the tail feathers. This "convulsive phase" last about 1.25 second, and apparently consistently consists of thirteen melodic figures. At the end of the song the bill is opened and air is expelled, producing a bellowslike sound. The song is often uttered repeatedly for rather

25. Male displays of the black grouse (after Hjorth 1970 and Glutz 1973), including (1) side and front views of rookooing male, (2) bowing duel of two territorial males, showing ground pecking by male on right, and (3) fighting by two males.

long periods, producing an almost continuous murmur of sound. While singing the male may either stand still or walk about, and in the latter case he usually moves with pompous steps, the body held low and the breast feathers nearly touching the ground. Occasionally the birds will also display from trees.

Territorial Defense and Fighting

When walking through an arena toward his territory, the male black grouse assumes a wide-necked upright posture similar to that of the male capercaillie. At this time he may hiss, and males also at times emit a nasal note, *kok*, which is sometimes uttered in a series. Often this note is extended into a series of calls that Hjorth terms the ''nasal whinny,'' and this vocalization occurs in a variety of agonistic situations, such as while two males are walking parallel along territorial boundaries or facing each other in ''bowing duels.'' As a male approaches a rival in the wide-necked upright posture, he gradually lowers his primaries and extends his neck farther forward. Upon meeting, two males may begin an actual fight or may perform various types of ritualized fighting.

Most such agonistic encounters occur at the edges of territories, and they usually take the form of bowing duels. While facing one another, the two males jerk their heads up and down while keeping their bodies parallel with one another and just beyond pecking distance of each other. Sometimes, as during the height of the display season, the two birds will perform rookooing displays during such confrontations, and at these times actual fights are likely to ensue. Fights often begin with pecking toward the opponent's red combs, while the attacked bird tends to jump backward with wing flapping, in turn attacking his rival. When one bird manages to grasp the other with his beak, battles involving wing beating follow, but even during such encounters the beak is the major weapon, and the wings are used only when the birds are clinging together. Sharp hissing sounds are uttered during these fights.

Precopulatory and Copulatory Behavior

As soon as females appear on the arena, the resident males begin to rookoo even more rapidly, and they spread their primaries downward so that the tips of the feathers drag on the ground. At this time the head is held lower than during normal rookooing, with the combs at about the same level as the conspicuous white shoulder-spot markings. In this posture the males attempt to approach the hens as closely as possible, and when a female is well within a male's territory he usually will walk to and fro in front of her or circle her. While circling, the male will often tilt his tail toward the female at a very slight but perceptible angle. When a male has courted a female for long periods but she fails to remain squatted at his approach, he will sometimes sink to the ground and rookoo from a crouched posture, while lifting his wings slightly and keeping his primaries more folded. This posture, which certainly corresponds to the ''nuptial bow'' of prairie chickens, appears to simply be a special modification of rookooing.

Females often pay little obvious attention to the displaying males, and their first indication of interest in mating is agitated running, pecking-intention movements, sitting down, and rapidly turning the head. The actual precopulatory display is a squatting posture, with the wings slightly spread and lifted and the head somewhat raised. A male will continue to circle a squatting female for a time, then he finally approaches from behind. During treading the male seizes the female's nape and strongly flaps his wings. Treading lasts only 2 to 4 seconds, and as soon as it is completed the female runs a distances, shakes, and either preens or flies away. The male immediately begins rookooing again. Hjorth noted that females sometimes selectively choose males for mating that are imperfectly plumaged and lack many tail feathers; will at times apparently momentarily squat before one cock before leaving him and allowing a male in an adjacent territory to mount; and will sometimes allow a male to mount that has just finished fighting with a rival. Rarely, a female will mate with two males during the same morning (Kruijt and Hogan 1964). As noted earlier, appropriate mating strategies on the part of experienced males seem to be an important aspect of attaining successful copulations. About three-fourths of the females copulate within 15 minutes of their arrival in the arena, and likewise 75 percent of the copulations occur during a 10-day period at the peak of the display season (Kruijt and Hogan 1967).

Nesting and Brooding Behavior

After being fertilized, females leave the lek area, and they probably begin laying their eggs about 10 days later. Nests are scrapes in the ground among fairly tall vegetation or in low scrub; rarely, the old tree nest of another species will be used, as far as 6 meters above the ground. The nest may be placed in the young growth of coniferous or deciduous vegetation, among the roots of a tree, under low branches or brush, or in grainfields (Glutz 1973; Bernard 1982). Clutch size appears to vary geographically, and in Finland the average size is largest in central areas, where population density is also the greatest. The average of 491 clutches in the zone between 60° and 62° N latitude was 8.1 eggs, while 637 clutches from there to about 64° 30′ N averaged 8.3 eggs and 185 clutches from still farther north averaged 7.7 eggs (Helminen 1963). The total range of clutch sizes is about 6 to 11, with clutches of 15 or more probably laid by two females. Although the birds are single-brooded, replacement clutches are often laid after clutch loss.

Eggs are laid 36 to 48 hours apart, and incubation begins with the last or penultimate egg. Females leave their nests to feed during the day, usually during morning, early afternoon, and evening, with the evening break averaging the longest and the total average time away from the nest about 200 minutes per 24-hour period (Robel 1969). Linden (1981) estimated nest mortality at 29 percent, with a hatching success of about 92 percent in successful nests.

Hatching is essentially simultaneous, and the birds soon leave the nest. The female broods the young for about the first 10 days after hatching, especially at night or during rain. The chicks may make short flights when they are only 10 to 14 days old, but they are not fully independent until they are about 3 months old. They reach sexual maturity the

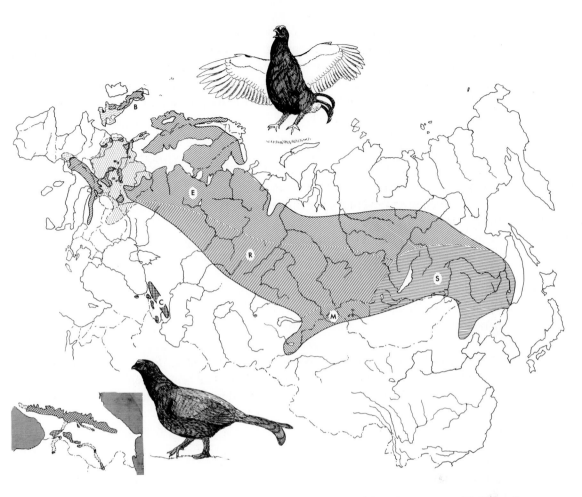

11. Current distribution of the Caucasian black grouse (C), and of the British (B), European (E), Mongolian (M), Russian (R), and Siberian (S) races of the black grouse. Probable historic distributions of the black grouse in Great Britain and Europe are indicated by stippling.

first year, but males probably do not become territorial until they are nearly 2 years old, and they rarely mate even then (Cramp and Simmons 1980).

In Finland an estimated average of 62 percent of hatched young survived until late August, with yearly averages ranging from 56.6 to 66.8 percent between 1963 and 1966 (Rajala 1974). First-year mortality (September to spring) averages about 64 percent (Linden 1981).

EVOLUTIONARY RELATIONSHIPS

As noted in the account of the Caucasian black grouse, these two species constitute an obvious superspecies, and they have probably been isolated from each other since early or mid-Pleistocene times.

The other important evolutionary and taxonomic question concerns the generic allocation of the two black grouse species, which have long been placed in the genus *Lyrurus* on the basis of the unusually shaped tail in males, the almost wholly iridescent plumage of males, and the less elongated neck feathers of males compared with those of *Tetrao* (Short 1968). As Short has already emphasized, these sex-associated traits do not warrant generic distinction, and furthermore the relative frequency and fertility of hybrids between capercaillies and black grouse suggest that reproductive isolation between these morphologically rather different types is actually very weak. Short also pointed out that the black grouse and capercaillies tend to be ecological replacement forms, as might be expected in close relatives, and their territorial and mating behavior patterns also strongly suggest a close genetic relation between them.

Caucasian Black Grouse

Tetrao mlokosiewiczi (Taczanowski) 1875

Other Vernacular Names

Caucasian blackcock; Kaukasisches Birkhuhn (German); tétras du Caucase (French); kavkazskiy tetrev (Russian).

Range

Resident in the alpine zones of the Greater and Lesser Caucasus from the basin of the Pshekha River in the west to the Samur basin in the east in the Greater Caucasus, the Iori and Alazan basins on the southern slopes of the same range, and in the Lesser Caucasus from the Gurisk Range east to the Karabakh Range (Vaurie 1965). Also in the upper levels of the Black Sea coastal range in northeastern Turkey (Cramp and Simmons 1980).

Subspecies

None recognized.

MEASUREMENTS

Folded wing: Males 180–220 mm, average of 12, 206 mm; females 196–211, average of 7, 187 mm (Cramp and Simmons 1980).

Tail: Males 151–71mm, average of 12, 162 mm; female 144 mm. Outer tail feathers of adult males average 48 mm longer, those of immatures, 10 mm (Cramp and Simmons 1980).

IDENTIFICATION

Length, 16 inches (females) to 20 inches (males). Adult males are uniformly glossy black on the head, neck, breast, and back; the remiges are dark brown, and the only white areas are a small carpal mark at the wrist and the axillaries. The beak is black, and there is a red comb above the eye. Females are narrowly and darkly barred on the top of the head and body, the bars sometimes becoming vermiculations, and the breast is heavily barred with dark brown and ochre, while the underparts are dark brown (Dementiev and Gladkov 1967).

FIELD MARKS

This species inhabits alpine meadows, subalpine forests of birch and rhododendrons, and coniferous forests near timberline. The range does not overlap with that of the black grouse, and thus separation of these two species in the field is not a problem. However, males of this species have longer and less lyre-shaped tail feathers and lack white wing bars and white under tail coverts. Females of the two species are very similar, but those of the Caucasian species are more uniformly and finely vermiculated on the breast and sides; they are more ochre in overall cast, and the tail is not so distinctly forked.

AGE AND SEX CRITERIA

Adult males are easily distinguished from females by their uniformly glossy black coloration. Evidently this plumage is not attained until the second year of life, and thus first-winter males look very much like females but apparently have the white carpal markings of adult males. Their rectrices are russet brown, with light vermiculations, and the body feathers are generally ochre, with a grayish tinge and narrow dark vermiculations (Dementiev and Gladkov 1967).

First-winter females are very similar to adult females but more deeply rufous; the upperparts are less regularly barred, and the tail feathers have broader dark bars, among other differences (Noska, cited in Cramp and Simmons 1980).

Juveniles resemble adult females but are smaller and more tinged with rufous, with the female usually paler than the male (Cramp and Simmons 1980).

Downy young are apparently almost identical to those of the black grouse but perhaps are lighter in color, with the brown parts more copper red and the crown cap less distinctly bordered with black laterally (Fjeldså 1977).

DISTRIBUTION AND HABITAT

In the Greater and Lesser Caucasus, this species occurs at elevations of 1,500 to 3,000 meters above sea level during much of the year, but it may descend to 700 to 800 meters during snowy winters. The usual summer habitats are timberline areas and subalpine meadows where there is an abundance of rich vegetation, rhododendron thickets, and low birches. In winter it uses the upper subalpine forests and higher fir forests and the sunny portions of the lower alpine zone. According to Averin (1938), the majority (56 percent) of the winter sightings are in the upper limit of the forests and the subalpine meadows, and more than a quarter are in the fir-beech forests. With spring, the use of the fir-beech forests declines and the use of alpine meadows increases, so that by summer the birds are gone from the fir-beech forests and most are found in alpine meadows. With fall there is again a downward movement into the subalpine meadows (65 percent of sightings) and a gradual reinvasion of the fir-beech forests. Tachenko (1966) found that most birds live within a 300 to 500 meter vertical range centering on the timberline, with spring and fall concentrations on south-facing slopes at about 2,400 meters that support subalpine meadow vegetation.

In Iran the species is restricted to the upper deciduous forest and associated alpine meadow zones between 1,800 and 2,000 meters elevation in the Kalibar Mountains northwest of Ahar, Azerbaijan (Scott 1976).

In Turkey the birds are probably fairly well distributed throughout the Black Sea coastal range, although in small numbers; there is only one recent record for that area (Cramp and Simmons 1980).

POPULATION DENSITY

The population in Iran is extremely low; perhaps 200 to 300 birds exist within a wildlife refuge of 38,320 hectares (Scott 1976). In Armenia the total population numbers about 300 to 500 birds. In southern Georgia the number of males sometimes reaches 3.5 per square kilometer in April and May, and on a preserve the density varied from 0.9 to 1.4 birds per square kilometer during the period between 1959 and 1960 (Flint 1978). Grazing reduces populations. Where cattle grazing is prohibited, as in a sanctuary in the Caucasus, two dozen birds may be flushed during a walk of a few hours (Dementiev and Gladkov 1967).

HABITAT REQUIREMENTS

During the winter, the primary factors influencing the birds are the distribution and depth of snow cover and associated availability of vegetation for food. This determines the birds' altitudinal distribution and their activities. Some birds remain on the northern exposures, where snow is deeper and nighttime burrowing into the snow is possible. Others remain on the southern slopes at night, roosting in bushes or under rock cover. Flocking is strongest in winter, when groups of as many as 15 to 20 individuals may sometimes be seen, or rarely even as many as 50. However, the usual situation even during winter is for the birds to remain as singles or in groups of 2 to 5 (Dementiev and Gladkov 1967). Of 128 winter flocks counted by Averin (1938), only 4 included more than 8 males, and in spring groups of more than 3 were rare except on display grounds.

Snow roosting is regular at night, and during bad weather the birds may remain in their holes during the day as well. The birds generally fly directly into their burrows rather than scratching them out. The hole is 15 to 25 centimeters deep and about 30 centimeters long. When snow crusting makes burrowing impossible, the birds will spend the night under rocks or snowdrifts or in shrubbery (Dementiev and Gladkov 1967).

The major spring requirement is certainly for suitable display areas. These occur in two general situations. The first comprises small areas closely associated with forests, to which the birds can readily walk from forest cover. These grounds are about 60 by 100 meters in area, in alpine vegetation bordered on one side by forest edge and on the other by tall herbaceous subalpine vegetation. The second comprises larger areas to which the birds often fly and which may be used only temporarily. These occur on alpine meadows on steep slopes about a half a kilometer above timberline. On these, groups of 1 to 4 birds sometimes gather to display and chase females but apparently do not defend permanent territories (Averin 1938).

FOOD AND FORAGING BEHAVIOR

These birds gather most of their food from the ground—primarily vegetable matter. Animal foods (insects and spiders) are eaten very rarely by adults. During the winter, the birds principally consume the buds and catkins of birches, juniper needles and berries, rose fruits, and the shoots and buds of willow. The most important foods are birch catkins and juniper needles, with rhododendron leaves, fir needles, and the leaves of whortleberries and foxberries next most important. In spring, birches, roses and junipers continue to be the most important foods, but as trees begin to bloom the greenery of herbaceous vegetation and shrub shoots and buds become more important. By late May the stalks, flowers, and ripening seeds and pods of alpine vegetation are major foods. In August the mature seeds of alpine plants and the berries of mezereon, whortleberries, and strawberries are the chief items. Finally, as the birds descend into the forest they begin to eat fir needles, the leaves of crabapples, acorns, and the like (Dementiev and Gladkov 1967).

The food of young birds differs from those of adults only during June and July, in the period of active growth. For the first 10 to 15 days of life the young eat insects almost exclusively, with little or no selectivity as to species. The items include primarily beetles (especially weevils and leaf miners), but also such forms as sawflies, flies, geometrid moths, butterflies, snails, and spiders. By August the birds have moved to berry patches, where they begin to eat the same foods as adults (Dementiev and Gladkov 1967).

At least during winter, the daily cycle of feeding is very uniform. The birds leave their nocturnal roosts by daybreak and disperse to forage. They spend much of the day basking in juniper shrubbery. After midday the birds begin to leave for evening browsing, and at dusk they fly as a group to nighttime roosts near or in forests (Dementiev and Gladkov 1967).

32. Black-billed capercaillie, male displaying. Photo by A. Andreev.

33. Black-billed capercaillie. Photo by A. Andreev.

34. Black-billed capercaillie, dominant male. Photo by A. Andreev.

35. Female Caucasian black grouse. Photo by W. Grummt.

36. Male Caucasian black grouse foraging. Photo by O. Vitovitsch.

37. Male Caucasian black grouse immediately after flutter-jump. Photo by O. Vitovitsch.

38. Male Caucasian black grouse displaying near female. Photo by O. Vitovitsch.

39. Female black grouse. Photo by Hans Aschenbrenner.

40. Male black grouse. Photo by Hans Aschenbrenner.

41. Male black grouse displaying. Photo by author.

42. Male black grouse. Photo by author.

43. Hazel grouse, male. Photo by Hans Aschenbrenner.

44. Hazel grouse, male in intense threat display. Photo by Hans Aschenbrenner.

45. Hazel grouse, fifteen-day-old chick. Photo by Hans Aschenbrenner.

46. Hazel grouse, male whistling. Photo by Hans Aschenbrenner.

47. Ruffed grouse, male. Photo by F. Jenson, courtesy G. A. Allen, Jr.

48. Ruffed grouse, male. Photo by F. Jenson, courtesy G. A. Allen, Jr.

49. Ruffed grouse, female defending brood. Photo by author.

50. Ruffed grouse, male. Photo by F. Jenson, courtesy G. A. Allen, Jr.

51. Ruffed grouse, male on drumming site. Photo by Ed Bry.

52. Ruffed grouse, male drumming. Photo by Ed Bry.

53. Ruffed grouse, male drumming. Photo by Ed Bry.

54. Lesser prairie chicken, male. Photo by author.

55. Lesser prairie chicken, male. Photo by author.

56. Greater prairie chicken, male and female. Photo by author.

57. Greater prairie chicken, male and female. Photo by author.

58. Greater prairie chicken, male and female. Photo by author.

59. Greater prairie chicken, male booming. Photo by author.

MOBILITY AND MOVEMENTS

Judging from Averin's (1938) observations, these birds have restricted home ranges; likewise, Tachenko (1966) found that even during winter their movements did not exceed 2 kilometers. The only recorded movements of this species are altitudinal ones, and the total maximum reported altitudinal range seems to be from 700 meters to somewhat more than 3,000 meters.

REPRODUCTIVE BEHAVIOR

Territorial Establishment

Territories are established and advertised during both spring and fall. Some males may begin displays in mid-April, but the peak period for display is in the first 3 weeks of May. Depending on the weather, some males may continue to display until the end of June, but warming weather causes them to cease (Tachenko 1966). Occasionally birds will display on snow cover or on thawed areas of snow-covered slopes, with each patch occupied by a single male, but normally the display areas are on south-facing grassy slopes, in troughlike depressions, or in the upper parts of ravines. Typically the same areas are used year after year, with some occupied for at least ten years (Potapov and Pavlova 1977).

The number of males on an arena obviously varies with density, but as many as 30 have been reported in a single location. However, there more often are 10 to 15 (Noska 1895) or 6 to 10 (Averin 1938). On one arena containing 18 males, 8 were nonterritorial yearlings (Potapov and Pavlova 1977). An arena containing 6 or 7 territorial males studied by Averin was about 60 by 100 meters; thus individual territories averaged no more than 1,000 square meters. A central territory was estimated by Potapov and Pavlova to be about 4,000 square meters.

Territories that are central and presumably dominated by master cocks tend to be adjacent to one another, but those on the periphery of the arena may be widely separated from all others. Beyond these well-established territories, there are other areas where some displaying might occur, but territorial boundaries apparently are absent. Thus Averin (1938) found that in late May grouse were observed displaying during evening hours in an alpine meadow well above timberline, where scattered groups of up to 4 individuals occurred and were sometimes separated by as much as 500 meters. However, this well may simply reflect the late seasonal conditions, where the males had abandoned their regular territories and were primarily engaged in foraging but still retained some sexual tendencies (Hjorth 1970).

There are probably about three categories of territorial occupation. First-year males, still not in adult plumage, often wander around the arena and may display at a low intensity, but they do not defend specific sites. Males in adult plumage, but perhaps only 2 years old, also may not have specific territorial attachment but nevertheless display at a

271

high intensity. Finally, other adult males, presumably at least 3 years old, have specific territories that they hold for long periods and on which they display very intensively. These males show considerable tolerance for younger males, at least those not yet in adult plumage, and sometimes allow them to enter their territories (Potapov and Pavlova 1977). Perhaps these immature males are so femalelike that the adult males fail to recognize them as potential rivals.

Territorial advertisement occurs during both morning and evening in spring but only during morning hours in autumn. During spring the morning and evening activity is of equal intensity. Early in the season, it begins while it is still dark and ceases at about sunrise, but later the greatest intensity occurs shortly after sunrise and may continue until midmorning. The birds gradually gather for the evening period, and late in the season activity is most intense between 6 and 7 P.M. (Potapov and Pavolva 1977).

Male Aerial Advertisement Behavior

Unlike the Eurasian black grouse, advertisement behavior in this species tends to be primarily aerial, and additionally it is largely or entirely lacking in vocalization. There is evidently only a single type of aerial display, flutter-jumping (terminology of Hjorth 1970). In this display the male begins in a resting posture, with the tail nearly horizontal, the body fairly erect, the breast and neck feathers greatly fluffed, and the combs maximally raised. From this posture the male may take a few steps forward or squat slightly, then he makes a nearly vertical leap into the air. He flies steeply upward for a meter or more, with 3 or 4 strong wingbeats, swivels sharply up to 180° as he reaches the apex of his flight, and makes a gliding descent to the point of takeoff. There is no associated vocalization, but the wings produce a whistling noise during both the ascending and the descending phases. This sound may carry about 150 meters, and additionally the white underwing surfaces are especially conspicuous at the apex of the flight as he veers around. The male keeps his ventral surface away from the slope, thus maximizing the visual effect, and at times may even somersault or make spiral movements in the air. The behavior is distinctly contagious; it is particularly common when females are present on the arena and declines when they are not in evidence (Averin 1938; Potapov and Pavlova 1977). The incidence of flutter-jumping also varies with weather and seasons. Thus counts in early May indicated an average of 135 jumps per male on a display ground, but in early June the average was only a single jump per male (Averin 1938).

Male Terrestrial Advertisement Behavior

When the males initially arrive on their territories, they are very alert, running up and down the slope with their necks stretched upward. From this stage the male assumes his basic display posture, that mentioned above as anticipatory of the flutter-jump. The breast area is swollen and projects forward of the head, the head is retracted until the eyes are directly above the carpal joints, the tail is slightly raised (23–26°) but not spread, and the wings are slightly drooped, revealing the white shoulder patch typical of most grouse. The

272

26. Male displays of the Caucasian black grouse (mainly after Potapov and Pavolova 1977), including (A) aggressive posture of male with neck ruffling, (B) stage in flutter-jump display, (C) display posture as illustrated by T. Lorenz, and comparisons of tails of male Caucasian black grouse (D) and black grouse (E).

body is held fairly erect, and the entire posture is reminiscent of that of a cooing rock dove (*Columba livia*). However, no hissing, cooing, or other notes have yet been heard, though it seems likely (as in the spruce grouse) that these might be so soft that they carry only a few yards and could so far have been overlooked by observers. The neck swelling may be only the result of feather ruffling, as Noska (1895) believed, or may in part be the result of esophageal inflation associated with vocalizations.

Male Antagonistic Behavior

Although observers differ in their ideas on the frequency of actual fighting, there is no doubt that hostile confrontations between males must be quite common. The usual threat posture during such confrontations is an approach with the neck extended diagonally forward, the neck feathers somewhat ruffled, and the tail held more or less horizontally. In more intensive threats the tail may be partially fanned and somewhat elevated as the birds perform bill snapping or ground pecking or engage in a bowing duel, uttering *chr-chr* notes. When one of the rivals retreats, the other follows with his tail raised and his wings drooping.

Sometimes two males meet at the border of their respective territories and walk in line with one another. At such times the tail is raised to about 45°, the head is drawn back, and the neck feathers are slightly swollen. Such behavior, with the two males keeping about 1.5 to 2 meters apart, may continue until the birds are well outside their territories, for up to 100 meters. Actual fighting includes breast-to-breast striking, pecking at the combs or throat, kicking, and wing thrashing. Submissiveness is indicated by a lowering of the orange combs. When a female appears, the male's posture is similar to that used when he is threatening another male, but the tail is more strongly erected and is opened to form a semicircle (Potapov and Pavlova 1977).

Courtship Behavior

Behavior associated with copulation is still unreported, but the presence of one or more females on the arena results in a strong increase in the incidence of display, particularly flutter-jumping. Males closely follow females, raising and spreading their tails, maintaining a thin-necked attitude, and sometimes performing flutter-jumps. If the female stops the male does the same, and he may also chase her up and down the slopes, even beyond the limits of the arena. Some observers have observed pairs moving away from the arena and out of sight, suggesting that perhaps at least some copulations occur away from the arena.

During periods of most intensive display, females may appear on the grounds during both morning and evening sessions. In the morning they arrive before daylight and rarely remain after sunrise, and during the evening they usually appear at dusk, during the peak evening activity between 6:30 and 7:30 P.M. (Averin 1938).

Nesting Behavior

Nesting begins in May, the nest site usually being a shallow scrape concealed in thickets of rhododendron, junipers, or other subalpine vegetation, or occasionally under a boulder. The nest, lined with grass and feathers, is about 21 centimeters in diameter. It is built by the female, and there is no evidence that the male participates in reproduction after fertilization. Ten clutches of eggs reported by Dementiev and Gladkov ranged from 2 to 10 eggs, averaging 6.0, while 13 clutches reported by Vitovitch (cited by Cramp and Simmons 1980) averaged 6.1. The eggs average slightly smaller than those of the Eurasian black grouse but are very similar in coloration. Not only are the eggs smaller, the clutch size is also somewhat smaller, based on available data, and adult females are approximately 85 percent the weight of the more widespread species.

The incubation period has not been firmly established but has been estimated as 24 to 25 days by Flint (1978). There is still only limited data on egg mortality, but of 80 eggs studied, 60 successfully hatched (75 percent hatching success), while 14 failed as a result of chilling and 6 were lost to predation (Vitovitch, cited in Cramp and Simmons 1980).

Incubation begins with the last egg, resulting in synchronous hatching. In the USSR this begins in mid-June, with most of the young hatching in late June or early July. The young are rapidly led away from the nest site to open alpine meadows. Young 2 to 3 days old weigh from 21.8 to 23.5 grams, but there is apparently very high chick mortality, probably well over 50 percent. Six broods counted in late June averaged 3.8 young, 31 broods seen during July averaged 3.7 young, and 6 August broods averaged 2.8 young. Assuming an average clutch of 6 eggs, this represents an average survival of 48 percent through August. Likely causes of chick mortality include severe weather (rain, fog, frost, hail) and probably also raptors. By late August the young birds are about the size of adults, and the broods disintegrate (Dementiev and Gladkov 1967).

The young are said to be able to fly when only 10 to 14 days old, and they are full grown at about 2 months. It is believed that females become sexually mature in their first year, but males probably do not breed until at least 2 years old. Young birds apparently first participate in arena display when nearly a year old, in the spring following hatching. At this time they are less active than adults, but reportedly their testes are just as enlarged as those of adults (Noska 1895).

After the display period is over the males begin to molt, and at that time they hide in thick vegetation. Females begin to molt in late August, after their broods have been raised (Dementiev and Gladkov 1967).

EVOLUTIONARY RELATIONSHIPS

All recent authorities have considered this species congeneric with the Eurasian black grouse, though opinions differ on whether the two species should be placed in the genus *Tetrao* or whether *Lyrurus* should be retained. That question has been considered in the account of Eurasian species, and the only remaining problem is the closeness of the relationship between the two black grouse species.

Voous (1960) believed that these two forms represented distinct species but that the Caucasian black grouse was probably a Pleistocene relict population and thus a very close relative of the Eurasian form. Short (1967) judged the Caucasian population to be closer to the ancestral type of black grouse, mainly on the basis of its less highly specialized tail, the lack of a white wing bar in males, and plumage differences in females. He suggested that future research may prove the populations should be considered only subspecifically distinct, but certainly the behavioral distinctions between them argue against such a conclusion.

I believe Short was essentially correct in concluding that these two forms should be considered allopatric species and that probably the Caucasian species is the more generalized. However, it is not in contact with any other grouse species, and thus its more generalized male plumage and displays may reflect this geographic situation and the absence of requirements for reproductive isolating mechanisms. Janossy (1976) suggested that the species probably was isolated from others by at least the Middle Pleistocene, based on fossil findings from the Upper Pleistocene.

Ruffed Grouse

Bonasa umbellus (Linnaeus) 1776

Other Vernacular Names

Birch partridge, drummer, drumming grouse, long-tailed grouse, mountain pheasant, partridge, pine hen, pheasant, tippet, white-flesher, willow grouse, wood grouse, woods pheasant; gelinotte à fraise, gelinotte huppée (French); Kragenhuhn (German).

Range

Resident in the forested areas from central Alaska, central Yukon, southern Mackenzie, central Saskatchewan, central Manitoba, northern Ontario, southern Quebec, southern Labrador, New Brunswick, and Nova Scotia south to northern California, northeastern Oregon, central Idaho, central Utah, western Wyoming, western South Dakota, northern North Dakota, Minnesota, Missouri, central Arkansas, Tennessee, northern Georgia, western South Carolina, western North Carolina, northeastern Virginia, and western Maryland. Recently introduced in Nevada and Newfoundland (modified from A.O.U. *Check-list*).

Subspecies (ex Aldrich and Friedmann 1943).

B. u. umbellus (Linnaeus): Eastern ruffed grouse. Resident in wooded areas of two regions, from east central Minnesota, southern Wisconsin, and southwestern Michigan south to central Arkansas, extreme western Tennessee, western Kentucky, and central Indiana (this population sometimes separated as *B. u. mediana* Todd 1940), and from central New York and central Massachusetts south to eastern Pennsylvania, eastern Maryland (formerly), and New Jersey.

B. u. monticola Todd: Appalachian ruffed grouse. Resident from southeastern Michigan, northeastern Ohio, and the western half of Pennsylvania south to northern Georgia, northwestern South Carolina, western North Carolina, western Virginia, and western Maryland.

B. u. sabini (Douglas): Pacific ruffed grouse. Resident of southwestern British Columbia (except Vancouver Island and the adjacent mainland) southwest of the Cascade Range, through west-central Washington and Oregon to northwestern California.

B. u. castanea Aldrich and Friedmann: Olympic ruffed grouse. Resident of the Olympic Peninsula and the shores of Puget Sound south to western Oregon.

B. u. brunnescens Conover: Vancouver Island ruffed grouse. Resident of Vancouver Island and adjacent mainland south to Puget Sound and north at least to Lund.

B. u. togata (Linnaeus): Canadian ruffed grouse. Resident from northeastern Minnesota, southern Ontario, southern Quebec, New Brunswick, and Nova Scotia south to northern Wisconsin, central Michigan, southeastern Ontario, central New York, western and northern Massachusetts, and northwestern Connecticut.

B. u. affinis Aldrich and Friedmann: Columbian ruffed grouse. Resident from central Oregon northward, east of the Cascades through the interior of British Columbia to the vicinity of Juneau, Alaska (not recognized in A.O.U. *Check-list*).

B. u. phaia Aldrich and Friedmann: Idaho ruffed grouse. Resident from southeastern British Columbia, eastern Washington, and northern Idaho south to eastern Oregon and on the western slopes of the Rocky Mountains to south-central Idaho.

B. u. incana Aldrich and Friedmann: Hoary ruffed grouse. Resident from extreme southeastern Idaho, west-central Wyoming, and northeastern North Dakota south to central Utah and western South Dakota.

B. u. yukonensis Grinnell: Yukon ruffed grouse. Resident from western Alaska east, chiefly in the valleys of the Yukon and Kuskokwim rivers, across central Yukon to southern Mackenzie, northern Alberta, and northwestern Saskatchewan.

B. u. umbelloides (Douglas): Gray ruffed grouse. Resident from extreme southeastern Alaska, northern British Columbia, north-central Alberta, central Saskatchewan, central Manitoba, northern Ontario, and central Quebec south, east of the range of *affinis* and *phaia*, to western Montana, southeastern Idaho, extreme northwestern Wyoming, southern Saskatchewan, southern Manitoba, southern Ontario, and across south-central Quebec to the north shore of the Gulf of Saint Lawrence, probably to southeastern Labrador.

MEASUREMENTS

Folded wing: Adult males 171–93 mm; adult females 165–90 mm (males of all races average 178 mm or more; females usually average under 178 mm).

Tail: Adult males 130–81 mm; adult females 119–59 mm (males average more than 147 mm; females average less than 142 mm).

IDENTIFICATION

Adults, 16–19 inches long. Both sexes have relatively long, slightly rounded tails that are extensively barred above and have a conspicuous subterminal dark band. The neck lacks large areas of bare skin, but both sexes have dark ruffs. Feathering of the legs does not reach the base of the toes; the lower half of the tarsus is essentially nude. Both sexes are definitely crested, but the feathers are not distinctively colored. In addition, males have a small comb above the eyes that is orange red and most evident in spring. Most races (*castanea* is perhaps the only exception) exist in both gray and brown phases, which appear with the first-winter plumage. Otherwise little seasonal, sexual, or age variation occurs. The birds are generally wood brown above, with blackish ruffs (less conspicuous in females and immatures) on the sides of the neck and small eye-spot markings on the lower back and rump (less conspicuous in females). The tails of both sexes have seven to nine alternating narrow bands of black, brown, and buff, followed by a wider subterminal blackish band that is bordered on both sides with gray and is less perfect centrally in females and some (presumably first-year) males. In winter both sexes develop horny pectinations on the sides of their toes, which are more conspicuous than in most other species.

FIELD MARKS

The fan-shaped and distinctively banded tail and neck ruffs of both sexes make field identification easy. The birds usually take off with a conspicuous whirring of wings, and in spring males are heard drumming much more often than they are seen.

AGE AND SEX CRITERIA

Females have shorter tails than do males (see above), and their central tail feathers lack complete subterminal bands near the middle of the tail. A mottled pattern on the central tail feathers (which occurs in about 15 percent of the population) can indicate either sex, but a bird with this characteristic is twice as likely to be male as female (Hale, Wendt, and Halazon 1954). Females also have little or no color on the bare skin over the eye, whereas in males this area is orange to reddish orange (Taber, in Mosby 1963). Davis (1969a) reported that the length of the plucked and dried central rectrices provides a 99 percent effective means of determining sex of both adult and immature ruffed grouse, but specific separation points for these groups vary with populations. Roussel and Ouellet (1975) reported that in postjuvenile birds an easy and reliable method of sex determination is made possible by the fact that females exhibit a single whitish dot at the center of each rump feather (rarely a double dot), whereas males typically exhibit a double dot (rarely triple) on these feathers.

Immatures can be identified by the pointed condition of their two outer primaries, especially the outermost one. Davis (1969a) stated that during the hunting season the condition of the tenth primary was useful for determining age of nearly 60 percent of the birds, with only a 2 percent error. However, the presence of sheathing at the base of the outer two primaries (adults) or on the eighth but not the ninth or tenth primaries (immatures) separated 79 percent of the birds examined with a 3 percent error. Immature males can be distinguished from adults by their shorter central tail feathers (length of plucked feather, 159 mm or less, compared with at least 170 mm in adults) as well as various other criteria (Dorney and Holzer 1957). Ridgway and Friedmann (1946) report that the two outer primaries of immatures have outer webs that are pale fuscous and mottled or stippled with ligher buff instead of being buff or whitish with darker brown markings.

Juveniles resemble the adult female but have barred tail feathers that lack the heavy subterminal band and have the gray tips poorly developed (Ridgway and Friedmann 1946). Juveniles also have white rather than buff chins and primaries with more mottling on their outer webs (Dwight 1900).

Downy young of the ruffed grouse can readily be identified by the restriction of black on the head to an elongated ear patch narrowly connected to the eyes and a few midcrown spots. The crown is otherwise a uniform ochraceous tawny, gradually blending with the buffy face color. The back lacks definite patternings and varies from russet or dark brown dorsally to pale buff or yellow ventrally.

DISTRIBUTION AND HABITAT

The distribution of the ruffed grouse in North America covers a surprising variety of climax forest community types, from temperate coniferous rain forest to relatively arid deciduous forest types. The unifying criterion, however, is that successional or climax stages include deciduous trees, especially of the genera *Betula* and *Populus*. For example, the range of various aspens (*Populus* spp.) bears a surprising similarity to that of the ruffed grouse, as does that of the paper birch (*Betula papyrifera*). Aldrich (1963) correlated racial variation in the ruffed grouse with major plant formations. He indicated that *togata* occurs in northern hardwood-conifer ecotone area, *umbellus* and *monticola* in eastern deciduous forest, *mediana* in oak-savanna woodland, *umbelloides* in typical boreal forest, *yukonensis* in northern or "open" boreal areas, *incana* in drier montane woodlands and aspen parklands, *brunnescens, castanea,* and *sabini* in the Pacific coast rain forest, and *phaia* in the corresponding wet interior forest. The relatively drier montane woodlands of the Pacific northwest are occupied by *affinis*. Not only is there a correlation between the relative wetness or dryness of these general habitat types and associated darkness or paleness of the body plumage, but there are also some relations between climate or vegetation and color phases. The gray phase of ruffed grouse is typically associated with northern areas or higher altitudes, while the reddish brown color phase is more characteristic of southern and lower-altitude populations. Gullion and Marshall (1968) have discussed the ecological significance of color phases in ruffed grouse, and they suggest that gray-phase birds are perhaps physiologically better adapted to cold than are red-phase ones and predominate in conifers and aspen-birch forest of these colder areas. They also

12. Current distribution of the Appalachian (A), Canadian (C), Columbian (Co), eastern (E), gray (G), hoary (H), Idaho (I), Olympic (O), Pacific (P), Vancouver (V), and Yukon (Y) races of the ruffed grouse. Regions of uncertain racial designation are indicated by cross hatching.

suggest that gray-phase birds may be less conspicuous in boreal forests, while in the hardwood forests where raptors have poorer hunting conditions and mammalian predators are more important the color phase may not be significant. However, their data indicate that gray-phase birds survive relatively better in hardwoods than do red-phase ones, and both phases survive better in hardwoods than in conifers.

Gullion (1969) has pointed out that, continentwide, the areas of highest population density of ruffed grouse correspond to the distributional patterns of aspens (*Populus* spp.), which he related to winter as well as summer food use by adults, as well as to their value as brooding habitat. Weeden (1965*b*) reported that ruffed grouse habitat in Alaska typically contains large amounts of aspen and usually also contains white spruce (*Picea glauca*) and white birch (*Betula papyrifera*). Where ruffed and spruce grouse occur together in Alaska, the ruffed grouse are found in earlier stages of succession, frequenting edges, shrubby ravines, and similar openings. Likewise in southern Ontario I have noticed that both species may be found within 100 yards of one another, but ruffed grouse are always associated with birch or popular, while spruce grouse are usually to be found under coniferous cover such as jack pine.

Edminster (1947) has analyzed the general shelter requirements of the ruffed grouse in the northeastern states according to vegetational succession stages. Open land types dominated by herbaceous plants provide some food sources for grouse but are of secondary importance. Overgrown fields with shrubs and saplings include single-species stands of high-quality quaking aspen cover (*Populus tremuloides*), pin cherry (*Prunus*), scrub oak (*Quercus*), or alder (*Alnus*) cover of moderate quality, and low-quality gray birch or hornbeam cover. Other important cover types include mixed-species stands of hardwood shrubs and trees and mixtures of hardwood and coniferous species. Slashings following lumbering produce an early stage dominated by many shrubs and herbaceous species, especially blackberries and raspberries (*Rubus* spp.), of considerable value to grouse. A later, thicker stand of saplings and taller trees is of less value, especially for young birds.

Older forest stands in the northeast include hardwood types, mixed hardwoods and conifers, and predominantly coniferous forest types. Edminster reported that younger hardwood stands have better undercovers for grouse than older stands and that scattered openings improve the value of either age-class. Pasturing also may affect the undercover development. Edminster believes that hardwoods with about 20 percent coniferous species provides better cover than pure hardwood stands and that woodlands with from 20 to 70 percent conifers provide both food and cover at all seasons, though summer cover may be imperfect. Predominantly coniferous stands of trees may be food-deficient in younger stages, but in mature stands with a hardwood understory this is not the case.

A study by Dorney (1959) in Wisconsin provides some additional information on grouse-forest relation. Dorney also reported that mixtures of hardwoods and conifers have greater ruffed grouse use than do hardwoods alone, but Wisconsin grouse appear to be less dependent on conifers for cover than are grouse in New York. Grouse need a heavy shrub understory for drumming sites, and an absence of shrubs in young hardwood stands causes rapid loss of drumming territories.

Gullion (1969) reported that in Minnesota young aspen stands first become habitable by adult ruffed grouse about 4 to 12 years after regeneration following logging or fire, when the trees are 25 to 30 feet tall and the stem densities are less than 6,000 per acre. Grouse continue to use the habitat throughout the year for the next 10 to 15 years, until stem densities drop below about 2,000 per acre. Older stands of aspen supply important winter food in the form of male flower buds besides providing nesting habitats.

The importance of small clearings in deciduous forest, as found by Edminster, was proved by Sharp (1963), who established a number of small clearings ¼ to 1 acre in size in half of a 1,470-acre pole timber forest. These changes were initiated in 1950, and during the next five years from 7 to 21 broods used the managed area, while 2 to 3 used the unmanaged portion of the forest. After ten years the openings in the forest had filled in, and the value of the area for brood use had declined.

Probably the overall range of the ruffed grouse has not changed greatly in historical times. Slight additions to the range have occurred with introductions. Wild-trapped grouse from Nova Scotia, Wisconsin, and Maine have apparently been successfully introduced into Newfoundland (Tuck 1968), and they have also been successfully introduced in the Ruby Mountain range of northeastern Nevada (McColm 1970).

Restrictions in ranges have occurred in a number of states, as indicated by Aldrich (1963). Although it once occurred in northeastern Nebraska, the ruffed grouse is now completely extirpated from the state. It is also gone from northeastern Kansas and northeastern Alabama (A.O.U. *Check-list*, 1957). However, a specimen was recently collected in Jackson County, Alabama (*Audubon Field Notes* 21:15, 1967). The population in Missouri was probably never high and may have declined to fewer than 100 birds by the 1930s, though recent attempts at reintroduction have had some success (Lewis, McGowan, and Baskett 1968). By 1930 the once-extensive Iowa population was also nearly gone except for a remnant in northeastern Iowa. This population still persists in good numbers locally, and in 1968 hunting was allowed for the first time in forty-five years (Klonglan and Hlavka 1969). In Ohio, where grouse once ranged over the entire state, a low ebb was reached about 1900, and the species was protected for thirty-two of thirty-four years following 1902 (Davis 1969*b*). Remnant populations occur in southern Illinois, where the species is protected. The species is also protected where it occurs in northwestern South Carolina, which is at the extreme southern limit of its range. Although limited to a small area of southern Indiana, the grouse population there has been fairly stable for the past two decades and is distributed over about 1,100 square miles in five counties. In 1965 the first limited season was held since 1937.

POPULATION DENSITY

Grouse populations have been intensively studied in New York by Bump et al. (1947), who reported breeding densities of from 8 to 22 acres per bird near Ithaca and from 21 to 38 acres per bird in the Adirondacks. Maximum fall densities in the two areas ranged from 5 to 20 acres in various years. Gullion (1969) estimated that maximum breeding densities in Minnesota allowed by territorial behavior are 1 pair (i.e., 1 territorial male) per 8 to 10 acres, although normal areawide densities are more commonly 4 to 6 birds per 100 acres. Slightly lower breeding densities of 2 to 4 birds per 100 acres occur in Ohio (Davis 1968). Porath and Vohs (1972) estimated a spring breeding density of 5.6 birds per 100 acres (assuming a 1:1 sex ratio) in northeastern Iowa during spring, which rose to from 14 to 21 birds per 100 acres by mid-July during the brooding period. In Indiana, Thurman (1966) reported a spring density of 18 males per square mile, while in western Washington spring densities of 5 to 7 males per square kilometer (13 to 18 per square mile) are similarly not uncommon (Salo 1978). In central Alberta the spring grouse populations increased from 15 to 22 birds per 100 acres during 1966–68, with minor yearly fluctuations (Doerr 1973). In Minnesota, populations of displaying males peaked in 1971 at 3.9 per 40.5 hectares (100 acres) of forest, but they had declined to 1.1 males by 1974 (Little 1978). According to Gullion (1976), density variations of ruffed grouse in Minnesota range from a maximum of 1 drumming male per 6 acres during population "highs" to an average of 1 per 10 acres in best habitats during "low" periods.

Consideration of ruffed grouse densities is not complete without mention of the well-known cycles of population abundance that have been reported for several grouse species but are especially often attributed to the ruffed grouse. Keith (1963) has made an intensive survey of population fluctuations in a variety of birds and mammals in northern North America, and his conclusions appear to be well founded. He believed that the ruffed grouse has undergone fairly synchronous ten-year population cycles at local, regional, and continental levels over most of its North American range with the exception of the eastern United States and New Brunswick. His book summarizes population density figures from a variety of studies in Minnesota, Michigan, and Wisconsin that indicate peak-year fall densities of from 123 to 180 birds per square mile in Michigan and up to 353 birds per square mile in Minnesota. The average ratios between densities of peak years and those of the subsequent low ones range from a ratio of 3:1 to as much as 15:1, with twelve such estimates averaging about 8:1.

In seven studies of local grouse populations, the ruffed grouse had peak populations or initial decline the same year as prairie grouse and spruce grouse; in two cases the ruffed grouse peaked or declined a year before the others; and in four cases the other grouse peaked or began declines one to three years before the ruffed. Likewise, at state or provincial population levels, the ruffed grouse peaked or began declines the same year as the prairie grouse in six of fourteen cases, while in six cases the other grouse peaked or declined one to three years before the ruffed grouse, and in the remaining two cases the ruffed grouse peaked or began its decline a year before the others (Keith 1963).

According to Watson and Moss (1979), the ruffed grouse is perhaps the most regularly cycling species of grouse, in that it seems to exhibit a consistent ten-year cycle in most areas, and they suggest that the most promising general explanation for such cycles is that they may be caused by changes in spacing behavior that occur at periods of high density. They believed that variations in environmental factors such as weather or food do not necessarily account for these changes, although they might influence the timing of cycles in local populations.

HABITAT REQUIREMENTS

Wintering Requirements

Although the ruffed grouse is one of the most temperate-adapted of all North American grouse, as indicated by its distribution in the southeastern states, it is well able to withstand cold weather. Edminster (1947) indicates that cold weather alone, if not accompanied by snow or sleet, does not materially affect grouse survival. However, during stormy weather the grouse resort to coniferous trees or roost beneath snow, where they may remain several days. Although the birds are rarely if ever frozen into such snow roosts, they become highly vulnerable to predation by mammals such as foxes, and Edminster reported mortality rates 25 to 100 percent higher than normal during a year of unusually heavy snow-roosting.

Although conifers provide valuable winter roosting cover for ruffed grouse in New York, the birds continue to rely on hardwood trees for their food, particularly buds and twigs of such trees as poplars, apples (*Malus*), birches, oaks, and cherries (*Prunus*). When available, understory shrubs and vines such as grapes (*Vitis*), greenbrier (*Smilax*), laurel (*Kalmia*), blueberry (*Vaccinium*), and wintergreen (*Gaultheria*) also provide important sources of winter food and cover (Edminster 1947).

Spring Habitat Requirements

The spring habitat needs of ruffed grouse appear to be closely tied to ecological situations associated with suitable drumming sites, or "activity centers" (Gullion and Marshall 1968). Within a general activity center, a specific display site, or "drumming stage," must be present, and Gullion and Marshall believe that two factors govern the choice of such a site. These are the presence of a number of 40- to 50-year-old aspens near or within sight of a drumming log and also a tradition of occupancy of the site by male grouse. They concluded that the presence of aspens is the most important aspect of cover that regulates the choice of activity centers, and they found strong relations between cover types and male survival. Males survived best in hardwoods completely lacking evergreen conifers (in contrast to conclusions of Edminster mentioned earlier), but the presence of spruce and balsam fir (*Abies balsamea*) did not reduce survival. However, survival did decrease as the density of mature pines increased, and male grouse did not survive as well in edge situations as in uniform forest types.

Boag and Sumanik (1969) gathered evidence supporting the view that ruffed grouse do not select drumming sites at random but that the nature of the surrounding vegetation plays an important role. Comparing 80 drumming sites with 98 similar sites that were not used, they found shrub sizes greater at used than unused sites, and canopy coverage as well as the frequency of young white spruce trees was higher at used sites. Only at used sites was aspen the predominant species in the tree layer. They believe that selective pressure for the male to choose open and visually effective sites for drumming is counterbalanced by selection favoring sites protected from predators. The result has been selection favoring sites that give the males sufficient height above the ground from which to observe other grouse or large ground predators, sufficient openings in the shrub layer to see at least 20 yards in most directions, and sufficient canopy and stem coverage to screen the birds from aerial predators. These conditions are met in Alberta by those areas where the density of young hardwood trees and the density and canopy coverage of young spruce are the highest.

The specific drumming stage is usually but not always a log; thus the presence of logs in suitable habitats is an important component of spring ruffed grouse habitat. Palmer (1963) analyzed 40 drumming logs in Michigan that had been regularly used by male grouse. Of the total, 34 were old, decayed conifers, primarily pines. Males always drummed near the larger end of these, usually about 5 feet from the end. The logs ranged from 7 to 21 inches in height at the drumming position, and none was shorter than 5.5 feet. Vegetation more than 8 feet high was significantly denser near the logs than in the surrounding cover, and among the large shrubs speckled alder (*Alnus incana*) constituted about three-fourths of the sampled stems. In general, drumming sites were associated with ground vegetation less dense, and large shrub and tree cover more dense, than was typical of the surrounding general vegetation.

Several studies have indicated that a male grouse may drum on more than one log in his territory, but one is typically favored. Gullion (1967*a*) called this log the "primary log" and designated additional drumming sites "alternate logs." Logs and activity sites may also be classified as perennial if they are used through the lifetimes of a succession of grouse, or transient if they are used by one grouse and not used again by other birds for several years. Although perennial logs apparently supply the appropriate ecological conditions that attract male grouse, Gullion and Marshall (1968) have found that male grouse using such sites suffer higher mortality, apparently because predators learn the locations of favored display areas.

In Alberta, drumming logs are typically associated with a sparser canopy cover of low shrubs, a lower density of saplings, and a greater density of trees. However, there does not appear to be any correlation between vegetational characteristics of drumming logs and the relative longevity of resident males (Rusch and Keith 1971). Another Alberta study suggests that mortality rates of drumming males are appreciably higher than those of nonterritorial males (Doerr 1973).

Various recent studies suggest that the characteristics of the display log itself are probably secondary to other environmental features in the selection of display areas. Thus

Stoll et al. (1979) found that the density of shrubs and small trees tended to be greater around display logs having a history of continuous or recent use than around logs with a history of less regular use, or abandoned logs. In general, logs that provide a nearly level platform, that are of at least a moderate diameter (range 18–98 centimeters), and that occur on upper slopes or ridges of the forest are distinctly preferred as display sites. In Washington the displaying males appear to concentrate in mixed forest stands some 40 to 50 years old, and most display logs were on nearly level sites. About 70 percent of the logs were within 50 meters of cottonwoods (*Populus trichocarpa*), a species that perhaps substitutes for aspens as a food source in this area. Within each territory, from as few as 1 to as many as 7 display logs were used by individual males. Of 49 "secondary" logs, only 6 were more than 30 meters from the primary display site (Salo 1978).

Nesting and Brooding Requirements

Habitats selected by female grouse for nesting have been analyzed by Edminster (1947), based on the study of 1,270 nests in New York. Medium-aged stands of hardwoods, with a few conifers, were the most common nesting habitat, followed by medium-aged stands of mixed hardwoods and conifers. When relative cover availability is considered, slashings were also found to be important grouse nesting habitat in New York. Middle-aged stands of hardwoods or mixed stands were found to be considerably more valuable as nesting habitat than were mature forest habitats. As to specific nest sites, the bases of trees appeared to be the most favorable sites, being used about two-thirds of the time. Most of these trees were hardwoods, and nearly all were rather large. Most of the remaining nest sites were at the bases of tree stumps or under logs, bushes, or brush piles. Edminster concluded that nest sites are chosen to provide a combination of visibility, protection, an escape route, and proximity to edges and to satisfy an apparent desire for sunlight. The undergrowth nearby is usually open, and the canopy density is also relatively open. More than half the nests were within 50 feet of a forest opening, often the edge of a road. Slope evidently is not important, except that steep slopes are avoided.

Gullion (1967b, 1969), summarizing research done at Cloquet, Minnesota, reported that female grouse probably begin to search for a clone of male aspen trees after mating, near which they locate their nests. The incubating hens forage on these trees during incubation.

Brood habitat analyses have also been made by Edminster (1947). From studies of 1,515 broods in New York, it was clear that females with broods showed a preference for brushy habitats, especially overgrown land, followed by slashings. Hardwood stands that have been "spot lumbered" exhibited high brood usage, as was later confirmed by studies in Pennsylvania by Sharp (1963). At the same time, hardwood forests continue to receive heavy use from adult grouse (males and broodless females) during the summer, while mixed woods and coniferous forest types serve for escape from extreme heat and summer storms.

Korschgen (1966) analyzed the nutritional value of seasonal foods of ruffed grouse in Missouri and concluded that high-protein foods were taken in greatest amounts during summer, foods high in fat and carbohydrate were taken most during winter, and the largest amounts of mineral sources were taken during reproduction. Evidently grouse select food to fulfill seasonal nutritional needs. Korschgen summarized the principal ruffed grouse foods indicated by twenty-four published studies. Aspen and poplars were listed as principal foods in seventeen of these studies, birch was cited in eleven, and all other food sources were mentioned less often, with apple, grape, sumac, beech, and alder all listed in several studies. In analyses of foods from six areas in the eastern United States, Martin, Zim, and Nelson (1951) list aspen as being of first or second importance in five areas and lacking only in samples from the Virginia Alleghenies. Other plants listed in several studies are clover, greenbrier, hazelnut, and grape.

Winter foods of the ruffed grouse consist largely of buds and twigs of trees. Edminster (1954) lists the following major winter sources of such foods: birches (several species), apple, hop hornbeam (*Ostrya*), poplar, cherry, and blueberry. In the Cloquet area of Minnesota, aspens (*Populus tremuloides* and *P. grandidentata*) are usually the most important source of winter food, and with the appearance of the male catkins in late winter these trees provide the most nutritious food source available to ruffed grouse as long as snow is on the ground (Gullion 1969).

A study in Utah by Phillips (1967) indicated that chokecherry (*Prunus virginiana*) was the most preferred winter food, followed closely by aspen and maple (*Acer*). Aspen was also the second most important fall food, but hips from roses (*Rosa*) were used more. In Ohio, Gilfillan and Bezdek (1944) found that the fruit and leaves of greenbrier (*Smilaz*) had high winter use, as well as aspen buds, fruit of dogwood (*Cornus*), grape (*Vitis*), sumac (*Rhus*), beech (*Fagus*), and other plants. Winter food in Maine, as reported by Brown (1946), consisted primarily of buds of aspens, followed by buds and leaves of willows, catkins and buds of hazelnut (*Corylus*), and buds of wild cherry and apple.

In interior Alaska, winter foods consist primarily of the buds of willow and quaking aspen, with the fall shift from berries to buds occurring in October. However, even during winter, only 30 percent of the crops collected contained only a single food item (McGowan 1973). Winter food diversity probably reaches its extreme at the southeastern edge of the ruffed grouse's range, in the southern Appalachians. Here winter snow accumulation is slight, and aspens are both rare and highly localized. Thus much ground foraging is possible, and herbaceous annuals and forest-floor perennials composed 30 percent of the foods found in 311 birds. Leaves and fruit of vines accounted for another 35 percent and the foliage of woody shrubs some 20 percent. Food sources from overstory trees composed only 3 percent of the foods taken in this area (Stafford and Dimminck 1979).

After winter, as ground vegetation is exposed, the foods of ruffed grouse become more diversified, but at least in New York the buds of poplar, birch, cherry, hop hornbeam, and blueberry are still eaten well into May (Edminster 1947). Likewise in Maine the buds and catkins of poplar are a primary spring food, in addition to buds and catkins of birch, willow buds, and the leaves of strawberry (*Fragaria*) and wintergreen (*Gaultheria*). In Minnesota male grouse sometimes continue to feed almost entirely on the male catkins of aspens long after snow melt allows succulent evergreen herbaceous plants to become available (Gullion 1969). Quaking aspen in this region is preferred over big-toothed aspen more than two to one.

The diet of adult grouse changes drastically in early summer as berries and fruits become available (Edminster 1947). These fruits include strawberries, raspberries and related species of the genus *Rubus*, cherries, blueberries, and Juneberries (*Amelanchier*). Insects compose a small percentage of adult foods at this time, rarely if ever exceeding 10 percent.

In contrast, the basic food of ruffed grouse chicks for at least the first week or 10 days of life consists of insects. Bump et al. (1947) reported that 70 percent of the food taken in the first 2 weeks consists of insects, compared with 30 percent during the third and fourth weeks, dropping to 5 percent by the end of July. Ants are among the most frequent food items, but a variety of other insect types, including sawflies, ichneumon flies, beetles, spiders, grasshoppers, and various caterpillar species make up the remainder of chick foods from animal sources. As dependence on insects declines with age, the amount of plant foods, particularly sedge achenes and the fruits of strawberries, raspberries, blackberries, and cherries, increases correspondingly (Bump et al. 1947).

Fall foods for juvenile and adult birds include a variety of fruiting shrubs, such as viburnums, dogwoods, thorn apples, grapes, greenbriers, sumacs, and roses (Edminster 1954). The availability of many of these persists into winter, when they supplement the standard diet of buds, twigs, and catkins.

Gullion (1966) has emphasized that the abundance of data on fall food intake by game birds is often misleading in that the diversity of foraging during that time of year is not representative of the critical dietary sources needed for survival through the winter. Thus, the availability of a winter source of male catkins of birch, alder, hazel, and particularly aspen is probably the most important single factor influencing the wintering abilities of ruffed grouse. Gullion believed that quantitative or qualitative difference in these winter foods might account for major population fluctuations in Minnesota ruffed grouse. Lauckhart (1957) had earlier pointed out that periodic heavy seed crops in trees may sap the nutrients from buds and stems for several years between such crops, causing a deficiency for animals highly dependent on these trees. The usual cycle of aspen seed crops is four to five years; thus an interaction of this cycle and some other factor or factors might account for the ten-year grouse "cycle." Clearly this idea has great promise and should be investigated thoroughly.

In an effort to determine the possible relation between ruffed grouse populations and nutrient levels of the major food sources during winter, Huff (1973) measured nutrient contents of crop materials taken from grouse during winter. He found wide annual variations of nutrient concentrations in aspens, the most commonly encountered winter food. Nutrients seemed to vary independently of one another in different years. Nutrients in tree samples from different ecological sites also varied, and fire was found to have varied effects on nutrient concentrations, while thinning of stands appeared to have no measureable effects. Another possible effect of winter on grouse populations concerns snow roosting sites; crusted snow or other conditions that make winter roosting difficult may decrease reproductive success the following year (Kubisiak, Moulton, and McCaffery 1980).

The importance of water, in the form of standing water, dew, or succulent plants, also should not be overlooked for ruffed grouse. Bump et al. (1947) indicate that captive grouse can easily survive at least 12 days without food if they are provided with water, but that in the absence of both food and water they will live only a few days. Since most grouse foods contain considerable water, it is probable that the birds can normally survive indefinitely without standing water.

MOBILITY AND MOVEMENTS

Ruffed grouse do not perform any movements that might be considered migratory, though there are some seasonal variations in mobility. Ruffed grouse broods normally move little before they break up and disperse; Chambers and Sharp (1958) reported that the cruising radius of most marked broods was no more than ¼ mile. With the dispersal of the broods, over half of the juveniles moved more than a mile, in one case up to 7.5 miles. Similarly, Hale and Dorney (1963) reported that about one-fourth of the juveniles they banded had moved more than 1 mile from the banding site at the time of recovery. One grouse they banded as a 3-month-old juvenile was shot 31 days later some 12 miles from the banding site. Apparently these fall movements were independent of population densities and were unrelated to "crazy flight" behavior, during which young grouse may make long and erratic movements apparently related to inexperience and perhaps fright.

By winter, movements of both young and adult grouse decline, and by spring the birds become virtually sedentary. Hale and Dorney (1963) found that males banded on drumming sites were highly sedentary and normally returned to the same site each year. Chambers and Sharp (1958) likewise reported that grouse become sedentary as they mature, with males only rarely moving more than ¼ mile, while females sometimes moved more than a mile. Hale and Dorney likewise reported that, except during winter, females were consistently more mobile than males. Gullion and Marshall (1968) noted a high degree of fidelity by adult male ruffed grouse not only to a particular territory but also to a specific display site. Only about 36 percent of 168 males that lived at least 12 months moved to another log during their drumming lifetimes, and such movements averaged only about 300 feet. At least 20 males, however, moved to new activity centers.

The mobility of males during spring apparently is partly influenced by their age.

Archibald (1975) found that the mean weekly range of drumming adult males was greater than that of drumming immature males. Further, the ranges of all drumming males were about half as large as those of females during the same period, and male home ranges always contained central core areas (averaging 2.26 hectares) of intensive use. The ranges of females increased during the 6 weeks before incubation but became appreciably smaller during the incubation period.

In studying the daily movements of three female ruffed grouse during May, Brander (1967) found that the females moved from their established winter home ranges of 7 to 26 acres toward male drumming sites, apparently stimulated by the drumming behavior, particularly drumming sounds. One female was apparently attracted to three different males on different days before copulation occurred, and the pair remained together no more than a few hours. Since the male continued to drum after her departure, Brander concluded that the ruffed grouse mating pattern should be regarded as promiscuous. He estimated that the three females each remained receptive for only 4 days, ending the day before the first egg was laid. The hen located her nest in each case within the area of her movements of the previous week to 10 days. As I mentioned previously, the female usually seeks out a clone of aspen near which she establishes her nest (Gullion 1969).

In a detailed study of home ranges of female grouse tracked by radiotelemetry, Maxson (1978) found that their greatest movements occurred during the prelaying period, when 9 hens had home ranges averaging 12.1 hectares (range 5.8–22.9). During the egg-laying period their home ranges declined considerably, averaging 8.4 hectares (range 4.1–14.0), and during the incubation period the ranges averaged only 0.9 hectare (range 0.4–1.5). During the postincubation period the ranges of hens leading broods averaged larger than those of broodless females. In a study of brood movements, Godfrey (1975) reported that the average brood range of six broods was 12.9 hectares, with average daily movements of 376.8 meters. Lowland areas of moist soils dominated by mature alders (*Alnus rugosa*) were used about 60 percent of the time by these broods. However, in similar studies in central Wisconsin, most brood observations were in upland areas, especially those dominated by aspens. Evidently these habitats provide broods with a variety of foods, especially berry-producing shrubs, evergreen herbs, and other succulent vegetation, as well as offering good protective cover (Kubisiak, Moulton, and McCaffery 1980).

REPRODUCTIVE BEHAVIOR

Territorial Establishment and Advertisement

According to Bump et al. (1947), captive male grouse exhibit aggressiveness as early as the first of March, though they have sometimes been seen strutting on warm days in winter. Edminster (1947) reported that drumming has been heard in every month of the year and at every hour of the day and night, but the most intensive drumming in New York occurs in early spring during late March and April, tapering off in May.

27. Displays of ruffed grouse (after various sources), including (A) stages in male drumming display, (B) rotary head shaking, (C) strutting, (D) ruff erection, (E) final phase of ''rush,'' and (F) posture of female defending a brood.

The two basic aspects of male reproductive display are drumming ("wing beating" of Hjorth 1970) and strutting ("upright," "bowing," and "rush" sequence of Hjorth 1970). There is no doubt that drumming is primarily an acoustic display and serves to advertise the location of the male in fairly dense forest cover. Strutting, however, is a predominantly visual display and is probably not normally released except in the visual presence of another grouse or similar stimulus. Undoubtedly both displays are essentially agonistic or aggressive in origin, serving to proclaim territory and establish dominance. Since drumming is the basic means of territorial advertisement, it will be discussed first.

The motor patterns of the drumming display are well described in Bent (1932) and many other references and need little amplification here. The male typically stands on a small log, facing the same direction and at virtually the same location on each occasion. With his tail braced against the log and his claws firmly in the wood, he begins a series of strong wingstrokes. These strokes, which start slowly at about 1-second intervals, rapidly speed up, with a complete series lasting about 8 (Allen, in Bent 1932) to 11 seconds (Hjorth 1970). Hjorth found that in a sample of drumming displays from Alberta there were consistently 47 wingstrokes, while a sample from Ohio had 51. Aubin (1970) noted that among six ruffed grouse studied in southwestern Alberta the number of wingstrokes varied only from 44 to 49 in his samples and was even more consistent for individual birds.

Allen hypothesized that the muffled drumming sound produced by the wings resulted from the forward and upward thrust rather than the return stroke. This strong forward thrust produces a counterpressure that forces the bird backward, thus explaining the need for the brace provided by the tail and the importance of clutching the log with the claws. At the end of the last stroke this pressure is released, and the bird tips forward on his perch. As Allen noted, the wings do not touch each other during the drumming, and the noise results simply from air compression, which accounts for its dull throbbing nature. Recently Hjorth (1970) has advanced the idea that the downstroke rather than the upstroke may be responsible for this sound.

Drumming usually begins well before daylight and may continue until sometime after sunrise. It usually begins again about an hour before twilight and may continue until dark (Bump et al. 1947). The usual interval between drumming displays is 3 to 5 minutes, but this interval varies from a few seconds to much longer periods.

As noted earlier, most males drum on a single log, but some may use more than one. Bump et al. (1947) reported an average of 1.33 logs per male used by 1,173 grouse, Aubin (1970) found that from 1.5 to 1.7 logs per male were used in different years and independently of population densities, while, as noted earlier, Gullion and Marshall (1968) observed a certain amount of movement in display sites of male grouse.

Gullion (1967a) found that only a few male grouse establish drumming logs their first fall, and a few also fail to become established the following spring. Most birds occupying logs in his study area were full adults, at least 22 months old. He also found a hierarchy of dominance among males. An established male on a drumming log is a "dominant drummer," and within his activity center a second, or "alternate," drummer may occur and take over the site if the dominant drummer is killed. Nearby rivals on adjacent activity centers are call "satellite drummers," but these are fairly rare. However, other males are "nondrummers" and drum infrequently or not at all. These are presumably young grouse that have been unable to establish drumming sites.

Gullion (1967a) also found "activity clusters" of males, consisting of from about 4 to 8 males occupying sites in fairly close proximity. These seem to represent an expanded collective display ground, similar to those that have been described for blue grouse.

More recently, Gullion (1976) has reinterpreted his earlier data and concluded that true "activity clustering" does not occur in Minnesota. Instead, his apparent male clusters were the result of habitat resource distribution rather than social interaction. However, Aubin (1970) also reported male clusters of displaying birds in his Canadian study area and suggested that a leklike social organization exists in ruffed grouse.

Gullion reported that males remain closely associated with their display sites during the summer and that fall drumming may approach or even exceed spring drumming. At least a few young males no older than 17 to 20 weeks may become established at this time. More recently, Gullion (1981) has determined that a proportion of the adult male ruffed grouse population consists of "nondrummers." These are birds that cannot be identified as occupying a territory ("activity center") for one or more seasons. Gullion marshaled several lines of evidence to document the presence of such birds, which he found to be least numerous when the population of birds was lowest and greatest when the population was at its peak density. During one such period of peak abundance, he estimated that a nondrumming grouse existed for every 2.3 territorial birds (or 30 percent of the adult male population). He believes these nondrumming birds may provide a kind of "momentum" to the population upswing at a time when the annual survival of males is declining. Survival rates among this nondrumming component equaled or even exceeded those of established males. At least 19 nondrumming males who lived for 2 to 3 years before they occupied logs lived longer than many drumming males from the same cohort. Two of these 19 birds survived for at least 3 years after finally occupying territorial sites. Apparently some males never become drummers, and very rarely males that have occupied drumming logs will abandon them and become part of the nondrumming population.

Male Strutting Behavior

Presumably the normal releaser for strutting rather than drumming is the appearance of another grouse near the display log. Edminster (1947) indicates that the drumming male will than strut very slowly toward the intruder, with tail erect and spread. The ruffs on the side of the neck are raised ("upright cum ruff display" of Hjorth 1970), and the male begins to emit hissing sounds that parallel the tempo of the drumming display. With each hiss the head is lowered and shaken in a rotary fashion ("bowing cum head twisting and panted hissing" of Hjorth 1970), giving the impression of a locomotive getting under way (Bump et al. 1947). The display ends with a blur of head shaking and hissing, followed by a short, quick run toward the other bird as both wings are dragged along the ground ("rush cum prolonged hiss" of Hjorth 1970). Photographs of this display suggest that in the early stages it is oriented laterally, with the tail and upper part of the body tilted toward the object of the display and the head turned in the same direction. However, the short rush is in a shallow arc toward the other bird (Hjorth 1970). The similarities of this display to the short rushes of the blue grouse and the spruce grouse are clear. Unlike that of the spruce grouse, however, the tail is neither shaken nor fanned to produce sound.

Bump et al. (1947) described a "gentle phase" after the strutting phase, which in turn was followed by a "fighting phase" of males. However, their data do not support such a strict interpretation of male behavior patterns, nor does such a sequence seem biologically probable. The strutting behavior of males serves equally well as a preliminary threat display toward other males before fighting and as a preliminary to attempted copulation. The means by which males recognize the sex of intruders on their territories is still uncertain, but in all likelihood there is a differential sexual response of males and females to strutting in another bird. Hjorth (1970) gave the posture associated with this reputed "gentle phase" the name "slender upright cum head shaking."

Females apparently are receptive from 3 to 7 days (Bump et al. 1947; Brander 1967), probably only until a successful copulation is achieved. The typical receptive posture of grouse, with the wings drooped and slightly spread and the tail slightly raised while the body feathers are depressed, will stimulate copulation attempts by the male.

Surprisingly, there are few good observations of actual mating in ruffed grouse. Gladfelter and McBurney (1971) observed one case in Iowa. After a period of drumming on his log, the male bobbed his head several times, jumped off the log, and began to pursue a female that had evidently been attracted by his drumming. As he chased her, he held his neck extended and his tail fully erect and fanned. When he caught the female, she squatted, upon which he immediately mounted her. After copulation the female ruffled her feathers, walked away, and within a minute flew to the branches of a fallen tree, below which the male continued to strut. She then flew back down to the ground and the male strutted over to where she had landed. Finally, 24 minutes later, she flew away.

Vocal Signals

Vocalizations of this species are still poorly studied, but Samuel (1974) has described several male vocalizations associated with drumming, including a "*queet*" call that may precede running or flushing, a hissing note, a soft *psst* call, and a whining call. These three vocalizations were observed to be interspersed with one another during an encounter with an intruding male or female.

Hissing is performed by both sexes. Males hiss during their head shaking and short rush displays, and females hiss when defending a brood (Bump et al. 1947). Females also squeal during distraction display and quiet their hiding chicks with a downward-inflected scolding note. After danger is past, they call the brood together with a low humming call (Bump et al. 1947).

Chicks have four principal call notes, according to Bump et al. (1947). These include alarm calls, two different notes uttered by scattered chicks, and a warning signal of several descending notes uttered by older chicks.

Nesting and Brooding Behavior

Typical nest sites for the ruffed grouse have already been mentioned in the discussion of nesting requirements. Bump et al. (1947) report that the female lays her eggs at an average rate of 2 eggs every 3 days, thus taking 17 days to complete the average clutch of 11 eggs. The female's attachment to the nest increases with the clutch size, but incubation does not begin until the last egg is laid. Incubation takes from 23 to 24 days, but low environmental temperatures may delay hatching a few days beyond this time. Bump et al. (1947) report that during incubation the female will leave the nest to feed for 20 to 40 minutes, rarely longer. Evidently she may feed twice each day under normal conditions, but during stormy weather she may remain on the nest continuously. Much-enlarged "clocker" droppings are typical of incubating females; these are usually found in the vicinity of nests near the usual foraging areas.

During incubation females spend about 95 percent of their time on the nest and normally leave it only to feed (Maxson 1977). Feeding usually is done in aspen trees, the birds eating both leaves and male or female catkins. However, they seldom feed in the tree nearest the nest; instead they may fly 28 to 185 meters to a foraging tree, suggesting that they may be selecting specific trees or clones (Maxson 1978).

Bump et al. (1947) report that although the average clutch size for 1,473 first nests was 11.5 eggs, 149 renesting attempts averaged only 7.5 eggs. Since no cases of second renesting attempts were found, they estimated that the maximum number of eggs a female might lay in a single season is about 19. There is no evidence that second broods are ever raised by this or any other species of grouse or ptarmigan.

Female ruffed grouse exhibit strong nest and brood defense tendencies and will often resort to a disablement display, feigning a broken wing, especially before hatching time. After hatching the female more often stands her ground, spreads her tail, and assumes a stance similar to the male's strutting posture as she hisses or squeals. When the chicks can fly, after 10 to 12 days, both hen and chick usually fly when disturbed. By mid-September, when the chicks are 12 or more weeks old, the families start to break up and the juveniles begin to disperse.

In an Alberta study, Doerr (1973) found that 80 percent of the nests he studied hatched at least one chick, and that in successful nests 96 percent of all eggs hatched. An average of 31 percent of the females alive in May raised young to 12 weeks of age, and from 37 to 51 percent of the young birds survived at least 12 weeks. Additionally, 35 to 69 percent of these 3-month-old juveniles survived until the following spring, with predation, especially owl predation, accounting for most of the overwinter mortality.

EVOLUTIONARY RELATIONSHIPS

In his revision of grouse genera, Short (1967) merged the monotypic genus *Bonasa* with the Eurasian genus *Tetrastes*, which contains two species of ''hazel grouse.'' The two Eurasian species lack neck ruffs but otherwise are very similar to the ruffed grouse, and Short considered that, of the two, the European hazel grouse is nearest to the North American ruffed grouse. The habitat of this bird in Europe consists of mixed hill woodlands and thickets, and the species is especially prevalent in aspen and birch, which strongly suggests a common ecological niche.

In contrast to the ruffed grouse, the hazel grouse is monogamous and forms a pair bond that lasts at least until hatching and sometimes beyond. An additional behavioral difference is that the male display consists largely of whistling calls. There is no drumming display, but an aerial display involving the whirring of wings also occurs. It seems that the evolution of a promiscuous mating system, the development of nonvocal acoustical signals rather than reliance on vocal whistles, and the correlated ritualization of aerial display flights into a sedentary drumming display all occurred after the separation of ancestral ruffed grouse stock.

Short (1967) concluded that the nearest relationships of the genus *Bonasa* (in the broad sense) are with *Dendragapus* and that the former genus probably arose from pre-*Dendragapus* stock. I agree that modern species of *Dendragapus* or *Tetrao* probably represent the nearest living relatives of *Bonasa*.

Hazel Grouse

Bonasa bonasia (Linnaeus) 1758

Other Vernacular Names

Hazel hen; Haselhuhn (German); gelinotte des bois (French); ryabchik, rjabok (Russian).

Range

Northern Europe, from Scandinavia, Germany, Belgium, and France south to the Balkans and east through Russia and Siberia to the Kolyma basin and the Sea of Okhotsk, south to the southern Urals, Russian Altai, northern Mongolia, Manchuria, central Korea, Sakhalin, and Hokkaido (Vaurie 1965).

Subspecies (after Vaurie 1965)

B. b. rupestris (Brehm): Western hazel grouse. Resident in Germany, northeastern France, and the Alps and Carpathians south to northern Italy, northern Albania, Macedonia, Bulgaria, and Romania, formerly also the Pyrenees.

B. b. bonasia (Linnaeus): Scandinavian hazel grouse. Resident in eastern Norway, Sweden, Finland, the Kola Peninsula, and Russia north to timberline, and south to East Prussia, Poland, and in Russia to Volynia, western Orel, and the Zhigolivsk Hills west of Kuibyshev.

B. b. sibiricus (Buturlin): Siberian hazel grouse. Resident in the Pechora basin and the Urals (where it grades into *bonasia*) eastward through Siberia to the Kolyma basin and the Sea of Okhotsk north to the Taygonos Peninsula, Transbaikalia east to the Yablonovy Range, and the Stanovoi Mountains.

B. b. vicinitas (Riley): Oriental hazel grouse. Resident in southeastern Siberia (intergrading with *sibiricus*), from eastern Transbaikalia (?) and Amurland to Sakhalin, south to Manchuria and Ussuriland to central Korea, and on Hokkaido, Japan.

MEASUREMENTS

Folded wing: Males 165–95 mm; females 161–84 mm. Males of at least three races average only 1–4 mm longer than females (Glutz 1973; Dementiev and Gladkov 1967).

Tail: Males (of *rupestris*) 111–31 mm; females 108–18 mm, with males averaging 10 mm longer than females (Glutz 1973). Adults of *rupestris* average 124 for males and 112 mm for females (Cramp and Simons 1980).

IDENTIFICATION*

Length, 14 inches. This Eurasian grouse is easily recognized by its generally gray to wood brown appearance and its rather small size. Males are gray with barring of black and russet on the top of the head, neck, and anterior part of the back. The back and tail coverts are gray with narrow transverse barring, and the tail is grayish brown with a dark, broad subterminal band and a whitish tip except for the central pair of feathers, which are barred with black throughout. The flight feathers are brown, with whitish markings on the external webs, and the upper coverts are mottled with gray, ochre, and white. The neck is black, fringed with a white band that extends narrowly upward above the bill, and there is a small white mark behind the eye. The breast is marked with brown and ochre bars that enlarge on the lower breast and belly to become spots, and the posterior underparts are marked with white, ochre, and blackish. There is a small area of red skin above the eye; the bill is grayish black; and the legs are feathered to the toes. Females resemble males but have less pronounced neck and head markings and less apparent red markings above the eye. Juveniles resemble adult females, but their back feathers have narrow white shafts and almost lack dark spotting (Dementiev and Gladkov 1967).

FIELD MARKS

This is a small woodland grouse slightly larger than a gray partridge (*Perdix perdix*), with a relatively long tail that is somewhat rounded. Its call is a high-pitched whistle, and when excited it erects the long feathers of the crown to form a distinct crest. The birds flush rapidly; on taking off, the grayish tail, with its interrupted dark band near the tip, is distinctive, and the wings produce a humming sound. The birds often perch in trees. None of the other European woodland grouse have white behind the eye and outlining the chin and throat areas.

AGE AND SEX CRITERIA

Adult males can be distinguished from females by the white markings that extend from the base of the beak downward to frame a black chin and throat. *Females* have a whitish to brownish throat, and the white markings do not extend above the beak. These differences are more evident in the breeding season than at other times.

*This and subsequent sections of the hazel grouse account were prepared primarily by Dr. E. O. Höhn.

First-winter birds cannot be distinguished from adults in the field after about September, but in the hand individuals less than 14 months old exhibit more (8–11) pale spots on the outer web of the ninth primary; adults have only 4 to 7 such spots, and in nearly 10 percent of the adults these spots are fused (Gajdar and Zhitkov 1974). Apparently the outer two primaries can also be used to recognize young birds, as can the length of the first primary (Stenman and Helminen 1974). These authors found that over 90 percent of the birds can be aged by the facts that the light area at the tip of the innermost primary is narrower (not over 2 mm) in adults than in juveniles (at least 4 mm) and that the number of crossbars in the outer vanes of the two outermost primaries is always less than 8 in adults but is 8 to 11 in juveniles. These criteria are more useful than the shape of the outermost primaries.

Juveniles of both sexes closely resemble adult hens but exhibit dorsal white shaft streaks, as do most other grouse, and lack a dark subterminal band on the tail.

Downy young have a distinctive pattern, with a brown crown cap that is not bordered by black, a very dark chestnut brown midline, and a heavy black eye stripe. The upperparts are rather uniformly reddish brown, with two buff bands extending down the back, and the underparts are buffy yellow to straw yellow (Fjeldså 1977).

DISTRIBUTION AND HABITAT

This species requires fairly large forests with a rich undergrowth that vary from place to place in density and variety of plant species and contain occasional clearings. At least in Europe, it requires a certain porportion of deciduous trees among the conifers and completely avoids well-tended pure coniferous forests. Conifers, although used for concealment and nocturnal roosting, are not essential, for the hazel grouse occurs in areas lacking evergreen conifers in its most eastern range as well as in parts of West Germany. In the Alps its breeding range extends up to about 1,600 meters, and there is no significant seasonal migration there (Glutz 1973).

A detailed account of hazel grouse habitats in representative regions extending from Siberia to West Germany was provided by Bergmann et al. (1978). I used their monograph extensively in preparing this entire account, and their summary on habitats can be briefly summarized here.

In far eastern Siberia the hazel grouse inhabits willow-alder thickets along the shores of watercourses; the dominant trees here are larches and poplars. Farther west in the Siberian taiga spruces and larches dominate, and there is an admixture of birches. The taiga provides good supplies of ericaceous berries, which also makes it attractive to other species of grouse such as black grouse and capercaillie.

The bare areas left by forest fires or destruction of trees by insects or storms, or by avalanches in the Alps, are soon colonized by berry plants and somewhat later by saplings that provide cover. Such areas, and the vicinity of streams and rivers with their more varied vegetation, are also favored by the hazel grouse. A supply of catkins and buds of deciduous trees is required for overwintering.

In the largely primeval lowland forest of Bielowicza in Poland, the habitats are spruce-oak and pine-oak forests with rich undergrowth; here too clearings with berries and newly grown deciduous hardwoods are favored, with birch, alder, and hazel all well represented. In this region too, areas along streams or rivers, or even moist depressions, have a good density of the birds. In Poland the birds are usually found in transitional areas between plant associations, and also in habitats showing a transition between young and fully grown trees. A horizontal structuring of vegetation is an important habitat component, and a well developed brush layer is associated with high spring densities (Wiesner et al. 1977).

The mountain forests of Bavaria and western Czechoslovakia are among the most important areas for the species in central Europe. Here valleys of streams with alder, willow, birch, and spruce are good habitats. Proceeding up the mountains, there is a zone of pure conifers lacking undergrowth that the birds avoid. The edges of meadows at the next higher zone provide a fairly good habitat, as does the next higher zone of mixed spruce and birch, but the two remaining higher forest zones are less favored.

The "Hauberge" of Hesse (West Germany) consists of hills covered with oak-birch forest where the trees are cut in selected areas every twenty-five years. Here the only consistent habitat of the hazel grouse is along streams where alder and willows grow. Hazel grouse spread into the clearings about seven years after the trees have been cut, by which time new saplings and undergrowth have developed there. They use these areas until the trees are cut once more. In this region, as in some Far Eastern habitats, there are no evergreens.

POPULATION DENSITY AND STATUS

Hazel grouse populations are at a minimum during the breeding season and at their annual maximum when the young of the year are added to the adult component in the fall, doubling or even trebling the adult segment. Thus only counts made during the same season are readily comparable.

Fall populations in northern European coniferous-birch forest areas range from averages of 4.6 to 37 birds per 100 hectares (maximum 81.1); those from northern European mixed-forest areas are similar. Populations from middle European mixed montane forests average about 10 to 11 birds per 100 hectares, and one estimate for western European broadleaf forest was 4.1 birds per 100 hectares. Spring densities average far less than those for fall and generally range from fewer than 10 to as high as 35.2 individuals per 100 hectares. In a Polish study the highest spring densities, 30.6 birds per hectare, were found in a mixed-forest area (Wiesner et al. 1977). A few substantially higher estimates of hazel grouse populations have been made in Siberia and in the Ural Mountains and Petshora District of Russia. Except for these very high estimates, the densities are similar to those that have been reported for ruffed grouse in North America. In general there seems to be an inverse relation between hazel grouse and human populations, which is not surprising and not unlike the situation in capercaillies and black grouse.

13. Current distribution of the Oriental (O), Scandinavian (S), Siberian (Si), and western (W) races of hazel grouse, and of the black-breasted hazel grouse (B), including specific locality records.

60. Greater prairie chicken, male. Photo by author.

61. Greater prairie chicken, male. Photo by author.

62. Greater prairie chicken, males fighting. Photo by Ed Bry.

63. Greater prairie chicken, males fighting. Photo by Ed Bry.

64. Hybrid pinnated × sharp-tailed grouse, male. Photo by Ed Bry.

65. Hybrid pinnated × sharp-tailed grouse, male. Photo by Ed Bry.

66. Sharp-tailed grouse, male dancing. Photo by Ed Bry.

67. Sharp-tailed grouse, male dancing. Photo by author.

68. Sharp-tailed grouse, male dancing. Photo by Ed Bry.

69. Sharp-tailed grouse, male cooing. Photo by author.

70. Sharp-tailed grouse, male posing. Photo by Ed Bry.

71. Sharp-tailed grouse, male posing. Photo by Ed Bry.

72. Sharp-tailed grouse, male in flight. Photo by Ed Bry.

In most of central Europe the hazel grouse populations have declined for several decades, and the species has been extirpated from many parts of its former range. Thus it has disappeared from central France (Cévennes and Auvergne) and is now restricted to the eastern part of that country, where 1,000 to 10,000 pairs exist. In West Germany its former lowland habitats are now deserted, and the species is essentially confined to mountainous areas. In East Germany it has been virtually extirpated. It is also extremely rare in Italy, Belgium, and Luxembourg and is declining in Austria and Czechoslovakia. The western limits of the breeding range are thus gradually shrinking eastward, and the European range has become quite discontinuous. The largest remaining populations in Europe are in Sweden (about 150,000 pairs) and Finland (228,000 pairs in 1956) (Cramp and Simmons 1980).

Considering the causes of this decline, Bergmann et al. (1978) point out that the proximity of humans is not detrimental to the birds, but that human alteration of the habitat is. Nevertheless, logging or relatively small areas can be beneficial, so that over the years the forest becomes a mosaic of areas in different stages of regrowth and a certain proportion of deciduous trees is preserved to provide winter food. Control of deer populations at a level that does not prevent the natural forest rejuvenation and the fruiting of berry-bearing plants is also desirable.

The population density of this species is highly variable but apparently does not exhibit a ten-year (or any other) cyclic variation. Peaks evidently occur at shorter and irregular intervals; in Finland successive peaks occurred at intervals of 3, 3, 4, 6, 4, and 5 years between 1911 and 1942. Semenov-Tian-Schanski (1959) has attributed the development of peak populations to the beneficial effects of mild average temperatures during the breeding season.

HABITAT REQUIREMENTS

Wintering Requirements

Hazel grouse require areas that offer winter food in the form of catkins and buds of birch, alder, and willow, and they generally prefer to occupy the edges of stands of such trees. The shift from ground foraging to browsing is done only when snow makes ground herbage inaccessible. Because of its relatively sedentary nature, this grouse uses only winter habitats close to those it occupies during other seasons.

Hazel grouse in Russia lose considerable weight over the winter, suggesting that their energy reserves are strained. Like other grouse, they have pectinated toes that reduce sinking in snow, and their tarsal feathers extend to the bases of the toes, reducing heat loss. When winters are severe, hazel grouse may emerge from their snow holes only to feed for 2 to 4 hours each day (Andreev and Krechmar 1976). In their snow holes the temperature is appreciably higher than on the surface ($-10°$ C in the bird's "chamber" and $-48°$ C on the surface, according to these authors).

Hazel grouse typically start scratching out their snow holes immediately on alighting,

303

thus avoiding a scent track on the snow, but on occasion they first walk about somewhat, apparently searching for snow of suitable density. Grouse leave their overnight snow holes by pushing their heads through the snow above them and emerge on the surface at the side opposite where they scratched out the burrow. A sudden thaw followed by frost may prevent them from digging a hole or may impede their leaving it in the morning. When the snow cover is unusually light the birds cannot dig snow holes, for which they need a depth of about 17 centimeters, and they may die of exposure. Danilov (1975, cited in Bergmann et al. 1978) reported mass deaths of hazel grouse in the Urals during a winter with very scanty snow, when the temperature stayed below $-30°$ C for two months. The numerous frozen birds that were found had full crops and clearly had died from freezing rather than from starvation.

Spring Habitat Requirements

Spring habitat requirements have generally been indicated in the section on distribution and habitat. During the spring, hazel grouse live largely on the ground, but some display is done from trees. In general the species is more arboreal than are other European *Tetrao* species, perhaps because it is especially fond of habitats providing dense low overhead cover, where both visual and acoustic aspects of male territorial display are less effective than when performed arboreally. Additionally, roosting is done mainly in trees, often near the ground in the densest tangle of branches, which probably provides some protection from predators such as martens.

FOOD AND FORAGING BEHAVIOR

In their first month after hatching, young hazel grouse feed almost exclusively on arthropods and other invertebrates. During the first 10 days these are mostly ants, small flies, caterpillars, grasshoppers, and spiders. During the second 10-day period the birds begin to take some vegetable matter, such as berries and seeds, but arthropods still predominate. Thereafter their diet is much like that of the adults, except that until they are about 3 months old the proportion of animal food in their total diet is somewhat higher than that of adults.

In view of the very wide range of this grouse, the plant species from which it takes buds or berries vary from region to region according to the geographical distribution of the plants concerned. Thus, in eastern Siberia the red wortleberry (*Vaccinium vitis-idaea*), which is an important item farther west, is replaced by the bilberry (*V. myrtillus*) and bog bilberry (*V. uliginosum*). Although in general the species uses tree buds and catkins during the winter, near the northern limits of its range in Russian Lapland the birds have been reported to abandon tree-feeding only during August (Semenov-Tian-Schanski 1959).

In Sweden the winter foods of the hazel grouse are almost entirely the twigs, buds, and catkins of birch. In spring the twigs of bilberry, and later its berries, as well as blueberries and various leaves and seeds, constitute the diet. In late fall the birds gradually shift to browsing as the berry crop is covered by snow (Ahnlund and Helander 1975).

Similarly, in Finland the late-autumn foods are mostly berries, seeds, and green plant materials, especially of *Vaccinium* species. Browse, especially the catkins, buds, and twigs of birch and alder, makes up 15 to 20 percent of the food even when the ground is exposed, and with the onset of winter this increases to as much as 90 to 95 percent of the total. Ground feeding begins again as soon as the snow begins to melt (Salo 1971).

While feeding on the ground during the snow-free period these birds forage like other grouse, by individual pecking movements. Scratching with the feet is very rare, although it occurs when the birds are taking a sand bath, making a nest scrape, or digging a snow hole. When searching for insects the birds have been seen turning over leaves and clumps of mosses and searching the bark of rotting trees. Grit is also obtained within their territories, and sometimes during winter they may use piles of fine gravel put out for the benefit of capercaillies. Even day-old chicks swallow small pieces of grit, and adults carry about 2 grams of grit in their gizzards. Maximum amounts, about 4 grams, are present in late fall. Grit lost in the droppings cannot be replaced after the snow becomes deep; thus the birds evidently take more in the fall in anticipation of winter. Daily intake of food during the winter (December to March) is about 41 grams; during spring it averages about 58 grams, during summer 28 grams, and during autumn 27 grams.

Although hazel grouse will drink water in captivity, wild birds have not been observed doing so, although they are often near water during the hot season. They apparently meet their requirements from dewdrops and the moisture content of their foods.

MOBILITY AND MOVEMENTS

That this species is unusually sedentary has been well established; banding studies in Russia reported that more than half of 88 marked birds were recovered within 200 meters of the banding place, 88 percent within a 500-meter circle, and the remaining 12 percent within 500 to 1,500 meters (Gajdar 1973). As has been observed in other grouse species, dispersal distances are greatest in young hens, somewhat less in young males, and least in adult birds, with distances averaging less for males than females. Movements are greater in years of peak densities and mainly involve females, since males establish territories and become sedentary during autumn. The average movement of 20 birds recovered in Sweden and Finland was 1.2 kilometers, though one juvenile in Finland was recovered 10 kilometers from its release point. In one area where the species was expanding its range in the southern Urals, birds were observed 10 to 30 kilometers from the nearest known habitats (Kirikov, cited in Bergmann et al. 1978).

REPRODUCTIVE BEHAVIOR

Prenesting Activities

The male's territory and the pair's home range probably coincide. Even in forests where these birds are fairly numerous there are generally unoccupied areas between territories. In Finland the territory may vary in size from about 2 to 16 hectares (Pynnönen 1954), and the same is true in the Bohemian forests (Bergmann et al. 1978). Since the males cannot generally see their neighbors, sound signals probably play a greater role in territorial defense than do visual ones.

The male's song is the most important element in territorial advertisement. It is reminiscent of the alarm calls of titmice (*Parus* spp.) and has been described as *tsee-tsee-tsitseviksi-tsooitee* or *ta-tee-tee-tee-tee*. At least in most areas, the first phrase is the longest, lasting up to nearly a second; it is followed by a variably shorter second phrase and finally a highly variable sequence of usually short elements. The song is surprisingly high in pitch, from 6,500 to 8,500 Hz, or about the same pitch as kinglet (*Regulus*) songs. A single song lasts 2 to 3 seconds, and from 20 to 60 seconds intervene between songs. It is performed primarily during spring, but there is also a secondary fall period of singing, and rarely it may be heard at other times of the year.

Males typically sing from a prominence such as a tree branch or stump. They assume a characteristic hunched attitude before singing and maintain it throughout a period of songs and pauses. The head is withdrawn, the feathers are loosely fluffed, the tail is slightly spread, the bill is wide open, and the nictitating membrane may be drawn across the eye. There are individual differences in the song, and neighboring males can perhaps recognize one another in this way (Bergmann et al. 1978).

Young males begin to produce a version of their territorial song in August, and by October their delivery cannot be distinguished from that of adults. Bergmann et al. (1975) have demonstrated some regional differences in hazel grouse songs from various parts of Europe on the basis of sonograms, and Scherzinger (1981) reported a vocal repertoire of at least twenty-three well-defined calls for this species.

Males also perform territorial flights, during which they fly at about half the height of the trees, and they interrupt normal flight with short periods in which the wings produce a distinctive whirring sound. This same wing whirring sound is produced during flutter-jumping, in which the male flies up a meter or two, producing sound both on the rise and again during the descent. The sound may also be delivered while the bird is on the ground in a marked upright stance, apparently much like that of the ruffed grouse. It seems probable that the drumming of the ruffed grouse has its origin in a wing whirring display in the ancestor of both species. Wing whirring can be heard up to about 100 meters and is often performed in response to noises or calls from adjacent cocks or evoked by a special call uttered by females (Pynnönen 1954; Hjorth 1970). Scherzinger (1976) has illustrated flutter-jumping.

While walking on branches or otherwise displaying, males often assume an imposing tail cocking and tail spreading posture, with a general fluffing of the body feathers. When the neck feathers are erected and the wings drooped, this becomes an intense threat posture very much like that of the ruffed grouse. Such displays primarily are performed when another grouse is in sight, and at times they may be directed toward birds of other species as well. Fights, however, are rare by comparison with their frequence in the lekking grouse, probably because visual contact between rivals is usually rare. During fights the

28. Male displays of hazel grouse (after various sources), including (A) male head twisting, (B) tail fanning and wing drooping during high-intensity courtship, (C) singing by male, (D) threat by male, (E) threat run by male, and (F) precopulatory behavior including head turning by female.

males jump at one another, and each attempts to strike his opponent's head with his wrists, or less frequently attempts to peck it.

Since the birds are monogamous and probably remain together for much of the year, courtship is fairly simple. Pairs are evidently formed during fall with the onset of territorial advertisement, but during winter the females often wander, and thus there is no real evidence that the same pair bonds are reestablished in the spring. Several weeks are probably required for pair formation to occur (Scherzinger 1981).

The only detailed observations on copulation are from captive birds. A male placed into an enclosure with a strange female first assumed the imposing posture, performed flutter-jumps, and occasionally sang, but only after a few days did copulation occur. The male's precopulatory display consists of walking slowly around the hen while fanning his tail, trailing his primaries, and sometimes twisting his tail toward her. The female, which is generally crouching, rapidly shakes her raised head from side to side in a 90° to 180° arc. The male may also crouch and make similar head shaking movements, thereby exhibiting his black-and-white throat markings. Finally the male mounts while uttering soft calls and grasps the feathers at the back of the hen's neck (Bergmann et al. 1978). Compared with courtship, behavior associated with copulation is rather silent and is performed with sleeked plumage (Scherzinger 1981).

Nesting and Brooding Behavior

Nest sites are generally very close to or beneath cover provided by tree trunks growing at an angle, shrubs, or young spruce, or among the roots of fallen trees. Tree nests are very rare, but nests of crows or various birds of prey as high as 4 meters above the ground have reportedly been used.

The nest depression is usually about 20 centimeters across and 5 centimeters deep. It is apparently made by foot scratching, and it is lined with plant materials. On leaving a nest containing eggs, the female covers them with grass, leaves, and the like.

Eggs are laid at intervals of 26 to 30 hours, and an entire clutch of 10 may be laid in 13 to 14 days. Clutch sizes vary from 3 to 14 eggs, but generally there are between 7 and 11. Incubation probably begins with the last or penultimate egg. The eggs are similar to those of ruffed grouse, with large and small brown spots on a reddish cream or pale ochre background. Occasionally they are unspotted, and apparently they tend to be less heavily marked than those of the black-breasted hazel grouse of China. Hazel grouse are single-brooded, but they often replace clutches lost early in the season with somewhat smaller second clutches.

The incubation period is 23 to 27 days, probably averaging about 25 days. During this period the male commonly leaves the territory but rarely may return later to accompany the female and the brood. Semenov-Tian-Schanski (1959) followed the behavior of an incubating hen with an electrical contact recorder. Even before incubation began with the penultimate egg, she spent more time on the nest each day. Once incubation was under way she left the nest progressively less often, from five times a day initially to two or three times a day near the end of incubation. For almost 37 hours before hatching she did not leave the nest at all. She was off the nest for only 5 percent of the total incubation time, in periods averaging 22 minutes. While off the nest she produced the typical extralarge droppings, or "clockers." The hen began to turn the eggs up to forty-seven times a day during the last week of incubation.

The young call while still in the egg, a soft *ssrsr* note, and as a contact call with their mother they produce a soft, piping *ssee*. Downy young initially lack adequate temperature control and depend on the hen to keep them warm. When separated from their mother or siblings, the chicks utter a louder version of the contact call. At about 18 days of age they begin to produce a new call, a repeated *plit, plit*, which adults also produce when agitated. After about a month they develop a number of lower-pitched calls that gradually replace the calls they uttered as chicks. Among these is a warning call, *pseeo*, which elicits fleeing by companions.

Four-day-old chicks are capable of very short fluttering flights, and at 4 weeks old the chicks are fully fledged and able to fly about as well as adults. Newly hatched young still have a small supply of yolk, which is used up in a few days. Thereafter they obtain protein in the form of various arthropods, but soon they also eat plant foods. Hens warm their young by brooding them at night and during cold weather. At the hen's alarm call the chicks crouch and remain immobile until summoned by a high-pitched whistling call. If an enemy approaches the female may show defensive behavior, but more often she distracts the enemy by zigzagging, leading it away from the young. Her final role is to guide the young to areas rich in animal food, without regard to territorial boundaries.

When only 54 days old, the young may show elements of sexual behavior, making a humming sound with their wings, and they have been observed attempting to sing when only 61 days old. The adult weight is nearly attained by about 90 days, and sexual maturity occurs during the first fall of life. The young usually remain in a group with the female for up to about 85 days, after which they disperse. At this time the adult pairs may reform, and the home range of adults may be as small as 2 hectares during fall. However, females wander extensively during winter. Probably most of the birds (55 percent in one study) overwinter singly, somewhat less than half in pairs, and a small portion as trios (Volkov, cited in Bergmann et al. 1978). Territories are generally reestablished during spring, and at this time the yearling birds also attempt to establish breeding territories (Bergmann et al. 1978; Cramp and Simmons 1980).

Juvenile and Adult Mortality

Hazel grouse experience their greatest mortality as chicks, particularly if the weather is cold and wet during the hatching period. Some figures from the forests of Bielowicza are indicative (Gavrin 1969). In dry, warm summers chick losses in June were 18 percent, increasing to 28 percent by the end of July and 45 percent by the end of August. In another year when June was cold and rainy, 26 percent of the chicks were lost that month, but with better weather later the total loss was only 37 percent by the end of August. Scherzinger

(1976) reported that the average size of twenty-seven dependent broods was 4.7 young, and for eight independent broods it was 4.1.

Adult mortality rates are still undetermined but must approach 70 percent annually, since that is the approximate incidence of young birds in fall populations. The most important predator in most of the range of this grouse is the goshawk (*Accipiter gentilis*). In the Bielowicza area, hazel grouse formed 16 percent of goshawk prey in the breeding season and 25 percent during fall and winter (Gavrin 1969). It is estimated that a pair of goshawks may take about 40 hazel grouse per year. Only in eastern Siberia does the sable (*Martes zibellina*) become a more important predator than the goshawk. The European sparrow hawk (*Accipiter nisus*) generally only takes chicks or juveniles, yet this grouse may form 12 percent of sparrow hawks' prey in the Petshora area of Russia. There is one case of a male sparrow hawk that, after losing his mate, brought 36 juvenile hazel grouse into his nest within a few days.

The Ural owl (*Strix uralensis*) and the hawk owl (*Surnia ulula*) have been estimated to account for 12 percent of the winter losses of hazel grouse in eastern Siberia. Eagle owls (*Bubo bubo*) also occasionally take hazel grouse, and in Poland the tawny owl (*Strix aluco*) sometimes preys on full-grown grouse.

A number of mammalian predators are reported to be important in hazel grouse mortality. In Europe the pine marten (*Martes martes*) rivals the goshawk as a predator in winter, and the sable is said to account for about 80 percent of hazel grouse kills during winter in Siberia. Other mammals, including the red fox (*Vulpes vulpes*), polecat (*Putorius putorius*), and various weasels are also capable of killing hazel grouse, but predation is probably of slight significance.

Certain other predators that hunt by scent sometimes endanger the eggs; they include the badger (*Meles meles*), raccoon dog (*Nyctereutes procyonoides*), wild pig (*Sus scrofa*), squirrels (*Sciurus vulgaris*), and additionally a few species of birds. These accounted for most of the egg losses in a Russian study area, which amounted to 12 to 18 percent.

In spite of the number of these predators, they are not the cause of population losses and the final local disappearance of hazel grouse. Several of the predators just named occur in the primeval forest sanctuary of Bielowicza, yet this area probably has the highest population of hazel grouse in Europe. The major reason for the decline or disappearance of this species from much of its European range is human alteration of habitats, the result of recent intensive forms of forest utilization.

EVOLUTIONARY RELATIONSHIPS

Short (1967) has already argued for the merger of *Tetrastes* and *Bonasa*, primarily on morphological grounds, and behavioral evidence also supports the idea of a close relation between the hazel grouse and the ruffed grouse. Short has suggested that the black-breasted hazel grouse closely approaches the condition of a generalized ancestral grouse, and Janossy (1976) has described from the middle Pleistocene of Europe a fossil species of hazel grouse (*Bonasa praebonasi*) that he considered the ancestral form of all three of the modern species of this genus. It seems probable that the ancestral hazel grouse initially evolved in Europe or western Asia and later gave rise to the two Eurasian populations and the North American population. Nothing is known of the behavior of the Chinese population of hazel grouse, but it would be interesting to learn whether it is more generalized behaviorally than *bonasia* and whether it approaches *Dendragapus* as Short has postulated.

Black-breasted Hazel Grouse

Bonasa sewerzowi (Przewalski) 1876

Other Vernacular Names

Severtzov's hazel grouse, Schwarzbrust Haselhuhn (German); gelinotte de Severtzow (French).

Range (after Vaurie 1965)

Resident in the mountains of southern Tsinghai and central Kansu, south to central and eastern Sikang and northern Szechwan. See distribution map 13.

Subspecies

None recognized here, but birds from Sikang are sometimes recognized as a separate form *secunda* (*Auk* 42:423).

MEASUREMENTS

Folded wing: Males 169–83 mm (average of 8, 175.5), females 167–76 mm (average of 9, 171.6).

Tail: Males 115–53 mm (average of 4, 126.2 mm), females 88–136 mm (average of 3, 109 mm).

IDENTIFICATION

Length, 14 inches. Males are very similar to males of the more widespread species of *Bonasa*, but the dark spotting on the breast is more extensive; the hindneck and upper back are more broadly barred. The central four rectrices are brown, vermiculated and narrowly

29. Comparative male plumage features of black-breasted hazel grouse (*left*) and hazel grouse (*right*).

barred, while the rest of the rectrices are strongly barred with black and white and have white tips. The throat of the male is black, with a narrow white border, as in *bonasia*, while that of the female is barred with black and white (Short 1967).

FIELD MARKS

This species doubtless is very much like the hazel grouse in its general appearance, posture, and behavior, but since these two species do not overlap in range there should be no problems in field identification. No other remotely similar species of grouse occurs within the range of this species.

AGE AND SEX CRITERIA

Age and sex criteria have not been described, but probably the statements made relative to the hazel grouse can be applied to this species with little or no modification. The downy young, not previously described in detail, are illustrated in plate 51, based on two specimens observed by Jon Fjeldså (pers. comm.). Short (1967) indicates that the head markings (loral mark, orbital lines, frontal mark traces, traces of black border of chestnut patch) are more evident in this species than in the other species of *Bonasa*, approaching the pattern found in *Dendragapus*.

DISTRIBUTION AND HABITAT

This species occupies dense coniferous forests, particularly thickets associated with small meadows and along brooks, and it breeds as high as 4,000 meters in Sikang but probably generally occurs at somewhat lower elevations (3,200–3,800 meters).

According to Stresemann, Meise, and Schönwetter (1938), the species is resident in the mixed woods and juniper woods of the northern Himalayas. During fall it has been observed in areas having abundant berry growth, and during the winter it primarily occurs in pairs, inhabiting mixed woods or juniper-dominated woods. Cheng (1979a) states that it is generally found above 1,000 to 1,200 meters, in the same range as such temperate-forest European species as goldcrests (*Regulus regulus*) and tree creepers (*Certhia familiaris*).

POPULATION DENSITY AND STATUS

No information is available on population density and status.

HABITAT REQUIREMENTS

Habitat requirements presumably are similar to those of the Eurasian hazel grouse, but no information is available.

FOOD AND FORAGING BEHAVIOR

The only specific note on food and foraging seems to be an observation of four birds seen in February, perched on the braches of a sea buckhorn (*Hippophae rhamnoides*) and eating its berries (Stresemann, Meise, and Schönwetter 1938).

MOBILITY AND MOVEMENTS

No information is available on mobility and movements.

314

REPRODUCTIVE BEHAVIOR

Nothing has yet been written on this species' pairing behavior, other than that it apparently remains in pairs throughout the winter and like the more common Eurasian species of hazel grouse is presumably monogamous.

Only a very few nests of the species have been discovered thus far. The first to be described were found by Beick (1927), who located a nest with 7 strongly incubated eggs on June 25, 1927, in the coniferous forest zone of the south Tungjen Mountains, near a tributary of the Hwang Ho River. The nest was in a steep rock area overgrown with pines and birches. The nest site was lined with pine needles and partly bordered with green moss. The eggs lay on a few feathers, and the nest was protected from above by a small birch. He found a second nest on July 2, on a rocky ledge. Each of these nests contained 7 well-incubated eggs. A third nest, with three unincubated eggs, was located in juniper woods on June 13. The associated eggs were all measured, and they proved to average very slightly larger than those of the Eurasian hazel grouse but were otherwise quite similar to them. Schönwetter (1929) later reported on the second of these clutches and stated that the eggs had the largest and darkest fleckings of tetraonids, the marking being without definable form, of dark chestnut brown, on a background of pale yellowish brown with a reddish tinge.

On the July 14, 1928, Beick observed a female leading three well-grown young. The mother uttered a normal alarm call that sounded something like *ze, ze, ze—dackdack*. On the July 16 a female with a brood of 7 or 8 young was encountered, but in this case as in the earlier one the male was not seen attending the brood. The habitat consisted of low birch woods mixed with shrubs such as *Rhododendron przewalskii*.

The collective average of the 17 eggs found was 44.0 by 30.8 mm, with extremes of 41.9 to 46.7 mm by 29.7 to 32.0 mm (Stresemann, Meise, and Schönwetter 1938).

EVOLUTIONARY RELATIONSHIPS

According to Short (1967), the ruffed grouse and the two species of hazel grouse form a morphological series in the sequence *sewerzowi-bonasia-umbellus*, with *umbellus* the most specialized in plumage characteristics, such as possessing a ruff. The barring on *sewerzowi* is somewhat like that of *Dendragapus falcipennis* and shows a less complicated pattern of vermiculations, streaks, and blotches than that present in *umbellus*. Thus, Short concluded that *sewerzowi* more closely approaches the ancestral condition than either of the other species.

I find no difficulty with this interpretation. The somewhat darker coloration of this species (and also its eggs) compared with its Eurasian counterpart is perhaps associated with a somewhat more humid environment, since it seems to particularly inhabit dense and moist montaine woods.

Pinnated Grouse

Tympanuchus cupido (Linnaeus) 1758

Other Vernacular Names

Prairie chicken, prairie cock, prairie grouse, prairie hen; cupidon des prairies, poule des prairies (French); Eigentliche Präriehuhn (German).

Range

Currently resident in remnant prairie areas of Michigan, Wisconsin, and Illinois and from eastern North Dakota and northwestern Minnesota southward to western Missouri and Oklahoma and portions of the coastal plain of Texas. Also (*pallidicinctus*) from southeastern Colorado and adjacent Kansas south to eastern New Mexico and northwestern Texas.

Subspecies

T. c. cupido (Linnaeus): Heath hen or eastern pinnated grouse. Extinct since 1932. Formerly along the East Coast from Massachusetts south to Maryland and north-central Tennessee.

T. c. pinnatus (Brewster): Greater prairie chicken. Currently limited to several small isolated populations in Michigan (nearly extirpated), Wisconsin, and Illinois and to the grasslands of northwestern Minnesota, eastern North Dakota, South Dakota, Nebraska, Kansas, western Missouri, and northern Oklahoma.

T. c. attwateri Bendire: Attwater prairie chicken. Currently limited to a few isolated populations along the coast of Texas from Aransas and Refugio counties to Galveston County, and inland to Colorado and Austin counties.

T. c. pallidicinctus (Ridgway): Lesser prairie chicken. Currently limited to arid grasslands of southeastern Colorado and southwestern Kansas southward through Oklahoma to extreme eastern New Mexico and northwestern Texas. Recognized by the A.O.U. *Check-list* (1957) as a separate species.

14. Current distribution of the Attwater (A), greater (G), and lesser (L) races of the prairie chicken, and the original distribution of the heath hen (H). The inset map shows the current (1979) Texas range of the Attwater prairie chicken, and the stippling indicates the historic ranges of all forms of pinnated grouse.

MEASUREMENTS

Folded wing (greater prairie chicken): males 217–41 mm (average 226 mm); females 208–20 mm (average 219 mm).

Folded wing (lesser prairie chicken): males 207–20 mm (average 212 mm); females 195–201 mm (average 198 mm).

Tail (greater prairie chicken): males 90–103 mm (average 96 mm); females 87–93 mm (average 90 mm).

Tail (lesser prairie chicken): males 88–95 mm (average 92 mm); females 81–87 mm (average 84 mm).

IDENTIFICATION (Greater Prairie Chicken)

Adults, 16–18.8 inches long. The two sexes are nearly identical in plumage. The tail is short and somewhat rounded and the longer under (but not upper) tail coverts extend to its tip. The necks of both sexes have elongated "pinnae" made up of about ten graduated feathers that may be relatively pointed (in *cupido*) or somewhat truncated (other races) in shape and are much longer in males than in females. Males have a conspicuous yellow comb above the eyes and bare areas of yellowish skin below the pinnae that are exposed and expanded during sexual display. The upperparts are extensively barred with brown, buffy, and blackish, while the underparts are more extensively buffy on the abdomen and whitish under the tail. Transverse barring of the feathers is much more regular in this species than in the sharp-tailed grouse, which has V-shaped darker markings and relatively more white exposed ventrally.

IDENTIFICATION (Lesser Prairie Chicken)

Adults, 15–16 inches long. In general like the greater prairie chicken, but the darker, blackish bars of the back and rump typical of greater prairie chickens are replaced by brown bars (the black forming narrow margins), the breast feathers are more extensively barred with brown and white, and the flank feathers are barred with brown and dusky instead of only brown. Males have reddish rather than yellowish skin in the area of the gular sacs, and during display their yellow combs are more conspicuously enlarged than those of greater prairie chickens. As in that form, females have relatively shorter pinnae and are more extensively barred on the tail.

FIELD MARKS

The only species easily confused with either the greater or the lesser prairie chicken is the sharp-tailed grouse, which often occurs in the same areas where greater prairie chickens are found. Sharp-tailed grouse can readily be recognized by their pointed tails, which except for the central pair of feathers are buffy white, and by their whiter underparts as well as a more "frosty" upper plumage pattern, which results from white spotting that is lacking in the pinnated grouse.

AGE AND SEX CRITERIA (Greater Prairie Chicken)

Females may readily be recognized by their shorter pinnae (females of *pinnatus* average 38 mm, maximum 44 mm, males average 70 mm, minimum 63 mm) and their extensively barred outer (rather than only central) tail feathers. The central crown feathers of females are marked with alternating buffy and darker crossbars, whereas males have dark crown feathers with only a narrow buffy edging (Henderson et al. 1967). In the Attwater prairie chicken the pinnae of females are about 9/16 inch (14 mm) long, while those of males are over 2 inches (53 mm), according to Lehmann (1941).

Immatures may be recognized by the pointed, faded, and frayed condition of the outer two pairs of primaries (see sharp-tailed grouse account). The pinna length of first-autumn males is not correlated with age (Petrides 1942).

Juveniles may be recognized by the prominent white shaft streaks, which widen toward the tip, present in such areas as the scapulars and interscapulars.

Downy young of the greater prairie chicken are scarcely separable from those of the lesser prairie chicken (see that account) and also resemble young sharp-tailed grouse. However, prairie chickens have a somewhat more rusty tone on the crown and the upperparts of the body and richer colors throughout. There are usually three (one small and two large) dark spots between the eye and the ear region and several small dark spots on the crown and forehead. Short (1967) mentions, however, that at least some downy specimens of *attwateri* have only one or two tiny postocular black markings, which thus would closely approach the markings of downy sharp-tailed grouse.

AGE AND SEX CRITERIA (Lesser Prairie Chicken)

Females may be identified by their lack of a comb over the eyes and their brown barred under tail coverts, which in males are black with a white "eye" near the tip (Davison, in Ammann 1957). Males have blackish tails, with only the central feathers mottled or barred, while the tails of females are extensively barred (Copelin 1963).

Immatures can usually be identified by the pointed condition of the two outer pairs of primaries. The outermost primary of young birds is spotted to its tip, while that of adults is spotted only to within an inch or so of the tip. In addition, the upper covert of the outer primary is white in the distal portion of the shaft, whereas in adults the shafts of these feathers are entirely dark (Copelin 1963).

Juveniles are more rufescent than the corresponding stage of the greater prairie chicken or the adults. The tail feathers are bright tawny olive and have terminal tear-shaped pale shaft streaks (Ridgway and Friedmann 1946).

Downy young are nearly identical to those of the greater prairie chicken (Short 1967) but are slightly paler and less brownish on the underparts. On the upperparts the brown spotting is less rufescent and paler, lacking a definite middorsal streak (Sutton 1968).

DISTRIBUTION AND HABITAT

The original distribution of the pinnated grouse differs markedly from recent distribution patterns; without doubt it is the grouse species most affected by human activities in North America. Aldrich (1963) identified the habitat of the now extinct eastern race of pinnated grouse, the heath hen, as fire-created "prairies" or blueberry barrens associated with sandy soils from Maryland to New Hampshire or Maine. The presence of oak "barrens" or parklands may have also been an integral part of the heath hen's habitat, particularly by providing acorns as winter food (Sharpe 1968). The range of the coastal Texas race, the Attwater prairie chicken, once extended over much of the Gulf coastal prairie from Rockport, Texas, northward as far as Abbeville, Louisiana, an area of more than 6 million acres (Lehmann and Mauermann 1963). The lesser prairie chicken once occupied a large area of arid grasslands, with interspersed dwarf oak and shrubs or half-shrub vegetation (Aldrich 1963; Jones 1963). The birds occurred over an extensive area from eastern New Mexico and the panhandle of Texas northward across western Oklahoma, southwestern Kansas, and southeastern Colorado. Over this area they were found on two major habitat and soil types, the sand sage–bluestem (*Artemisia filifolia–Andropogon*) shrub grasslands of sandy areas and the similarly sand-associated shin oak–bluestem (*Quercus havardi–Andropogon*) community (Jones 1963; Sharpe 1968). The greater prairie chicken originally occurred in the moister and taller climax grasslands of the eastern Great Plains from approximately the 100th meridan eastward to Kentucky, Ohio, and Tennessee, and northward to Michigan, Wisconsin, Minnesota, and South Dakota (Sharpe 1968). Sharpe suggested that the presence of oak woodlands or gallery forests throughout much of this range, and the more extensive oak-hickory forests to the east of it, may have been an important part of the greater prairie chicken's habitat. There absence in the western and northwestern grasslands may have made those areas originally unsuitable for prairie chickens. Probably a winter movement of no more than 250 miles to woody cover was typical, according to Sharpe.

With the breaking of the virgin prairies in the central part of North America and their conversion to small-grain cultivation, the prairie chickens responded strongly and moved into regions previously inhabited only by the sharp-tailed grouse (Johnsgard and Wood 1968). Thus they moved into northern Michigan and southern Ontario, into northern Wisconsin and much of Minnesota, into the three prairie provinces of Manitoba, Saskatchewan, and Alberta, and westward through all or nearly all of North Dakota, South Dakota, and Kansas to the eastern limits of Montana, Wyoming, and Colorado. At the same time the lesser prairie chicken may have undergone a temporary extension northward into western Kansas, northeastern Colorado, and extreme southwestern Nebraska, where it may have been geographically sympatric for a relatively few years with the greater prairie chicken (Sharpe 1968). However, their habitat requirements are quite different (Jones 1963), and no natural hybrids between these forms have ever been reported.

During several decades the greater prairie chicken survived extremely well in these interior grasslands, where remaining native vegetation provided the spring and summer habitat requirements and the availability of cultivated grains allowed for winter survival. Eventually, however, the percentage of land in native grassland cover was reduced to the point where these habitat needs could no longer be met, and the species began to recede from much of its acquired range and to seriously decline or become eliminated from virtually all of its original range. The sad history of this range restriction and population diminution has been recounted in various places and by many writers (Johnsgard and Wood 1968). Space does not allow a detailed review of these changes, and all I will attempt here is a statement of the current range and status of the three extant subspecies.

Of the three races, the Attwater prairie chicken is clearly in the greatest danger of extinction. The race was extirpated from Louisiana by about 1919, and between 1937 and 1963 the Texas population declined from about 8,700 to 1,335 birds (Lehmann and Mauermann 1963). The remaining populations suffered from a badly distorted sex ratio, intensified farming practices, predators, fire exclusion, pesticides, and bad drainage practices, and realtively little area was set aside specifically for their protection. The purchase of 3,420 acres of land in Colorado County by the World Wildlife Fund in the mid-1960s was a critical step toward retaining viable population. By 1965, when the total Texas population was estimated to be from 750 to 1,000 birds, the estimated refuge population was 100 birds. Lehmann (1968) provided a summary of the status of this bird in the late 1960s. As of 1967 an estimated 1,070 birds occupied some 234,000 acres, which represents a habitat loss of 50 percent since 1937 and a population reduction of 85 percent during the same time. By the late 1970s the population was estimated at about 2,000 birds, with the largest populations in Aransas, Refugio, and Goliad counties, a secondary population in Austin and Colorado counties (Attwater's Prairie Chicken National Wildlife Refuge), and a third population on private rangelands in Victoria County (Kessler 1979). As of 1980 the total population was estimated by Lawrence and Silvy (1980) at 1,584 birds, or up 48 percent from 1967. These authors indicated that the population is now centered on about 120,400 hectares (75 square miles) of potential range, with 53 percent in Aransas, Goliad, and Refugio counties, 32 percent in Austin and Colorado counties, 6 percent in Galveston County, 4 percent in Victoria County, and the remaining 5 percent in Fort Bend and Brazoria counties. They judged that the total population should remain stable for the next decade, given no unpredictable catastrophes, though remnant populations in Galveston and Brazoria counties are likely to disappear. About 10 percent of the population is on refuge lands in Colorado and Aransas counties, and the rest is on private lands.

The present range of the lesser prairie chicken centers in the panhandles of northern Texas and northwestern Oklahoma, but it also includes parts of southwestern Kansas, southeastern Colorado, and eastern New Mexico. In Oklahoma the present range includes several isolated populations spread over about 1,090 square miles (2,800 square kilometers), and contained about 7,500 birds in the spring of 1979. There has been a loss of more than half of the population's range in twenty years, and an approximate reduction of the population by half since 1944 estimates. The largest remaining populations are in Beaver, Ellis, and Woodward counties, with much smaller numbers in Beckham, Harper, Roger Mills, Texas, and Woods counties (Cannon and Knopf 1980). Hunting is still allowed in the state, and the 1979 harvest was estimated at 134 birds.

In Texas lesser prairie chickens have likewise been in a long-term declining trend, and the estimated fall population in 1979 was between 11,000 and 18,000 birds. After thirty years of protection, hunting was first allowed again in 1967, and since that time a few hundred birds have been taken annually. The 1979 harvest was probably about 600 birds (Crawford 1980). The occupied range is about 4,625 square kilometers (Taylor and Guthery 1980).

The third of four states where the lesser prairie chicken is sufficiently abundant to be legally hunted is Kansas, where the estimated fall 1979 population was 17,000 to 18,000 birds and the harvest was about 2,900. There too the population is declining (Crawford 1980). Its habitat is diminishing at the rate of 1.5 to 6 percent annually, largely because of the effects of center-pivot irrigation (Waddell and Hanzlick 1978). The occupied range is about 778 square kilometers (Taylor and Guthery 1980).

New Mexico is the other only state that currently allows the hunting of lesser prairie chickens. Its estimated fall 1979 population was some 10,000 birds, and about 1,200 birds were legally shot that year. Since 1960 the birds have been legally hunted, and harvests have generally been about 1,100 birds per year. Its population in that state is apparently stable and is essentially limited to five counties, centering on Roosevelt County (Sands 1968). The occupied range is about 18,898 square kilometers (Taylor and Guthery 1980).

Besides these four states, there is a small population in Colorado that probably numbered 400 to 500 in the fall of 1979. It is considered a threatened but stable population and is protected from hunting (Crawford 1980). It has probably benefited from such protection; Hoffman (1963) reported a substantial increase of males on censused display grounds between 1959 and 1962. The occupied range is about 1,634 square kilometers (Taylor and Guthery 1980).

Beyond these states, the lesser prairie chicken has been extirpated from Missouri (where it is known only from nearly century-old winter records) and is considered hypothetical in Nebraska (where it may have once occurred in the extreme southwestern corner of the state). It was released on the island of Niihau, Hawaii, during the 1930s, but its present status there is unknown (Crawford 1980).

Crawford (1980) estimated the total fall 1979 continental population as 44,400 to 52,900 birds, and generally it is either stable or declining. The species' range of about 27,300 square kilometers has decreased by more than 90 percent in the past century, and there has been a 78 percent decrease in range since 1963 (Taylor and Guthery 1980). These authors estimated a current population of 46,700 to 55,330 birds.

The status of the greater prairie chicken is almost as alarming as that of the lesser. It now may be regarded as virtually extirpated from all of its prior range in the Canadian provinces (Hamerstrom and Hamerstrom 1961). Westemeier (1980) has provided a useful summary of the bird's status in the United States. Considering the form's probable original range, it has been extirpated as a breeding species from Iowa, Ohio, Kentucky, Texas, and Arkansas. The birds were gone from Ohio before 1930 and from Kentucky, Texas, and Arkansas at even earlier dates. The last nesting prairie chickens in Iowa were seen as late as 1952, and stray birds were seen as late as 1960 (Stempel and Rogers 1961). The estimated population in Indiana diminished from more than 400 males occupying 33 booming grounds in 1942 to 4 males on a single booming ground by 1966. Christisen (1969) indicated a current estimated total Indiana population of only 10 birds in 1968, but by 1979 the population was considered extirpated.

Similarly, although Michigan may have had a population of about 200 birds in 1968, by 1979 there were fewer than 50 (Westemeier 1980). In 1980 the estimated number was only 20 birds, which were limited to a section and a half of publicly owned grassland (*Nebraskaland,* September 1980, p. 35).

In Illinois the estimated population dropped from 300 to 230 birds between 1968 and 1979, but in 1980 the spring count was up to 334 birds. These occur on some 1,640 acres of grassland that is being specifically managed for prairie chickens, so their population is probably secure. About 1,000 of these acres are in Jasper County, and the remainder are in Marion County. Additional very small remnant flocks occur in Washington and Wayne counties, not on protected lands (Westemeier 1980; in litt.).

In Wisconsin the population shifted from about 1,000 birds in 1968 to an estimated 1,842 in 1979 and appears to be increasing. In central Wisconsin about 100 square kilometers (39 square miles) are being managed for the species, and a reintroduction attempt will be made on the Crex Meadows Wildlife Area (47 square miles) of northwestern Wisconsin (Westemeier 1980). Hunting was last permitted in 1951.

In Minnesota, however, the estimated populations declined from 5,000 birds in 1968 to about 2,000 in 1979. These birds occur on about 78 square miles (about 200 square kilometers) of habitat, and in 1977 they were apparently successfully reintroduced in the Lac Qui Parle Wildlife Management Area, which would increase their available habitat somewhat (Westemeier 1980). Hunting was last permitted in 1942.

In Missouri the prairie chicken population was approximately 9,600 birds in 1979, or nearly the same as the same as the 10,000-bird estimate of 1968. These are distributed on some 900 square miles of range, most of which is in public ownership and is being managed specifically for prairie chickens. Hunting has not been allowed since 1906.

In North Dakota the total 1979 population was estimated at some 1,000 birds, which occur on about 220 square miles of habitat. In 1969 the estimated state population was about 1,800 birds, but locally the populations may have increased recently because of better management of federally owned grasslands by the United States Forest Service (Westemeier 1980). The last legal hunting season was in 1945.

In Colorado the population has declined from an estimated 7,600 in 1968 to between

300 and 3,000 in 1979. The total range of greater prairie chickens in the state is about 430 square miles and is currently declining as a result of grassland losses to center-pivot irrigation and row-crop farming (*Nebraskaland,* September 1980, p. 35). Plans are now under way to restore the South Platte Management Area to conditions suitable for prairie chickens, and to reintroduce them there (Westemeier 1980). Hunting was last allowed in 1937.

These seven states must all be considered relatively marginal, and they collectively support fewer than 20,000 birds at present. Most of the greater prairie chicken's population is to be found in Oklahoma, Kansas, Nebraska, and South Dakota.

In Oklahoma the total greater prairie chicken population was estimated at about 8,400 birds in the spring of 1979. These were distributed over 2,400 square miles (6,100 square kilometers) in thirteen northeastern counties, with the largest populations in Osage and Craig counties. Since 1943 there has been a 42 percent decrease in occupied range and a 34 percent decline in actual numbers. The Oklahoma population includes a fairly stable western component and a rapidly declining eastern one (Martin and Knopf 1980). The population is currently hunted, with an estimated 1979 harvest of 4,971 birds (Mark Byard, in litt.).

The South Dakota population of prairie chickens is in the vicinity of 40,000 birds, or half of that estimated to be present in 1968 (Westemeier 1980). The birds are distributed over an area of about 14,000 square miles, and the population is probably declining as a result of recent droughts, plowing of rangelands, and heavier grazing pressure. The state has a regular hunting season lasting approximately 80 days, and in 1978 an estimated 5,233 birds were legally taken. The highest populations occur in Jones County, where native grasslands occupy about 68 percent of the land area and cultivated lands about 30 percent; woody cover in South Dakota's prairie chicken range covers less than 1 percent of the total area (Janson 1953).

In Nebraska the greater prairie chicken probably originally occurred in the eastern part of the state, but it is now largely limited to the central portion, where it occurs along the eastern and southern edges of the sandhills, where native grasses and grain crops are close by and provide both summer and winter habitat needs (Johnsgard and Wood 1968). The state's population is relatively static, and both this species and the more common sharp-tailed grouse have been regularly hunted, except in the case of the small and isolated population of this species in southeastern Nebraska, which is an extension of the large Flint Hills population of eastern Kansas. In 1967 the Nebraska harvest was estimated at 15,000 birds, and the state's total population in 1968 was estimated at 100,000 birds (Christisen 1969). In 1980 the total state population was placed as 75,000 to 100,000 birds, and the occupied range at approximately 7,000 to 10,000 square miles. The population is thus apparently stable, but increasing row-crop agriculture may bring about future declines. Between 1965 and 1968 the area of irrigated croplands in sixteen sandhills counties increased from about 26,000 hectares to 214,000 hectares, and there was a concurrent decline in the number of active prairie chicken leks (Robertson 1980). The usual annual harvest is between 10,000 and 30,000 birds, taken over an approximate fifty-day season (*Nebraskaland,* September 1980, p. 35).

The heart of the greater prairie chicken's present range is in eastern Kansas, amid the bluestem (*Andropogon*) prairies that extend from the Oklahoma border in Chautauqua and Cowley counties to near the Nebraska border in Marshall County (Baker 1953). This area includes an easternmost zone of interspersed natural grassland and croplands, a zone of sandy soils associated with natural grasslands and wooded hilltops, a zone of flinty, calcareous hills and associated native grasslands, and a transition zone between these hills and the cultivated lands to the west. In the best areas for prairie chickens, the ratio of natural grasslands to cultivated feed crops is roughly two to one (Baker 1953). Prairie chickens have been given protection in Kansas periodically since 1903. The population apparently underwent a marked decline in the early 1940s, followed by an increase to the end of that decade, when 50,000 birds were conservatively estimated to be present in the state (Baker 1953). In 1967 some 46,000 birds were harvested, and an estimated 750,000 were believed present in the late 1960s (Christisen 1969), suggesting that the Kansas population was by far the most secure of any state's. Yet by 1979 the estimated population had declined to 200,000 birds (Westemeier 1980). Although rural mail carrier surveys between 1963 and 1980 indicate recent downward population trends, booming-ground counts have indicated a generally upward recent trend, and the latter seem to correlate better with hunter harvest data (Rogers Wells, Prairie Chicken Population and Harvest Summary, 1979, Federal Aid Project no. W-23-R18). The 1979 estimated hunter kill was 88,400 birds, the highest since 1959 and well above the recent state average of about 34,000 birds.

In summary, it seems that the total collective fall populations for the three extant prairie chicken forms might be about 1,500 for the Attwater, about 50,000 for the lesser, and about 500,000 for the greater. Of the three, the endangered Attwater has remained stable or increased somewhat, the lesser has likewise remained stable in its most important population centers but has declined locally, and the greater has exhibited a probable population decline of about 50 percent in about a decade. However, the areas of apparently greatest decline (South Dakota, Oklahoma, and Kansas) are the ones where the population estimates have often been based on limited information, and in Kansas and Nebraska the population may actually be fairly stable. Correlated with the population decline there has been an approximate 28 percent reduction in hunter kill since the fall of 1967 (Westemeier 1980). During the same period the species was extirpated from at least one state (Indiana) and is nearly gone from another (Michigan). Its survival in Wyoming (where it had been limited to Goshen County) is dubious, and it is probably also now gone from Manitoulin Island, the mainland of southern Ontario, and southern Manitoba. Only in Wisconsin, were a great deal of money is being invested in land purchase and management, does the species seem to be increasing at present.

POPULATION DENSITY

Population density estimates for prairie chickens vary greatly for different areas and in general probably reflect the deteriorating status of the species, with declining populations being studied more intensively than the relatively few healthy or increasing populations. Grange (1948) estimated a spring prairie chicken population in Wisconsin of 1 prairie chicken per 110 acres in 1941 and 1 per 138 acres in 1942, or between 4 and 6 birds per square mile. In 1943 the prairie chicken range in Missouri likewise averaged 4.8 birds per square mile. In South Dakota's best remaining prairie chicken habitat of six counties, spring population densities of from 2 to 4 birds per square mile occur (Janson 1953).

In contrast, Baker (1953) studied several flocks of prairie chickens in high-quality Kansas range on a study area covering about 3½ square miles. Two flocks used this area exclusively, while two other flocks used it in part. Spring numbers of one flock varied over a three-year period from 15 to 104 birds, while a second flock varied from 15 to 43 birds during these three springs. A third flock consisted of about 20 birds. Using conservative figures, an average spring population of at least 50 birds must have been dependent on the area, or at least 14 birds per square mile. During population "highs," the spring density may have reached about 50 birds per square mile for the study area as a whole, and even more if only the composite home range areas are considered.

Data on male spring densities for the lesser prairie chicken are available from Oklahoma (Copelin 1963). Over a six-year period on four different study areas having display grounds, the densities of males per square mile varied from 1.5 to 18.31 and averaged 7.4 males. Earlier figures aviable from one of these study areas for the 1930s indicated densities of from about 15 to nearly 40 males per square mile. Hoffman (1963) reported that male densities on three areas in Colorado increased from 0.8 to 5.8 males per square mile over a four-year period in this marginal part of the species' range. In Texas, Jackson and DeArment (1963) noted that numbers of males on a 100,000-acre area reached as high as 600 birds in 1942 (about 4 birds per square mile) but more recently have averaged about 200 males. These data collectively indicate that spring densities of males in favorable habitats may exceed 30 per square mile but probably average less than 10. Similarly, Lehmann (1941) reported spring densities of about 10 birds per square mile for the Attwater prairie chicken in Texas for the late 1930s. A 1967 survey of this population indicated that 645 birds were present on about 136,000 acres, or a density of 210 acres per bird (3 birds per square mile).

HABITAT REQUIREMENTS

Wintering Requirements

The winter requirements for pinnated grouse seem to center on the availability of a staple source of winter food, rather than protective cover or shelter from the elements. Lehmann (1941) reports that Attwater prairie chickens moved into lightly grazed natural grassland pastures by mid-November and remained there until spring. In Oklahoma Copelin (1963) found that the lesser prairie chickens used cultivated grains, especially sorghum, extensively during two winters. In the following winter, when production in the shin oak grassland pastures was apparently high, the birds remained in this pastureland area. During the following two winters use of cultivated grains increased, particularly in late winter when snow was nearly a foot deep for a week or longer, and shocked grain sorghum was then extensively utilized.

Edminster (1954) concluded that grainfields are an important part of present-day prairie chicken habitat, with corn providing the best winter habitat, provided it is either shocked or left uncut. Sorghum, like corn, stands above snow during the winter and thus is almost as valuable. Robel et al. (1970) confirmed the importance of sorghum in winter for Kansas prairie chickens. Other small grains such as wheat and rye are utilized whenever the birds can reach them during winter.

In contrast to the sharp-tailed grouse and nonprairie grouse, there is little evidence that the pinnated grouse ever resorts to buds as primary foods during winter. Martin, Zim, and Newlson (1951) list the buds and flowers of birch as a minor source of winter food for pinnated grouse from the northern prairies but found them of far less importance than cultivated grains or wild rose (presumably rose hips). Edminster (1954) lists the buds of birch, aspen, elm, and hazelnut among items used in the northern range during winter, but as long as grain or other seeds are available this does not appear to be critical to winter survival. Mohler (1963) reported that the best winter habitats for prairie chickens in the Nebraska sandhills were areas where cornfields were near the extensive and lightly grazed grasslands of the larger cattle ranches, providing a combination of available food and grassy roosting cover.

According to a summary by Taylor and Guthery (1980), the lesser prairie chicken uses grass cover for nearly all winter activities, with dwarf half-shrubs also being used as roosting cover. Sites having Harvard oak (*Quercus harvardii*) are also used during winter, perhaps for acorns, and crop plants such as sunflowers and sorghum are sometimes also important fall and winter food sources. In the Attwater prairie chicken the clumped midgrass cover type is apparently a preferred type throughout the year (Horkel 1979).

Spring Habitat Requirements

The habitat requirements of the lesser prairie chicken for display grounds have been summarized by Copelin (1963). He reported that the males always selected areas with fairly short grass and that the grounds were usually on ridges or other elevations. In sand sagebrush habitat, on the other hand, display grounds were in valleys on short-grass meadows if the sagebrush on adjacent ridges was tall and dense. A variety of studies has indicated that this subspecies favors display areas that are nearly devoid of vegetation and provide excellent visibility. In west Texas 12 of 14 leks were on open areas created by humans, while the other 2 were on slightly elevated terrain where Harvard oak was 10 to 20 centimeters tall (Taylor and Guthery 1980).

Of several hundred Attwater prairie chicken booming grounds, most were on level ground, but they typically consisted of a short-grass flat, about an acre in extent, surrounded by heavier grassy cover (Lehmann 1941). Horkel (1979) reported that, of 24 sites he studied, 20 were in artifically maintained areas such as pipeline rights-of-way or roads.

Ammann (1957) has provided similar observations for the greater prairie chicken in Michigan. He noted that of 65 prairie chicken and 95 sharptail display grounds observed, 47 percent were on elevated sites and only 4 were in depressions. Of 97 Michigan prairie chicken grounds studied in 1941, 27 contained some woody growth other than sweet fern or leather leaf, while of 65 grounds studied since 1950 only 2 contained a sparse stocking of woody cover. Prairie chickens evidently will not tolerate as much woody cover on their booming grounds as will sharp-tailed grouse.

Robel et al. (1970) found that booming grounds in Kansas were associated with clay pan soil types, and the birds remained on these sites for some time after display activities ceased, feeding on succulent green vegetation, especially forbs. With the coming of hot summer weather, the steep limestone hillsides received greater use, probably because of the availability of shade for loafing. Lehmann (1941) likewise reported that heavy shrub cover provides shade for hot summer days, protection against predators and severe weather, and a source of fall food.

Comparing habitat requirements of greater and lesser prairie chickens, Jones (1963) found that both forms preferred level or elevated sites with short grasses. Plant cover differences were not significant, but the greater prairie chickens tolerated somewhat taller vegetation than did the lesser (a mean of 15.1 cm versus 10.4 cm). Anderson (1969) reported that greater prairie chickens preferred grass cover less than 6 inches tall for their booming grounds, the combination of short cover and wide horizons apparently being far more important than specific cover type.

Nesting and Brooding Requirements

Ammann (1957) indicated that of 13 prairie chicken nests found in Michigan, 8 were in hayfields, 1 was in sweet clover, 3 were in wild land openings, and one was on an airport. All the nests were in fairly open situations. Hamerstrom (1939) has similarly reported on 23 prairie chicken nests in Wisconsin. Eleven of these were in grass meadows near drainage ditches, 3 were in dry marshes or marsh edges, 3 were in openings of edges of jack pine–scrub oak woods, 3 were in scattered mixtures of brush, small trees, and grass, 2 were in small openings in light stands of brushy aspen or willow, and 1 was in rather dense mixed hardwoods. Both of these studies indicate the importance of grassy, open habitats for prairie chicken nests. Hamerstrom, Mattson, and Hamerstrom (1957) and Yeatter (1963) have emphasized the importance of mixed natural grasslands or substitutes in the form of redtop (*Agrostis alba*) plantings as nesting and rearing cover types for prairie chickens. Yeatter (1963) correlated a decline in redtop production and prairie chicken populations in Illinois and found that birds nesting in redtop had a nesting success as high or higher than those using pastures, idle fields, or waste grasslands.

Schwartz (1945) also provided information on nest site preferences in greater prairie chickens and noted that, of 57 nest locations, 56 percent were in ungrazed meadows. Half of the remainder were in lightly grazed pastures, while the others were in sweet clover, fencerows, sumac, old cornfields, or barnyard grass. The usual proximity of nests to booming grounds has led Schwartz (1945), Hamerstrom (1939), and Jones (1963) to comment on this relation. However, Robel et al. (1970) found considerable movement between booming grounds by females and questioned whether the location of booming ground has any major influence on female nesting behavior. He found that 19 nest sites average 0.68 mile from display grounds and ranged up to 1.13 miles away. Jones (1963) noted that all of the 9 greater prairie chicken nests he found were near pastures or old fields that had a large number of forbs into which the broods were taken after hatching.

Lehmann (1941) reported that of 19 Attwater prairie chicken nests found 17 were in long-grass prairie, one was in a hay meadow, and one was in a fallow field. All of them were in the previous year's grass growth, and 15 were in well-drained situations, often on or near mounds or ridges. Twelve were near well-marked trails, such as those made by cattle. All the nests were roofed over with grassy vegetation, and most had good to excellent concealment. Copelin (1963) reported on 9 lesser prairie chicken nests in Oklahoma and Kansas. None of these were among shrubs more than 15 inches high, and 7 were between grass clumps, particularly little bluestem (*Andropogon scoparius*). Two were under bunches of sage, and 1 was under tumbleweed. Shin oak shrubs from 12 to 15 inches tall were associated with 5 of the nests.

After hatching, females with broods typically moved to somewhat heavier cover than they used for nesting. Copelin (1963) noted that only 1 brood of lesser prairie chickens was found in the low shinneries of oak, but 27 were seen in oak motts, which are clumps of oak 4 to 20 feet tall in stands up to 100 feet in diameter. Oak motts provide better shade than do oak shinneries. In the absence of oak, the birds moved into cover provided by sagebrush or other bushy plants. Lehmann (1941) likewise found a movement of both young and old Attwater prairie chickens toward cover that provided a combination of shade and water. The importance of free water for prairie grouse is questionable (Ammann 1957), but certainly in moister habitats the availability of succulent plants, insects, and shade all contribute to the value of the area as rearing cover.

Yeatter's (1943, 1963) studies in Illinois indicated that females with newly hatched young feed mainly in redtop fields and to some extent in small grain or grassy fallow fields. They also move along ditch banks and field borders, where there is heavier cover. In Missouri, females take their young to swales that provide cover in the form of slough grass, which gives a combination of shade, protection, and easy movement. As the birds grow older, they gradually move to higher feeding grounds such as grainfields or stubble but still return in the heat of the day to rest in the shade provided by shrubs, large herbs, or trees.

Winter foods of the prairie chicken are virtually all from plant sources (Judd 1905; Schwartz 1945). Judd indicated that the prairie chicken eats only about half as much mast as does the ruffed grouse, the mast consisting mostly of the buds of poplar, elm, pine, apple, and birches. It also consumes some hazelnuts (*Corylus*) and acorns, which it swallows whole. In most parts of the bird's present range, however, grain is much more important than buds as winter food. As I noted earlier, corn and sorghum are major winter foods for the species, with corn more important in northern areas and sorghum increasing in importance farther south.

Korschgen (1962) found that in Missouri corn kernels and sorghum seeds are the primary winter foods, with corn remaining important well into spring. In late spring soybeans (*Glycine*) exceed corn in usage, with the leaves eaten first and later the seeds and seed pods. Sedge (*Carex*) flower heads are also important in the spring diet, as are grass leaves. Two cultivated grasses, oats and wheat, are heavily depended on in summer, first for their leaves and later for their grains. Korean lespedeza (*Lespedeza*) foliage is used almost throughout the year, but especially from July through September. In September ragweed (*Ambrosia*) seeds begin to appear in the diet and are used to a limited extent until February.

Year-round, Judd (1905) reported that animal foods (mostly grasshoppers) constitute about 14 percent and plant foods 86 percent of the greater prairie chicken's diet. Martin, Zim, and Nelson (1951) stated that the animal portion may reach 30 percent during summer but in winter and spring is as little as 1 to 3 percent. Lehmann (1941) found that adults of the Attwater prairie chicken consume about 88 percent plant material and 12 percent insect food, with seeds and seed pods alone constituting more than 50 percent of the materials eaten. In contrast to the high percentage of cultivated grains noted in most studies of the greater prairie chicken, native plants found in lightly grazed pastures provided the major food items listed by Lehmann. These included ruelli (*Ruellia*), stargrass (*Hypoxis*), bedstraw (*Galium*), doveweed (*Croton*), and perennial ragweed (*Ambrosia*) as well as many other less important species.

Jones's study (1963) of the greater and lesser prairie chickens in Oklahoma brought out some striking differences in foods taken in study areas about 250 miles apart. The percentage of insects eaten was much higher for the lesser prairie chicken (41.8 and 48.6 percent average yearly volume in two habitats) that for the greater prairie chicken (8.2 and 20.8 percent average volume in two habitats). The rest of the food of both species consisted of seeds and green vegetation, with the latter usually greater in volume than the former. Both species fed in grassy cover, but whereas the lesser prairie chickens preferred midlength grasses for foraging, the greater were found feeding more frequently in short grasses. Jones also reported (1964*b*) that during the six-month period when plants were important food items, the half-shrub cover type (associated with sandy soils) was used for foraging for five months, and the short-grass cover type (associated with clay soils and used for display purposes) was heavily used only during April. Copelin (1963) reported that the relative use of sorghum in winter was closely related to the amount of snow cover, with large flocks moving to grainfields when snow was about a foot deep for a week or more. When such snow is present, lesser prairie chickens regularly make snow roosts (Jones 1963), suggesting a fairly recent climatic adaptation to the warmer climates typical of the bird's present range.

More recent studies of the lesser prairie chicken in New Mexico tend to confirm the results of Jones's study in Oklahoma (Davis et al. 1979, 1980). In New Mexico insects make up most of the summer diet, with chicks and young juveniles eating almost nothing but insects and even adults using mostly insects for their summer food. Spring foods in New Mexico are mainly green vegetation, especially the catkins of shinnery oaks or their acorns. Shinnery oak alone provides the most heavily utilized food source for the species throughout the year, and it is also the preferred concealment cover for foraging birds of all age-classes during the summer.

MOBILITY AND MOVEMENTS

An early analysis of greater prairie chicken seasonal movements was made by Hamerstrom and Hamerstrom (1949) for the Wisconsin population. They suspected that there was little movement during summer, especially during the brood-rearing period. However, during autumn considerable movement does occur, and some slight migratory movements may exist. Autumn movements of up to 29 miles were established using banded birds, which perhaps correspond to the "fall shuffle" of quail or the general fall dispersion of young birds known for other grouse. Most of the longer movements were by females; 6 of the 8 females recovered had moved at least 3 miles, while 18 of 30 males had moved less than 3 miles.

During winter, prairie chickens typically occur in large packs formed by mergers of the fall packs. In Wisconsin these can consist of 100 to 200 birds, which become progressively less mobile in the most severe weather. During very bad weather the birds move very little and may scarcely leave their winter roosts. Roosting sites in the Hamerstroms' study area were often from ¼ to ½ mile from feeding fields and were seldom more than 1¼ mile away.

By February the winter packs begin to break up and the males start returning to their booming grounds. The Hamerstroms found that most of 56 banded males moved less than 2 miles from their winter feeding grounds to their booming grounds (50 birds), while the remaining males moved from 2 to 8 miles. Apparently many males winter at the feeding sites nearest their booming grounds, and in late winter some daily movements between these locations may occur. During spring males move little; they may roost on their territories or within a few hundred yards of them. Sources of water, shade, dusting places, and loafing sites are often within ½ mile. After display is over, the males may remain close to their booming grounds for much of the summer.

More recent studies of movements of greater prairie chickens have been made by Robel et al. (1970) in Kansas, using radio telemetry. They established monthly ranges for 39 adult males, 37 adult females, and 31 juveniles. Movements of adult males were greatest in February, as the birds began to visit their booming grounds and also had to search somewhat harder for food. Flights of a mile or more between feeding areas and display grounds were sometimes seen, and there was also some movement between display grounds. Immature males, however, exhibited their greatest movements in late February and March, with the later flights largely between display grounds as the birds unsuccessfully attempted to establish territories at various grounds. During April and May both adults and immatures moved less, the birds remaining closely associated with specific booming grounds. Females moved most in April, during the time of peak male display. Females often visited several booming grounds, with movements of up to 4.8 miles recorded. One female that attempted to nest three times was fertilized at a different booming ground before each nesting attempt. Summer movements by both sexes were minimal, as the birds molted and females were rearing broods. However, during fall longer movements again became typical, especially among juveniles. Three juvenile males moved from 2.7 to 6.7 miles during October and November, but comparable data for females are not available. However, daily movements of females during that time averaged farther than those of males (808 yards versus 660 yards).

Monthly movements of the prairie chickens studied by Robel et al. (1970) reflect this seasonal behavior pattern. Summer monthly ranges of adult males were greatest in June (262 acres), fairly small in July (132 acres), and smallest in August (79 acres). In fall and winter the monthly ranges increased from 700 to almost 900 acres from November to February and reached 1,267 acres in March, then decreased sharply and were at a minimum of 91 acres in May. Data for juvenile males indicated a similar monthly mobility pattern for the year. On a daily basis, adult males were most highly mobile in February (with an average daily movement of 1,121 yards), and they decreased their daily mobility through August (320 yards per day). The movements increased again in fall and through the winter averaged from 600 to 700 yards per day until February. During the period of February through September, adult females had average daily movements of from 332 to 928 yards. Juveniles of both sexes had daily movements rather similar to those of adult males, least extensive in August and increasing to a peak in March.

Comparable data for the lesser prairie chicken are not available, but Copelin (1963) does provide some observations on mobility. He also found that movements were most limited in summer and most extensive in winter. The summer range of a female and her brood was estimated to be from 160 to 256 acres, or somewhat less than the estimates of monthly summer mobility in greater prairie chicken females. Of 114 banded birds retrieved, 79 percent were found within 2 miles of their point of capture, and 97.4 percent were within 4 miles. The maximum known movement was 10 miles. In common with the Hamerstroms' study, Copelin found that juveniles often moved considerable distances between their brood ranges and display grounds used the following spring, with all of 14 birds moving at least 0.5 mile and 2 moving nearly 3 miles. Considering birds captured in fall and winter and observed the following spring on display grounds, he found that juvenile birds tended to move farther than adults during this time and that juvenile hens moved farther than juvenile males. Forty juvenile males moved an average distance of 0.93 mile, and 20 adult males moved an average of 0.46 mile; 6 juvenile hens moved an average distance of 2.12 miles, and 1 adult hen moved 3.75 miles.

Lehmann (1941) provided some observations on seasonal movements in the Attwater prairie chicken that in general support the studies already discussed. He noted a summer movement of adults and fairly well grown young from nesting areas into heavier summer cover that provided shade and water, followed by a sedentary state until fall. At this time, from September onward, the birds moved out of some pasturelands and into others that provided winter food and cover. During this time large concentrations of up to 250 to 300 individuals were sometimes seen, in addition to many smaller flocks of 8 or fewer birds. These winter packs broke up late in January, when males began to display.

More recent radio telemetry studies of the Attwater prairie chicken (Horkel 1979) indicate that average monthly ranges of males of this population varied from 28 to 211 hectares, with the largest ranges associated with the fall and winter booming periods and the minimum ranges associated with the nesting and brooding periods. Average monthly ranges of females varied from 35 to 267 hectares, with maximum ranges associated with the winter booming period and minimum ranges associated with the brooding season. Daily movements were estimated to range from 0.12 to 0.56 kilometer for males and from 0.13 to 0.46 kilometer for females.

In a similar radio telemetry study of the greater prairie chicken, Svedarsky (1979) found that females tended to return each spring to the booming ground on which they were trapped the preceding year, and that two females returned to nest within 30 meters of their previous year's nest site. During the first 2 weeks after hatching, broods moved approximately 2,000 meters per week.

Copelin (1963) summarized numbers of male lesser prairie chickens on display grounds in Oklahoma from 1932 to 1951. For a total of 64 grounds studied over varying periods of years, the number of males present averaged 13.7 and reached as high as 43. These grounds occurred on a study area of 16 square miles, and in different years from as few as 8 to as many as 40 display grounds were found there. The average figure of 24 display grounds indicates that good lesser prairie chicken habitat might support about 1.5 active display grounds per square mile. Taylor (1980) found 14 leks in an area of 5,200 hectares (20 square miles), or an average of 0.7 lek per square mile. These leks were separated by an average of 1.2 kilometers (0.74 mile). By comparison, Baker (1953) indicated that 6 greater prairie chicken booming grounds were present on a study area of 3.5 square miles of excellent range in Kansas, or 1.7 grounds per square mile. However, studies such as that of Sisson (1976) in Nebraska suggest considerably lower display ground densities of from 0.04 to a maximum of 0.53 per square mile. In general there thus seems to be a greater scattering of display grounds for the greater prairie chicken, which may in part reflect the effective acoustical distances of the male vocal displays. The lower-pitched booming calls of the greater prairie chicken presumably are effective over greater distances than are the homologous "gobbling" calls of the less prairie chicken, and this might affect spacing of display grounds.

REPRODUCTIVE BEHAVIOR

Territorial Establishment

As in the sharp-tailed grouse, fall establishment of territories and associated display occurs regularly in the pinnated grouse. Copelin (1963) noted that during the fall old male lesser prairie chickens reestablish territories they held during the spring, and although young males visit the booming grounds they are apparently not territorial. In the greater prairie chicken an active period of fall display is likewise usual, at least in Missouri (Schwartz 1945), Michigan (Ammann 1957), and various other states, although Hamerstrom and Hamerstrom (1949) did not regard it as typical in Wisconsin. Whether or not the females regularly visit the grounds during fall is not so important as the fact that territorial boundaries are reestablished by mature and experienced males, and that young males learn the locations of these display grounds. During the following spring some shifting about may occur as winter deaths among the males remove some territory holders, but the basic structure of the booming ground is probably formed during fall display.

The average size of the lek, in terms of participating males, is similar to that in sharp-tailed grouse. Lehmann (1941) indicated that for 5 Attwater prairie chicken grounds studied over a three-year period, the average yearly numbers of participating males ranged from 7.2 to 8.4. Grange (1948) found that on 17 display grounds in Wisconsin in 1942, an average of 6.9 males were present. In Nebraska an average of about 9 male prairie chickens is typical of booming grounds (Johnsgard and Wood 1968). Generally similar figures have been indicated for Missouri (Schwartz 1945) and Illinois (Yeatter 1943). The largest reported booming grounds were those noted by Baker (1953) for Kansas; he observed one ground containing approximately 100 males.

Of 610 booming grounds observed in Wisconsin between 1950 and 1971, 36 percent had between 1 and 5 birds, 34 percent between 5 and 10 birds, 19 percent between 10 and 15 birds, 8 percent between 15 and 20 birds, 2 percent between 20 and 25 birds, and 1 percent more than 25 birds (Hamerstrom and Hamerstrom 1973). Before experimentally manipulating lek composition, Ballard and Robel (1974) found that the average size of 13 territories was 152 square meters.

Male Display Behavior

Since the basic sexual and agonistic behavioral patterns of the greater, lesser, and Attwater prairie chickens are virtually alike, a single description of motor patterns will be given, with comments on any differences, based on Sharpe's comparative analysis of the three forms (1968).

334

30. Male displays of pinnated grouse (after various sources), including booming by (A) greater prairie chicken and (B) lesser prairie chicken, (C) flutter-jumping of greater prairie chicken, (D) bowing by greater prairie chicken, and (E) fighting by greater prairie chickens.

Booming is the collective term given to the sequence of vocalizations and posturing that greater prairie chicken males use both to announce territorial residence to other males and to attract females. During booming, the tail is elevated, the pinnae are variably raised until they may be almost parallel with the ground, the wings are lowered while held close to the body, and the primaries are somewhat spread. The bird then begins a series of foot stamping movements (about twenty per second according to Hjorth 1970), during which he moves forward a relatively short distance, then he snaps his tail in three rapid fanning movements. At the same time as the tail is initially clicked open and shut, a three-syllable vocalization ("tooting" of Hjorth 1970) begins, lasting almost 2 seconds and sounding like *whoom-ah-oom*, with the middle note of reduced amplitude. During the second note a rapid and partial tail fanning also occurs, and the "air sacs" are partially deflated. During the third note the esophageal tube is again inflated, and the lateral apteria or "air sacs" are maximally exposed. Simultaneously, the tail is rather slowly fanned open and again closed. Sharpe (1968) indicated that in the lesser prairie chicken a single exaggerated tail spreading movement occurs during the first phase of booming and the later tail-spreading elements are lacking. He estimated that maximum amplitude of the fundamental harmonic during booming at about 300 cycles per second (Hz) in the greater and Attwater prairie chicken and about 750 Hz in the lesser prairie chicken. In addition, the vocalization phase of the lesser prairie chicken lasts about 0.6 second, as opposed to nearly 2 seconds in the greater. The associated call ("yodeling" of Hjorth 1970) sounds more like a "gobble" and has two definite syllables plus a terminal humming sound. However, "low-intensity" booming may have up to four syllables. Hjorth (1970) has distinguished a variant of the lesser prairie chicken's gobbling call that he called "bubbling," but it appears to be an incomplete and less stereotyped version of the more typical call and posture and probably corresponds to Sharpe's "low intensity booming." In contrast to the greater prairie chicken, male lesser prairie chickens frequently utter their booming displays antiphonally ("duetting" of Hjorth 1970), performing up to ten displays in fairly rapid sequence. An additional visual difference between the displays of the two forms is that the exposed gular sac of the lesser prairie chicken is mostly red, whereas those of the greater and Attwater prairie chickens are yellow to orange (Jones 1964a; Lehmann 1941).

A second major display of prairie chickens is flutter-jumping. It is performed in the same fashion by this group as by sharp-tailed grouse and no doubt serves a similar advertisement function. Unlike those of the sharptail, however, most prairie chicken flutter-jumps have associated cackling calls ("jump-cackle" of Hjorth 1970). Sharpe (1968) found that calls occurred during 27 of 30 flutter-jumps in Attwater prairie chickens, 16 of 20 in lesser prairie chickens, and 17 of 20 in greater prairie chickens. He noted that flutter-jumping is especially typical of peripheral males when hens are present near the middle of the display ground.

When defending territories against other males, males typically use several display postures and calls. Ritualized and actual fighting, such as Lumsden (1965) described for the sharp-tailed grouse, is commonly seen, often including short jumps into the air and striking with the feet, beak, and wings. Between active fights the males will commonly "face off," lying prone a foot or two apart and calling aggressively. Associated calls during facing off include a whining call much like that of sharptails, and a similar, more nasal "quarreling" note (Sharpe 1968) that sounds like *nyah-ah-ah-ah*. Grange (1948) describes the "fight call" as a very loud, raucous *hoo'-wuk*. Apparent displacement sleeping, displacement feeding, and "running parallel" displays have also been noted by Sharpe at territorial boundaries. A white shoulder spot is often evident in such situations, and Hjorth (1970) noted that in both sexes of lesser prairie chickens this may frequently be observed.

When a female enters a male's territory, his behavior changes greatly. He performs booming with high frequency as well as extreme posturing, particularly pinnae erection and eyecomb enlargement. The eyecombs of all three forms are a bright yellow, but those of the lesser prairie chicken are relatively larger than those of either the greater or the Attwater prairie chicken. Between booming displays, the male will sometimes stop and "pose" facing the female, but most booming displays are not oriented specifically toward the hen. Rather, the male circles her, and all aspects of his plumage are visible to her.

In the presence of females, either nearby or at some distance, a characteristic *pwoik* call ("whoop" of Hjorth 1970) is frequently uttered (Lehmann 1941). Sharpe reports that this call is very similar in both the greater and the Attwater prairie chickens, but in the lesser it is higher pitched and sounds like *pike* ("squeak" of Hjorth 1970). It is shorter (0.23 second compared with about 0.4 second in the larger forms), and the greatest sound amplitude occurs at about 1,000 Hz rather than 550 to 600 Hz.

All three forms of prairie chickens perform the "nuptial bow" ("prostrate" of Hjorth 1970), which Hamerstrom and Hamerstrom (1960) originally described for the greater prairie chicken. They regarded it as a sexual display that often precedes copulation and yet is not a prerequisite for it. Sharpe (1968) found that the same applies to the Attwater and lesser prairie chickens, and in all three the display has the same form. The male, while actively booming and circling about a nearby female, suddenly stops, spreads his wings, and lowers his bill almost to the ground while keeping his pinnae erect. He may remain in this posture for several seconds as he faces the female.

When females are ready for copulation they squat in the typical galliform manner, with wings slightly spread, head raised, and neck outstretched. When mounting, the male grasps the female's nape, lowers his wings on both sides of her, and quickly completes copulation. After copulation, females usually quickly run forward a few feet, then stop to shake. Males lack any specific postcopulatory displays and often begin booming again within a few seconds.

Ballard and Robel (1974) found that, of 132 attempted copulations they observed before manipulating lek composition, 92 percent were successful, and 89 percent of these were performed by dominant (alpha or beta) males. After the dominant males had been eliminated, only 13 percent of 39 attempted copulations were successful. These were performed by males of lower social status that had moved into the center of the lek after the dominant birds were removed. Not only did removing these dominant males disrupt the social organization of the lek during the year when it was done, but similar effects carried over to the following spring. Thus, during the next year there were fewer males on the lek, fewer females visited it, and fewer successful copulations were observed (Robel and Ballard 1974). Not only did these authors confirm a well-developed and experience-related social hierarchy among males, but they also noted that females varied in dominance status as well. One undesirable result of such dominance behavior among females may have been to delay fertilization, and thus the onset of nesting, in some low-ranking females, thereby affecting nesting success rates.

Although Robel and Ballard thus effectively proved the importance of social structuring and the role of experienced males in regulating reproductive efficiency, it should be noted that Hamerstrom and Hamerstrom (1973) reported that 18 percent of 2,264 copulations were performed by first-year males, and that these young birds were apparently as successful in their copulation efforts as were adults (84 and 76 percent respectively).

Vocal Signals

In addition to the booming, whining, quarreling, and *pwoik* calls already mentioned, pinnated grouse have several other vocal signals, including many cackling sounds. Sharpe (1968) recognized a "long cackle" that consists of several individual notes spaced about 0.2 second apart and sometimes lasting several seconds. The notes uttered during flutter-jumping are essentially the same as these individual long cackle sounds. Lehmann (1941) has listed several variants of these cackling calls and combinations of *pwoik* and cackling notes, and he also mentions several other notes. These include calls sounding like *kwiee, kwerr, kliee, kwoo,* and *kwah.* In the absence of comparative study and analysis, their possible functions cannot be guessed. Hjorth (1970) has noted that between bouts of flutter-jumping or booming the male often utters an indefinite staccato cackle, and during territorial confrontations he may produce cackling sounds that range from whinnies to whining cackles and explosive cackles. Sparling (1983) has provided a detailed recent sonagraphic analysis of the vocalizations of greater prairie chickens, sharp-tailed grouse, and their hybrids.

Nesting and Brooding Behavior

After mating, the female almost immediately begins to lay a clutch; indeed, it is probable that she has already established a nest scrape before successful copulation. She may move a considerable distance away from the display ground to her nest site and may actually nest nearer to another booming ground than to that at which copulation occurred (Robel et al. 1970). Robel et al. found that females had to visit a ground for an average of 3 consecutive days before copulation occurred but did not return thereafter except perhaps for renesting attempts. Lehmann (1941) and Robel et al. found that renesting birds laid progressively smaller clutches, and sometimes two such attempts were made. The average clutch size of first clutches is about 12 to 14 eggs for the lesser (Copelin 1963), Attwater (Lehmann 1941), and greater prairie chickens (Hamerstrom 1939; Robel et al. 1970). Later clutches, probably the result of renesting, often have only 7 to 10 eggs. Svedarsky (1979) found an average clutch size of 14.6 eggs for initial nests, and an egg-laying rate of 1 per day, for greater prairie chickens in Minnesota. He determined that egg laying began an average of 3.8 days after copulation, and that six females began second nesting efforts an average of 6.4 days after losing their initial clutches.

Nest sites that provide dense, vertically oriented cover may be important, as are habitats that have remained undisturbed for a year or longer (Svedarsky 1979). Incubation may begin the day before or several days after the last egg is laid, according to Lehmann (1941). Apart from two feeding and resting periods in early morning and late afternoon, the female incubates constantly. The incubation period is probably 23 to 26 days in all three forms (Lehmann 1941; Schwartz 1945; Coats 1955; Svedarsky 1979). Svedarsky found an average period of 25.5 days, with slightly longer periods typical of early nests.

Judging from the work of Ballard and Robel (1974), nests that are initiated later in the season (May 5 or later in their study) not only have lower clutch sizes than earlier clutches (10.2 vs. 12.1 eggs), but also have a lower probability of successful hatching (10 percent vs. 44 percent). Somewhat higher nesting success rates were reported by Sisson (1976) for Nebraska (50 percent of 29 nests) and by Svedarsky (1979) for Minnesota (62.4 percent of 36 nests).

Pipping may require up to 48 hours, during which the female appears highly nervous and the nest apparently is extremely vulnerable because of the noises made by the chicks and the odors of the nest (Lehmann 1941). Normally the nest is deserted within 24 hours after the last chick is out of the shell. Females with young chicks typically perform decoying behavior with heads held low and wings drooping and nearly touching the ground, uttering a low *kwerr, kwerr, kwerr* (Lehmann 1941). After the young are able to fly well, both the hen and the brood typically flush when disturbed.

Chicks less than a week old may be brooded much of the time, possibly up to half the daylight hours (Lehmann 1941). However, older chicks are brooded only at night, during early morning hours, and in inclement weather. Broods typically remain with females for 6 to 8 weeks, after which families gradually disintegrate. There is also considerable brood mixing, as when separated chicks join the broods of other females, even if the young are of different ages.

EVOLUTIONARY RELATIONSHIPS

The close and clearly congeneric relation of the pinnated grouse to the sharptail has already been mentioned in the account of that species. Thus, comments here will be restricted to the relation among the four forms of pinnated grouse. Short (1967) has already dealt extensively with the criteria advanced by Jones (1964a) for considering the lesser prairie chicken specifically distinct from the greater prairie chicken. Since then, Sharpe (1968) has found some male behavioral differences between the lesser prairie chicken and the two surviving races of *cupido*. These consist of acoustic differences (higher frequencies in the lesser), time differences (more rapid and shorter displays in the lesser), and some motor differences (one versus two tail movements during booming in the lesser). A few other contextual and orientational differences were also found, but Sharpe admitted that these may be attributed largely to size differences in the birds and possible selection related to aggressive behavior patterns rather than being the result of reinforcement for species differences during some past period of sympatry. He concluded that the lesser should be considered an ''allospecies'' to emphasize that it is more unlike *T. c. pinnatus* than is *T. c. attwateri*. This may well be the most effective way of handling questionable allopatric populations, but it is not used elsewhere in this book and has not been generally adopted.

It appears that the living forms of pinnated grouse and those that have recently become extinct were all derived from some ancestral grouse associated with deciduous forest or its edge, since the original ranges of the lesser and greater prairie chickens as well as the extinct heath hen all had affinities with oak woodlands or oak-grassland combinations. The Attwater prairie chicken, on the other hand, is apparently associated with pure grassland vegetation. The separation of the ancestral stock of the lesser prairie chicken probably occurred during an early glacial period, and subsequent adaptation during postglacial times to an unusually warm and dry grassland habitat in the southwestern states has accounted for its smaller size and generally lighter coloration. More recent separation of gene pools no doubt brought about the separation of the East Coast (heath hen) and Gulf Coast (Attwater) populations from the interior form, but the behavioral and morphological differences among these are minimal.

Sharp-tailed Grouse

Tympanuchus phasianellus (Linnaeus) 1858
(*Pedioecetes phasianellus* in A.O.U. *Check-list,* 1957)

Other Vernacular Names

Brush grouse, pintail grouse, prairie grouse, prairie pheasant, sharptail, speckle-belly, spiketail, sprigtail, white-belly, white-breasted grouse; cupidon phasanelle, gelinotte à queue fine (French); Schwief-Waldhuhn (German).

Range

Currently from north-central Alaska, Yukon, northern Mackenzie, northern Manitoba, northern Ontario, and central Quebec south to eastern Washington, northeastern Utah, Wyoming, and Colorado, and in the Great Plains from eastern Colorado and eastern Wyoming across Nebraska, the Dakotas, northern Minnesota, northern Wisconsin, and northern Michigan.

Subspecies

T. p. phasianellus (Linnaeus): Northern sharp-tailed grouse. Breeds in northern Manitoba, northern Ontario, and central Quebec. Partially migratory.

T. p. kennicotti (Suckley): Northwestern sharp-tailed grouse. Resident in Mackenzie from the Mackenzie River to Great Slave Lake.

T. p. caurus (Friedmann): Alaska sharp-tailed grouse. Resident in north-central Alaska east to the southern Yukon, northern British Columbia, and northern Alberta.

T. p. columbianus (Ord): Columbian sharp-tailed grouse. Resident from north-central British Columbia and western Montana south to eastern Washington, northern Utah, and western Colorado. Formerly extended to Oregon, Nevada, and perhaps New Mexico, although the racial identification of the now-extirpated New Mexican population is uncertain and may have been *jamesi* (Miller and Graul 1980).

T. p. campestris (Ridgway): Prairie sharp-tailed grouse. Resident from southeastern Manitoba, southwestern Ontario, and the Upper Peninsula of Michigan to northern Minnesota and northern Wisconsin. Formerly extended to northern Illinois.

T. p. jamesi (Lincoln): Plains sharp-tailed grouse. Resident from north-central Alberta and central Saskatchewan south to Montana (except the extreme west), northeastern Wyoming, northeastern Colorado, and western portions of Nebraska, South Dakota, and North Dakota. Formerly extended to Kansas and Oklahoma, and perhaps New Mexico.

MEASUREMENTS

Folded wing: Adult males 194–223 mm; adult females 186–221 mm (males of all races average 202 mm or more; females, 201 mm or less).

Tail: Adult males 110–35 mm; adult females 92–126 mm (males average 4 mm longer than females).

IDENTIFICATION

Adults, 16.4–18.5 inches long. The sexes are nearly identical in plumage. The tail is strongly graduated in both sexes, with the central pair of feathers extending far beyond the others, but the tips are not pointed. Both sexes are feathered to the base of the toes, and males have an inconspicuous yellow comb (somewhat enlarged during display) and pinkish to pale violet areas of bare neck skin that are also expanded during display, though not to the degree found in prairie chickens. Both sexes have inconspicuous crests, and the head and upperparts are extensively patterned with barring and spotting of white, buffy, tawny brown, and blackish. White spotting is conspicuous on the wings, and the relative amount of white increases toward the breast and abdomen, which are immaculate. The middle pair of tail feathers is elaborately patterned with brown and black, but the others are mostly white. The breast and flanks are intricately marked with V-shaped brown markings on a white or buffy background.

FIELD MARKS

The grassland, edge, or scrub forest habitat of this species varies considerably throughout its range, but the bird is basically to be found in fairly open country, where its pale, mottled plumage blends well with the surroundings. In flight the white underparts are conspicuous, as is the whitish and elongated tail. On the ground the birds have a much more "frosty" appearance than do prairie chickens, which are generally darker and lack definite white spotting.

AGE AND SEX CRITERIA

Females may be identified with about 90 percent reliability by a transverse barring pattern on the central tail feathers, compared with the more linear markings of males. Also, the crown feathers of females have alternating buff and dark brown crossbars, whereas the male crown feathers are dark with buffy edging (Henderson et al. 1967).

Immatures are identified by the usual character of pointed outer primaries. Ammann (1944) suggested that a comparison of relative amounts of wear on the eighth and ninth primaries (equal or little wear on both in adults, greater wear on the ninth in immatures) is the most suitable method of judging age in prairie grouse.

Juveniles have white rather than buffy throats and have shorter median tail feathers than do adults. The lateral tail feathers of juveniles are more buffy, mottled and speckled with brown, while the median two pairs have broad, buffy central stripes (Ridgway and Friedmann 1946). White shaft streaks are conspicuous on the upperparts as well. Pepper (1972) has provided a method for estimating the age of juveniles by weekly categories to 10 weeks of age by the lengths of the seventh and ninth juvenal primaries, and from weeks 10 to 16 by the seventh and eighth postjuvenal primaries.

Downy young of the sharp-tailed grouse are a clearer and paler mustard yellow overall than are prairie chickens of the same age and lack the rusty tints of that species. There is the trace of a median black crown line and a few small crown spots, but only one or two black spots between the eyes and the ear region are present.

DISTRIBUTION AND HABITAT

This species and the pinnated grouse together constitute the "prairie grouse" of North America. Such a designation for the sharp-tailed grouse is not wholly accurate, for its original distribution included not only grassland habitats but also sagebrush semidesert, brushy mountain subclimax communities, oak savannas and successional stages of deciduous and mixed deciduous-coniferous forests of the eastern states (*T. p. campestris*), and brushy habitats of boreal forests from Canada through Alaska (*phasianellus, caurus,* and *kennicotti*), as summarized by Aldrich (1963).

Three of the races have suffered considerably from habitat changes associated with man's activities. One of these is the Columbian sharp-tailed grouse, which has been reduced in a remnant distribution pattern to the point that by 1960 it was wholly eliminated from California, virtually gone from New Mexico, rare in Utah, Nevada, and Oregon, uncommon to rare in Oregon and Washington, and generally uncommon in Colorado, Wyoming, and Montana (Hamerstrom and Hamerstrom 1961).

A recent (1980) review by Miller and Graul allows for updating the status of the Columbian sharptail. It evidently was extirpated from Nevada by 1952 and from Oregon in 1968 or 1969. In Idaho, Montana, Utah, and Wyoming the birds now occupy less than

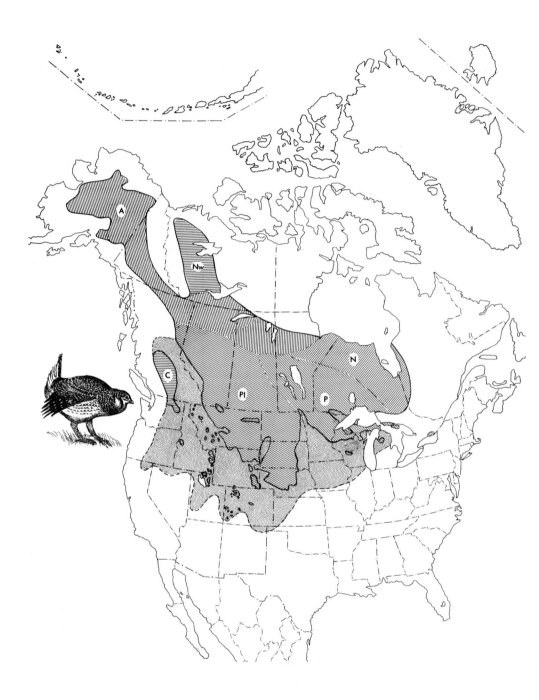

15. Current distribution of the Alaska (A), Columbian (C), northern (N), northwestern (Nw), plains (Pl), and prairie (P) races of the sharp-tailed grouse. The known historic range is indicated by stippling.

10 percent of their original range; they occupy from 10 to 50 percent of it in Colorado and Washington and at least 80 percent in British Columbia. Current populations were placed as 60,000 to 170,000, with 60 to 80 percent of these in British Columbia. Much of the prime habitat types (sagebrush steppe and fescue-wheatgrass) has been eliminated, primarily by conversion to cropland. The birds are still legally hunted in Colorado, Idaho, Utah, Washington, and British Columbia, and 13,000 to 15,000 are taken annually.

The prairie race of sharp-tailed grouse has similarly been extirpated from Illinois, Iowa, and southern portions of Wisconsin and Minnesota and is in danger of extirpation in the northern parts of these states (Hamerstrom and Hamerstrom 1961). In the Lower Peninsula of Michigan, introduced sharptails probably reached their greatest distribution by 1950 (Ammann 1957), and by the early 1960s only a few hundred birds could be counted on display grounds (Ammann 1963a). In the Upper Peninsula the sharptail population had decreased at least 9 percent between 1956 and the early 1960s, primarily through habitat losses (Ammann 1963a). In Minnesota the general population trend appears to be downward, as a result of changing farming practices as well as increased reforestation and tree-farming (Bremer 1967). Hamerstrom and Hamerstrom (1961) reported that the Wisconsin population is in greater danger than those in Minnesota and Michigan as a result of fire protection, forest succession, pine plantations, and modern farm practices. The Canadian populations of this race in Ontario, Manitoba, and eastern Saskatchewan then appeared to be in relativley good condition.

Miller and Graul's (1980) review indicates that the prairie sharptail has retracted to less than 10 percent of its historical range in Michigan and Wisconsin, to 30 percent in Minnesota, and to 50 to 90 percent in Manitoba and Saskatchewan. Habitat losses, especially of preferred oak-savanna habitats in Wisconsin and Minnesota, are evidently primarily responsible for these changes. The total population of this race probably numbers between 600,000 and 2,000,000 birds, exclusive of Ontario and Saskatchewan, but Michigan and Wisconsin probably have no more than 5,000 collectively. The birds are currently hunted in three states and three provinces, and the total annual harvest is perhaps 60,000 to 140,000, exclusive of Ontario and Saskatchewan.

The plains sharptail, with its extensive range from northern Alberta to North Dakota and central Colorado, apparently has suffered the least of the United States races, and it still maintains legally hunted populations in four provinces and five states. However, it is gone from northwestern Oklahoma and western Kansas, and its range in eastern Colorado has shrunk appreciably (Johnsgard and Wood 1968). It probably disappeared from its tiny range in northeastern New Mexico in the early 1960s. It now occupies less than 10 percent of its former range in Colorado, from 10 to 50 percent in North Dakota and Wyoming, and from 50 to 90 percent in Montana, Nebraska, Saskatchewan, and South Dakota. Intensive grazing and conversion of range to cropland are evidently major causes of this contraction. Several habitat types, including wheatgrass-needlegrass, grama-needlegrass-wheatgrass, sand sage–bluestem, and Nebraska sandhills prairie, have all decreased substantially in recent decades. The total population of this race may consist of 600,000 to 3,000,000 birds, exclusive of peripheral populations in Manitoba and British Columbia. The birds are hunted in all the states and provinces where they still occur, except for Colorado, where they are considered endangered (Miller and Graul 1980).

The remaining, predominantly Canadian, populations of sharp-tailed grouse are evidently in relatively satisfactory condition.

HABITAT REQUIREMENTS

General habitat characteristics of the prairie race of sharp-tailed grouse have been analyzed by Grange (1948) for Wisconsin and by Ammann (1957) for Michigan. Grange concluded that sharp-tailed grouse are abundant in areas 25 to 50 percent covered by wooded vegetation, and Ammann believed that 20 to 40 percent woody cover is ideal, preferably with the trees in clumps rather than widely scattered. Sparse or bare patches in the ground cover should not exceed half of the total, and the area of suitable open habitat in wooded vegetation should not be less than 1 square mile, in Ammann's opinion. According to him, ideal summer sharptail habitat on a square-mile unit should include an open portion of about 6 percent of the total area that would provide a display site, loafing and foraging habitat for adult males and broods, and roosting sites for displaying males. About half of the area should consist of scattered large shrubs and trees, especially aspens. Heavy ground cover serves for resting, dusting, and feeding, especially by broods. The remaining 44 percent of the cover should consist of an alternating series of small (10-acre) brushy clearings and heavier second-growth timber stands of mixed hardwoods and conifers, which serve as a source of winter browse and protection from severe weather as well as escape cover. The scattered small clearings provide additional nesting and brood-rearing habitat and winter roosting opportunities. Paper birch (*Betula papyrifera*) and aspen (*Populus tremuloides*), especially the former, are major winter food sources when snow cover prevents foraging on grains or similar foods.

Although these habitat needs may apply to the prairie sharp-tailed grouse, they clearly are not strongly applicable to the Columbian and plains races, which occur in simidesert scrub and relatively dry grasslands, respectively. For the Columbian race at least, shrubs and small trees are important habitat components only during late fall and winter, whereas during the rest of the year weed-grass cover types as well as cultivated crops such as wheat and alfalfa provide important food and cover (Marshall and Jensen 1937). Likewise, Hart, Lee, and Low (1952) list a variety of grasses and herbs as important components of Columbian sharptail habitat in Utah. Similarly, the plains sharp-tailed grouse inhabiting the sandhills of central Nebraska and the comparable sand dune areas of north-central North Dakota are relatively independent of extensive tree cover (Aldous 1943; Kobriger 1965). In the late fall and winter these birds resort to foraging on rose hips and willow buds in the North Dakota sandhills (Aldous 1943), and in Utah the buds of maples and chokecherries are major winter foods (Marshall and Jensen 1937). According to Edminster (1954), a minimum of 5 percent brush cover to total land surface is tolerable to sharptails in North Dakota.

Wintering Requirements

Grange (1948) reported that sharptails do not roost overnight in trees during winter; instead, they utilize snow burrows they scratch out among fairly dense marsh or swamp vegetation or sometimes in open stands of tamarack or spruce in northern Wisconsin. During snowless periods they usually roost in dense and fairly coarse marshy vegetation.

Ammann's observations (1957) for Michigan sharptails are similar. During fall the birds concentrate in "packs" on grain plantings near their summer habitat, and they may continue to use grain as long as it remains available. When the snow is deep and grain becomes unavailable, they prefer the catkins, twigs, and buds of trees such as paper birch, aspen, Juneberry, hazel, and bog birch, as well as the fruit of mountain ash, sumac, common juniper, rose, and black chokecherry. Of all these, the buds and catkins of birch and aspen are most important, particularly birch. A wide variety of grains are taken if they are available, including wheat, buckwheat, field peas, corn, barley, soybeans, millet, and rye. Thus the availability of grain or native food sources in the form of fruiting shrubs or deciduous trees is an important component of winter habitat.

Adequate snow during unusually severe weather conditions may be important to sharptails. Marshall and Jensen (1937) found that movement to maple-chokecherry cover in Utah was related to snow depth; there the birds could feed on buds and roost under the snow unless it crusted heavily, when they preferred to roost above the snow in brushy cover. Some deaths by freezing have been reported when strong winds were combined with low winter temperatures and no snow was available for roosting (Edminster 1954).

Spring Habitat Requirements

Ammann (1957) reported on the general cover characteristics of 95 sharptail dancing grounds in Michigan. Of these, 27 were on cultivated lands and 68 were on wild lands. Although most contained no woody cover, 35 percent had such cover, but rarely did it exceed 30 percent of the surface area. Favored sites for both sharptails and pinnated grouse appeared to have low, mottled, or sparse vegetation with good visibility, allowing for good footing and unrestricted movement. Elevated, rather than level or depressed, sites were preferred for both species; of 65 pinnated grouse and 95 sharptail display grounds, 47 percent were elevated and only 4 were in depressions.

In Wisconsin, Grange (1948) found that wild hay meadows and marshes were frequent display locations for pinnated grouse and sharptails, with sharptails apparently having greater preference than pinnated grouse for wet marshes. A variety of other cover types were also found to be used by both species, including abandoned fields, cultivated fields, and, less commonly, upland grassland, peat burns, and clover fields.

In Alberta, Rippin (1970) noted that of 36 display grounds he studied, 32 were on open, dry, and elevated sites, 3 were on level ground, and 1 was on an elevation with heavy shrub cover. In the Nebraska sandhills Kobriger (1965) found that three-fourths of all prairie grouse display grounds studied were on wet, mowed sites. Similarly, Sisson (1970) reported that 26 of 36 sharptail dancing grounds in the Nebraska sandhills were within ⅛ mile of a windmill, where the vegetation was fairly low as a result of grazing and trampling by cattle, and where visibility was good in all directions.

At least in Nebraska, not much long-term use of areas for display is evident. About half of 161 display grounds used by sharp-tailed and pinnated grouse had also been active the previous year, though several grounds were found to have been active every year for a twelve-year period (Sisson 1976). About a fifth of the display grounds studied in South Dakota remained active for at least ten years (Hillman and Jackson 1973). Changes in land use appear to be a major factor affecting the "lifetime" of a display ground. Changes that result in taller and denser vegetation may cause areas to be abandoned within two to five years (Sisson 1976).

Nesting and Brooding Habitat Requirements

Ammann (1957) has provided a fairly detailed analysis of nesting requirements for sharptails in Michigan. He reported that they choose a wider variety of sites with respect to woody cover than do pinnated grouse, site conditions varying from open to 75 percent shaded. Most nests either were protected by overhead cover or were within a few feet of such cover. Of 29 nests found, none was more than 10 feet from brushy or woody cover. Of 10 nests studied, 6 were in open aspen, 3 were in cutover pines, and 1 was in an open marsh. These sites averaged 43 percent shrub cover, from 3 to 6 feet high, and 4 percent tree cover more than 6 feet tall. Associated shrubs were chokecherry, willow, and alder, and associated trees were aspen, spruce, and Juneberry. Of seven additional nests, 4 were at the base of small trees or bushes, and there were one each in a hayfield, on an aspen-birch ridge, and in a heavy grass–sweet fern site.

Hamerstrom (1939) reported on cover sites for 17 sharptail nests in Wisconsin. Of these, 8 were at the edges of marshes, brush, or woods in brushy or woody (aspen, willow, etc.) cover. Three were in small openings of dense brush such as aspen or willow, 2 were in openings or edges of jack pine–scrub oak woods, 2 were in grass meadows, 1 was in a dry marsh, and 1 was in a mixture of scattered brush, trees, and grass. In this study as well as Ammann's, it appears that birds avoid nesting in cultivated areas.

Nesting preferences of the plains race of sharptail may be somewhat different from these. Of 78 nests found in North Dakota, 62 were in rolling grassland, 11 were in lowland draws, and most were more than 164 feet from woody cover (Kobriger 1980). Females select taller vegetation within a pasture for nest sites; plant height is evidently more important than species. However, brushy or woody areas apparently are used only when grassland quality is poor.

Since the males do not participate in nesting, they gradually move away from their display grounds to foraging and daytime resting sites that usually include brushy cover, aspen or willow thickets, or young conifer stands. In Utah summer daytime resting places gradually change from weeds and grass during June and early July to shrubs and bushes in late July and August (Hart, Lee, and Low 1952). For night roosting fairly open and upland cover with good ground cover is preferred by sharptails over marsh and bog vegetation (Ammann 1957).

Brooding habitat requirements have been analyzed by Hamerstrom (1963) in the Wisconsin pine barrens and by Ammann (1957) for Michigan. Ammann concluded that the birds tend to favor somewhat more woody cover than that chosen for nest sites but in general remain in areas that do not exceed 50 percent shading by woody cover. Peterle (cited by Ammann) estimated a higher (70 percent) average overall shading by woody cover, with shrubs covering 43 percent of the area and trees covering an average of 70 percent in locations where 15 broods were observed.

Hamerstrom's observations on about 190 broods confirm the importance of openings in forested areas as brood habitat. Of his brood habitat records, about 80 percent were in open situations, 14 percent were in edge situations, and only 5 percent were more than 50 yards inside woody habitats. He concluded that brood cover should be basically grassland, with some shrubs and trees, but the taller the woody plants present, the fewer there should be. Shrubs are more important than trees, since they provide not only cover but also food for chicks. Thus, berry-producing species such as blueberries, cherries, and Juneberries are valuable, as are catkin-bearing shrubs that are a source of winter foods. Aspens and willows, though valuable as sources of winter buds, are most useful in small thickets and young trees. Hamerstrom stressed the importance of distinguishing the open, predominantly herbaceous brooding habitat from the fall and winter woody cover that is also critical to sharptail survival.

Other studies of brood habitat characteristics generally suggest that favored brooding sites are those that are relatively dense, associated with a mixture of shrubs and forbs (Sisson 1976). Artmann (1970) reported that early in the brooding period brushland and grassland habitats are used, whereas later the birds use grassland and agricultural areas. Schiller (1973) likewise reported that denser brush was used during early brood stages than later on.

POPULATION DENSITY

Some of the best figures on spring population densities for sharp-tailed grouse come from the work of Grange (1948). Using spring dancing ground counts and assuming a 55 percent ratio of males in the total populations, he calculated an estimate of 235.2 acres per bird on 130,560 acres for 1941 and 186.7 acres per bird on the same areas in 1942. Considering only the occupied range in 1942, the average area per bird was calculated to be 138 acres. Ammann (cited by Edminster 1954) reported spring densities on 13 square miles of habitat on Drummond Island, Michigan, for a three-year period as averaging 1 bird per 45 acres, and the fall population of sharptails on the island was approximately 1 bird per 18 acres of occupied range over a seven-year period. This island represents prime Michigan sharptail habitat, and these figures must be regarded as reflecting unusually high densities that have not been maintained recently. Edminster (1954) summarizes a variety of other fall density estimates from various states that in general indicate that 27 to 125 acres per bird in summer or fall is probably typical. One other high-density figure has been reported for Saskatchewan, with Symington and Harper (1957) estimating late summer populations of between 25 and 40 birds per square mile (16 to 25.6 acres per bird) in the Sandhills area, where an ideal combination of native grasses, shrubs, and small trees occurs.

In the Nebraska sandhills, similar late-summer densities of sharp-tailed grouse have been estimated, with yearly averages ranging from 19 to 70 birds per square mile, of which about half are young birds. In the same area and during the same period, the density of displaying males per square mile ranged from 3.4 to 5.6 (Sisson 1976). The rather wide discrepancy in these numbers might cause one to question their value. Hillman and Jackson (1973) concluded that spring display ground counts provided more reliable indexes of population changes than did summer brood counts. However, Sisson (1976) doubted whether the average number of males on a lek necessarily reflected population levels, since after reaching a certain size leks cease to grow and new leks are formed.

FOOD AND FORAGING BEHAVIOR

Dependable and nutritious winter food sources are critical to the survival of all grouse, and the sharptail appears to be somewhat flexible in its winter diet compared with other grouse species. In central Wisconsin, paper birch (*Betula papyrifera*) buds and catkins are the primary winter foods, with aspen (*Populus tremuloides*) of secondary importance. Among shrubs, rose (*Rosa*) hips and hazel (*Corylus*) buds and catkins are important (Grange 1948). In Ontario the paper birch is also the primary winter food, supplemented by browse of willow, aspen, blueberry, and mountain ash (Snyder 1935). In North Dakota willow buds are the most important single winter food, but chokecherry, popular, and rose hips are also major supplementary species (Aldous 1943). During periods of heavy snow in Utah, sharptails move into thickets of maple, chokecherry, and serviceberry, where they feed on buds. In the Nebraska sandhills the sharp-tailed grouse appears to be more efficient than the pinnated grouse in finding winter foods and surviving the severe weather conditions, and it is much more common and more extensively distributed through that region (Kobriger 1965; Johnsgard and Wood 1968).

Throughout the range of the species, the percentage of woody mast foods sharply decreases in spring as herbaceous plants become available after periods of thawing. Such plants include cultivated grain species, clover, alfalfa, and native annuals and perennials. Jones (1966) found that during the spring and summer green materials composed the bulk of the diet in Washington, with grass blades alone (especially *Poa secunda*) totaling half of the spring diet and three-fourths of the summer diet. Flower parts made up the rest of the spring and summer foods, particularly those of dandelion (*Taraxacum*) and buttercup (*Ranunculus*). The importance of dandelion continued into fall, when its seeds and grass leaves were the leading foods. Apparently the sharptail relies less on animal foods during the summer than does the pinnated grouse (Jones 1966), although Grange (1948) reported that grasshoppers are a major summer food and Edminster (1954) estimated that from 10 to 20 percent of adult summer food is insects. Kobriger (1965) found that juveniles had increased the amount of vegetable food in their diets to more than 90 percent; he reported that in Nebraska such important food plants included clover, roses, cherry, and dandelion, the most important of which were favored by wetland mowing practices.

During fall, a diverse array of seeds and cultivated grains are eaten, especially in agricultural areas. Otherwise the fruits of shrubs such as roses, snowberry, wolfberry, bearberry, blueberry, mountain ash, and poison ivy are eaten, as well as seeds and green leaves of herbs, shrubs, and trees. Probably a superabundance of suitable foods is normally available during this time, and much local or yearly variation in foods taken might be expected. Grange (1948) has pointed out that in general the sharptail closely resembles the ruffed grouse in its food cycle, and differences occur only because of the sharptail's preference for more open habitats. Differences in foods taken are most pronounced in late summer and fall, but from late fall through spring the diets may be nearly identical. The primary differences noted between the sharptail and the pinnated grouse were that the pinnated grouse uses a greater amount of grains and weeds and more generally depends on food sources associated with cultivation. Pinnated grouse may also feed to a somewhat larger extent on insects than do sharptails, but recent studies in South Dakota (Hillman and Jackson 1973) show that sharp-tailed grouse eat many grasshoppers during the summer. Similar studies in Nebraska suggest that the two species consume very similar foods, and that at least differences simply reflect habitat preferences such as greater use of subirrigated meadows by prairie chickens (Sisson 1976).

MOBILITY AND MOVEMENTS

Seasonal Movements

One of the first and most complete summaries of sharptail movements was that of Hamerstrom and Hamerstrom (1951). They reported that evidence for a definite seasonal migration dates from fifty to one hundred years ago, when most or all of the original sharptail range was occupied. At that time marked seasonal movements evidently did occur, but there is no clear evidence on migratory distances or even the directions involved. In areas of mountains or hills where there was woody cover, an upward altitudinal migration apparently occurred, but few if any cases of downward movement have been reported. Much of what has been interpreted as migration has consisted simply of movements to woody cover for the winter period, with distances of such movements gradually being reduced as the birds were driven out of their grassland habitats to woody edges, ravines, and similar brushy or woody situations. Thus long-distance movements from prairies to wooded wintering habitats have in recent years been completely eliminated, although seasonal changes in habitat preferences still persist in local areas.

With the advent of agriculture the prairies were made relatively unsuitable as breeding grounds for sharptails, and the availability of fall and winter grain has also influenced their movements. However, the sharptail has not been as strongly influenced by this food source as has the pinnated grouse, and it is less likely to leave its brushy winter habitat to

351

obtain grain. Where sharptails have simply incorporated grain into their winter diets they have altered their winter behavior very little, but in some areas the availability of grain throughout the winter has enabled the birds to winter in relatively open situations.

During the period of habitat shift from open to relatively brushy habitats, fall "packing" occurs, as coveys or broods gather into small flocks, which in turn form packs of up to several hundred birds. To a smaller extent, packing may occur in late winter during movement back to breeding grounds.

The Hamerstroms presented data on mobility for 167 sharp-tailed grouse banded in Wisconsin. Of the 162 birds for which the point of return was known, 81 percent were retaken within 2 miles of the point of banding. Only 12 percent had moved more than 3 miles, and only 10 percent were retaken more than 5 miles away. The longest distance from the point of banding was 21 miles. Similarly, Aldous (1943) found that short-range movements were the rule, with the maximum distance for any return 58 miles. Judging from comparable data on Wisconsin pinnated grouse, the relative overall mobility of the two species appears to be about the same. By transplanting sharptails and plotting their later recoveries, the transplanted birds were found in general to move farther than nontransplanted birds but to show no tendency to return to the point of banding. The maximum mobility of these transplanted birds was found to be between 26 and 27 miles from the point of release.

The distances that sharptails move from their wintering quarters to spring display grounds doubtless vary greatly in different areas. Kobriger (1965) found that in the Nebraska sandhills the dispersal distance of 35 male sharptails from winter feeding stations to spring dancing grounds ranged from 0.2 to 3.3 miles and averaged 0.9 mile. The majority of these birds moved from their wintering areas to the nearest dancing ground. However, this probably implies that the birds picked the suitable wintering area nearest their dancing ground rather than vice versa, since Evans (1969) found a high degree of fidelity of male sharptails to specific leks between successive years. Similarly, most nests are within a mile of the nearest dancing ground (Hamerstrom 1939; Hamerstrom and Hamerstrom 1951). Similarly, in studies in North Dakota the dispersal distance of females from dancing grounds averaged less than a mile but ranged to a maximum of 2 miles (Kobriger 1980). In a radio-tracking study of 13 females, the birds moved as far as 2 miles from the dancing ground but later returned to nest within ½ mile of it (Artmann 1970).

On the basis of 576 recoveries of birds banded in South Dakota, it was found that females exhibited stronger dispersal tendencies than did males, and juvenile males tended to move farther than adult males (Robel et al. 1972; Hillman and Jackson 1973). Sisson (1976) also reported larger average movements (8.1 vs. 1.3 miles) for females than males but noted that the differences were not statistically significant. The longest movement Sisson reported was for an adult female, of 44 miles, while Hillman and Jackson (1973) reported a movement of 93 miles for a juvenile female.

Daily Movements and Home Ranges

Hamerstrom and Hamerstrom (1951) reported that in the fall sharptails had a rather large covey range that totaled about 100 to 200 acres in extent, with from 3 to 6 such coveys usually to be found in an area of 1,000 to 1,500 acres. They estimated that the usual winter daily cruising radius was about 1 mile.

Kobriger (1965) tracked a sharptail male by radio telemetry through the summer months, during which it moved about 2.5 miles from its dancing ground. Similarly, a female was tracked from a dancing ground to a nest site 2 miles away. In a Minnesota study, Artmann (1970) found that 13 females has spring and summer home ranges of from 33 to 263 acres, with the largest movements occurring before nesting and again after the chicks were 8 to 9 weeks old. Ramharter (1976) estimated that the average brood range size was 44.5 hectares (26.9 to 68.9 hectares) for 5 hens that were tracked for at least 4 weeks. A radio-telemetry study in North Dakota showed that during the first 2 days after hatching 11 broods remained within 0.25 mile of their nest sites, though maximum brood movements usually exceeded 0.5 mile (Kobriger 1980).

REPRODUCTIVE BEHAVIOR

Territorial Establishment

Sharp-tailed grouse probably establish territories as early as the first fall of life. Hamerstrom and Hamerstrom (1951) found that at least 3 of 18 males seen on a dancing ground in North Dakota during late September were young birds. Likewise, Rippin (1970) found that although only adult males were among those trapped or shot on a display ground in late August, by late September and early October several juvenile males were also present. This regular fall period of display, which is also typical of pinnated grouse but not of sage grouse, may be important in helping young birds learn traditional display sites. Rippin found that when he killed all the males using a dancing ground during the spring, there was no use of that display site the following fall, but on another area where he killed all but one of the displaying males the lone bird formed a nucleus for display behavior with several juvenile birds that next fall. Young probably begin trying to establish peripheral territorial areas their first fall of life, and these are held again the following spring. Rippin reported that on two control dancing grounds (on which he did not experimentally remove any males), the percentage of immature males was 43 percent in 1968 and 37 percent in 1969. On his experimental grounds, he first mapped the relative territorial positions of the participating males; in each he recognized one or more central males and approximately three outer rings of less dominant males defending peripheral territories. On one display ground that contained 18 males, a marginal male originally defending a peripheral territory gradually established himself centrally as Rippin progressively reduced the number of males on the dancing ground to 5 birds. When the ground was reduced to 4 participating males, no single bird was able to maintain a central dominant position. His studies clearly showed that a strong centripetal tendency was present in all the males, with each attempting to attain and defend a relatively central territory.

When such display ground social structures are not disrupted by the death or removal of males, they exhibit great stability. Evans (1969) found that of 10 males that were marked one spring, 5 returned to the same dancing ground the following spring, while the other 5 disappeared and apparently had died. The areas defended by the 5 returning males were virtually the same as those they had defended the previous spring, with a single minor exception. Hjorth (1970) analyzed Evans's data and concluded that on 2 grounds the average territorial size was about 90 square meters, ranging from 14 square meters in the central area to 170 square meters on the periphery. He also determined that the average territorial size for a Montana display ground was about 50 square meters, with the 4 central territories averaging 25 square meters.

The average numbers of territorial males present on display grounds probably vary with population density. Ammann (1957) provides numbers of birds of both sexes present on 10 different sharptail dancing grounds; they averaged 12.4 but ranged from 3 to 29 birds in different years and on different grounds. In the Nebraska sandhills, display grounds of both the sharptail and the pinnated grouse usually have between 9 and 10 males (Johnsgard and Wood 1968). Grange (1948) indicated that the average number of males on 14 sharptail grounds in Wisconsin was 6, while 7 pinnated grouse grounds averaged 7 males in attendance. In Utah, Hart, Lee, and Low (1952) reported the average number of birds present on 29 dancing grounds as 12, although as many as 50 were seen. Lumsden (1965) summarized data from several areas in Ontario that indicated from 2 to 24 males present on dancing grounds. In North Dakota the twelve-year average for 1,664 dancing grounds was 12.9 males (Johnson 1964). It seems that from 8 to 12 males represents a typical dancing ground for sharp-tailed grouse in most parts of their range.

Lumsden (1965) confirmed for sharptails the earlier observations of persons working with pinnated grouse and sage grouse as to the reproductive advantage of holding central territories in dancing grounds. He reported that such central positions were held by socially dominant birds that readily won disputes with neighbors. These central territories were often smaller than peripheral ones, and Lumsden thought that normally only fairly old males could successfully hold such territories. On one display ground Lumsden noted that the dominant male performed 76 percent (13) of the copulations or attempted copulations observed, which emphasizes the enormous selective value of occupying such central territories.

Moyles and Boag (1981) have found not only that central territories are superior in terms of reproductive efficiency, but also that birds holding them have lower mortality rates than do peripheral males. The annual rate of turnover in displaying males may be as high as 83 percent. Young males are able to obtain only peripheral territories and gradually work inward, according to these authors.

Territorial Advertisement and Defense

Lumsden (1965) has classified the social displays of sharptails as those that serve aggressive functions, those concerned with courtship and mating, and those that are specifically associated with advertising the location of the display grounds. In addition, several signals serve as a predator warning system. Lumsden's account is unusually complete, and his terms and descriptions will be used here. More recently, Hjorth (1970) has made an equally detailed analysis; his comparable terms will be noted and a few divergent observations briefly mentioned.

Signals that serve primarily to advertise the location of the dancing ground and of specific males include the flutter-jump and cackling calls. Both sexes perform cackling calls. Females usually cackle as they approach the dancing ground, and this stimulates strong responses by the males, especially flutter-jumping. Flutter-jumping was first described for the pinnated grouse, and it is virtually identical in both species. The male jumps a few feet into the air, sometimes uttering a *chilk* note as he takes off, flies a few feet forward, and lands again. In so doing, he clearly advertises his own presence, as well as the location of the dancing ground as a whole. Males may cackle between flutter-jumps or when others are flutter-jumping.

A large number of male sharptail displays are primarily aggressive and serve to establish and maintain territories. Secondary functions no doubt include attracting females and sexual recognition. These primarily aggressive signals include several calls and postures. The calls may be named the *lock-a-lock*, "cooing," the "cork" call, and the *chilk* and *cha* calls. Lumsden regards the last two as associated with courtship, since they are most often uttered when hens are present.

The *chilk* and *cha* calls are both loud, high-pitched notes that carry great distances. They are often given before or after flutter-jumping and during the "tail rattling" display, and both may be uttered very rapidly. They evidently grade into one another and probably serve similar functions.

The "cork" note is a squeaking sound like that made by pulling a cork from a bottle and is uttered only during the tail rattling display. It is most often heard when a female is near but may be elicited by another displaying male. A similarly aggressive call, called "whining," consists of drawn out and repeated singsong *kaaa-kaaaaa* notes. Such notes are usually associated with territorial defense and are often uttered by birds facing one another.

The *lock-a-lock* call is a gobbling note produced by males standing at rest. With head lowered slightly, a male may utter this call as he approaches his territory before dawn. It is not uttered in the presence of females and apparently serves only an aggressive function.

The "cooing" display is a combination of posturing ("oblique" posture of Hjorth 1970) and sound production that is clearly homologous with the "booming" of pinnated grouse. As in that display, the tail is partially cocked, the esophagus is inflated, and the head is distinctly lowered ("bowing" of Hjorth 1970) as a low-pitched cooing sound of one or two notes is uttered. However, the folded wings are not strongly lowered, and the throat skin is not as strongly distended as the pinnated grouse's during booming. The neck skin is usually pink to purple and thus is also different from that of the greater prairie chicken. Lumsden believes cooing does not serve as a sexual signal but rather is evoked in aggressive situations, thus also differing functionally from the booming display.

Several postures or movements are also closely associated with territorial defense. These include an "upright advance" ("wide-necked upright" of Hjorth 1970), which is

31. Male displays of sharp-tailed grouse (in part after Lumsden 1965), including (A) dancing, (B) cooing, (C) bowing, and (D) running parallel. Tracks made while dancing are also shown (E).

an aggressive approach posture of a male during which the tail is cocked and the neck feathers are erected to expose the apteria. "Walking or running parallel" consists of two males' moving along their territorial boundaries while threatening one another, often uttering the *lock-a-lock* call. During this display the head is usually held low, the eye combs are enlarged, and the tail is cocked. During "ritual fighting" the birds face one another, often squatting, and utter aggressive calls while periodically making short lunges toward each other. When not attacking, they usually hold their wings partly open and on the ground. During overt attacks the birds leap up into the air, flailing one another with their claws and beaks and sometimes striking with their wings. Between such attacks the birds watch each other intently, and Lumsden reports that "displacement sleeping" may occur when the attack intensity wanes to a certain point. Should a male attempt to withdraw from such an encounter, he typically lowers his tail, covers his neck skin, withdraws his eye combs, and sleeks his feathers. These submissive patterns give him the appearance of a female and tend to inhibit attack by other males. Lumsden reported that the sharptails he observed in Montana, but not those in Ontario, performed a shoulder spot display when fighting and also just before copulation, exposing the white under wing coverts in the region of this elbow. The shoulder spot display is a conspicuous feature of several grouse species, such as the pinnated grouse, and in several seems to indicate fear or submission. However, Hjorth (1970) did not observe this display in Montana sharptails, and I have not seen it in Nebraska. Lumsden (1970) has reviewed its relative occurrence in various grouse species and has concluded that in some species (such as black grouse and capercaillie) it serves as an aggressive signal among males while in females it is an expression of fear.

By far the most complex and interesting of the male displays is the "tail rattling" or "dancing" display of sharptails. Lumsden considers this a courtship display, but it is also closely associated with territorial defense and proclamation. It consists of a highly ritualized series of rapid stepping movements, performed with the tail erect, wings outstretched, head held forward and rather low, and neck feathers erected to exhibit the bare purple skin. With the cocking of the tail the white under tail coverts are exposed and appear to be somewhat expanded for maximum visibility. In this rigid posture the male begins a series of very short and rapid stepping movements (eighteen to nineteen per second according to Hjorth 1970), causing him to move forward in a generally curving direction ("aeroplane display" of Hjorth 1970).

In synchrony with the stepping movements, the male also performs a strong lateral vibration of his tail, producing a clicking or rattling frictional sound that is a combination of these pattering sounds and the scraping noises of the overlapping tail feathers. Hjorth (1970) has recently found that during tail rattling not only are the lateral rectrices alternately spread and shut, but the male also occasionally performs a rapid (0.08 second), symmetrical tail spreading while breaking his stamping rhythm momentarily.

Not only are the foot and tail movements of the male a highly coordinated series, but males tend to perform the tail rattling in highly synchronized fashion. Two or more closely adjacent males will start and stop their display almost simultaneously, and sometimes all

357

the males on a dancing ground will simultaneously become silent. At such times the birds appear to be highly attentive and sensitive to disturbance, whereas when they are all actively "dancing" they remain nearly oblivious to their surroundings.

When performing tail rattling in the presence of a female, the male often alternates this display with a stationary posture Lumsden has called "posing." In this posture the male usually faces or nearly faces a female, with wings slightly spread and drooped and eye combs greatly enlarged. Soft crooning notes may also be uttered. Typically the male moves from this into a crouching position or "nuptial bow" before the female, in which he lowers his body to the ground, fully spreads his wings to the sides, and almost touches the ground with his bill ("prostrate" of Hjorth 1970). His rear end is held high, so that the tail remains vertical, and in general the upper body surface and dorsal view of the tail appear to be presented to the female. In contrast to the comparable posture of the pinnated grouse, the male may perform several short and repeated bowing movements; in the pinnated grouse the male typically remains prostrate and motionless before the female for several seconds. Although this display is normally performed by a male that is beside a female and not being bothered by rival males, Lumsden noted that he observed it as a precopulatory display in one 1 of 19 copulation sequences.

Most copulations by sharp-tailed grouse occur before or approximately at sunrise. Preliminary postures may include the nuptial bow, posing, or tail rattling displays. The female squats in the usual manner and is immediately mounted by the male. Usually the hen runs forward rapidly immediately after copulation, then vigorously shakes her body and wing feathers. After a successful copulation the hen often leaves the display ground within a few minutes, and there is no evidence to date that more than one copulation is needed to fertilize all the eggs in a single clutch.

As noted earlier, nearly all copulations are apparently performed by dominant and central males, although Sexton (1979) observed a case of off-lek copulation. Rippin and Boag (1974b) found that, as dominant males were systematically eliminated from two arenas, their territories were filled in an orderly manner, first by other central males and then by more peripheral birds. The sizes of individual territories remained constant, as did the density of the males on the arena.

Vocal Signals

Although all the major vocalizations associated with lek behavior have already been noted in the previous section, a brief review of calls and other types of acoustic communication associated with lek activity is warranted. Kermott and Oring (1975) have provided a functional analysis of these acoustic signals, together with sonagrams and results of playback experiments. According to them, there are six major male calls. Cooing is a species-typical and stereotyped vocalization, used primarily for territorial advertisement. Playbacks of cooing usually elicited the same vocalizations from territorial males, at least until midseason. Gobbling or *lock-a-lock* calls are aggressive, differ acoustically among individuals, and apparently are largely associated with the establishment and defense of territories. The *chilk* and *cha* calls are evidently used to attract and stimulate females, as is the "cork" note. The mechanical sounds associated with dancing (tail rattling and foot stamping) likewise attract and stimulate females. However, unlike the calls just mentioned, the playback effects of dancing sounds consistently stimulate dancing in other males, suggesting that male-oriented communication functions may exist as well. Chattering and whining calls, typically uttered during territorial boundary disputes, are obviously aggressive in function, while a soft clucking sound sometimes is uttered during periods of inactivity and may serve to maintain contact among individuals.

In addition to the calls already mentioned, Lumsden has described several others. In a situation of uneasiness or slight disturbance, birds utter a *yur* note with a downward inflection. In flight they frequently utter a series of rapid calls *tuckle . . . tuckle . . . tuckle*, or *tuk . . . tuk . . . tuk*, and they may produce the same calls before flight.

One other vocalization that serves as a courtship signal, or at least is produced only when hens are on the display ground, is the *pow* call. When courting a hen, males will utter this call several times in rapid succession. Most probably, as Lumsden has suggested already, it is homologous to the loud *whoop* call of greater prairie chickens.

Lumsden has described several predator-response postures of sharp-tailed grouse, which include an "upright alert" posture, in which the bird stands upright to its fullest extent with its feathers sleeked and crest raised. A "prostrate alert" is performed in a similar situation, but the bird crouches in a "frozen" posture. "Alarm strutting" may be performed as the bird walks around or away from a source of possible danger, in a stiff gait and with occasional tail flicks that reveal the white outer tail feathers.

Nesting and Brooding Behavior

The female begins to make a nest scrape in a protected site at about the time she begins to visit the dancing grounds or possibly even before. After successful mating, she leaves the dancing ground and probably will not return to it again, except in the event of renesting. The eggs are laid approximately daily until the total clutch of about 12 is produced (Hamerstrom 1939; Ammann 1957). The female typically begins incubation about the time the last egg is laid, and the incubation period is 23 to 24 days. Renesting attempts evidently do sometimes occur but probably contribute no more than 10 percent of the offspring in an average season (Ammann 1957). Kobriger (1980) reported that, of 78 nests studied in North Dakota, 9 were renesting attempts. Schiller (1973) reported finding 4 second nesting attempts and 2 third nesting attempts during radio-telemetry studies in northern Minnesota.

During incubation, the female is on the nest constantly except for a short period in the morning and evening (Artmann 1970). After hatching, the female leads the young away from the nest fairly rapidly, usually toward areas where insects and green herbaceous foods are abundant. Areas of fairly heavy cover are used early in brood development, but the birds gradually move into more open habitats and begin to eat a smaller proportion of insect foods (Schiller 1973). Young sharp-tailed grouse feed to a large extent on insects during their first few weeks, with grasshoppers, spiders, ants, and weevils all contributing to their diet, but leaves and berries are also important foods (Grange 1948). Chicks are able to fly to a very limited degree by the time they are 10 days old, and from that time they become increasingly independent of their mother. By the time they are 6 to 8 weeks old they are virtually fully independent, and broods begin to gradually break up and the young birds disperse, often fairly long distances.

EVOLUTIONARY RELATIONSHIPS

There can be little doubt that the nearest living relative of the sharp-tailed grouse is the pinnated grouse, and I agree with Short (1967) that they are obviously congeneric. Similarities in their downy young as well as in their adult plumage patterns bear this out, as well as the frequency of hybridization under natural conditions (Johnsgard and Wood 1968). The two forms also share a number of display patterns, such as "booming" and "cooing," "foot stamping," the "nuptial bow," and "flutter-jumping." The sharptail's *pow* call no doubt is homologous to the *pwoik* of the pinnated grouse, and the whining and cackling calls of the two species are very similar. The sharptail's *lock-a-lock* aggressive call probably corresponds to the pinnated grouse's *hoo-wuk*; I have heard a hybrid male utter an intermediate call sounding like *wuk-a-wuk'*. However, the lateral tail rattling of the sharptails is replaced in pinnated grouse by symmetrical tail fanning movements, the forward "dancing" is represented by foot stamping almost in place, and cooing in the sharptail appears to have much less visual and acoustical importance than the homologous booming of the pinnated grouse.

Short (1967) suggests that the sharp-tailed grouse is probably closer to the ancestral prairie grouse type than is the pinnated grouse, on the basis of its less specialized neck feathers (rudimentary pinnae) and reduced esophageal sacs. However, its tail-feather structure is specialized for the tail rattling display (Lumsden 1968), and these differences largely reflect the relative importance of "booming" and "dancing" in the two species. I suggest that both species have diverged equally from a common forest-dwelling ancestral type, the pinnated grouse in a more easterly and southerly location (oak woodland or savanna habitat) and the sharptail in a more westerly and northerly location (grassland, coniferous forest edge habitat). There was probably little contact between these two forms until fairly recently, when human activities greatly altered the habitats of both species (Johnsgard and Wood 1968).

Appendixes

Name Derivations of Grouse and Ptarmigans

(This listing includes all generic and specific names, and most but not all subspecies.)

Bonasa—from Latin *bonasum*, a bison (the drumming of the male resembling the bellowing of a bull), or perhaps from Latin *bonus*, good, and *assum*, roast.

    affinis—Latin, related

    bonasia—see *Bonasa* above

    brunnescens—Latin, dark brown

    castanea—Latin, chestnut

    incana—Latin, hoary

    monticola—from the Latin *montis*, mountain, and *colo*, I inhabit

    phaia—Greek, dusky

    rupestris—Latin, existing among rocks

    sabini—after Joseph Sabine (1770–1837), brother of English astronomer Sir Edward Sabine

    severzowi—after Professor N. A. Severtzow (1827–85), naturalist of central Asia

    sibiricus—of Siberia

    togata—Latin, clad in a toga

    umbelloides—from the Latin *umbella*, umbrella, and *eidos*, resemblance

    umbellus—Latin, umbrella

    vicinitas—Latin, neighboring

    yukonensis—of the Yukon

Canachites—from the Greek *kanacheo*, make a noise, with formative suffix *-ites*.

Capercaillie—probably derived from the Gaelic *capull*, great horse, and *coille*, a wood; or perhaps from *cabher*, old man, and *coille*, meaning old man of the woods. Also spelled capercalze, capercailzie, etc.

Centrocercus—from the Greek *kentron*, spine, and *kerkos*, tail

    phaios—Greek, dusky

urophasianus—from the Greek *oura*, tail, and *phasianos*, pheasant
Dendragapus—from the Greek *dendron*, tree, and *agape*, love
    atratus—Latin, clothed in black
    canace—Latin, a proper name
    canadensis—of Canada
    falcipennis—from the Latin *falcis*, sickle, and *pinna*, feather
    franklinii—after Sir John Franklin (1786–1847), explorer of the Canadian arctic
    fuliginosus—Latin, sooty
    howardi—after O. W. Howard, California naturalist
    obscurus—Latin, dusky
    pallidus—Latin, pale
    richardsonii—after Sir John Richardson (1787–1865), English naturalist
    sierrae—of the Sierras
    sitkensis—of Sitka, Alaska
Galliformes—from the Latin *gallus*, cock, and *forma*, appearance
Gallinaceous—from the Latin *gallina*, hen, and adjective suffix *-aceus*, pertaining to
Grouse—probably from the French *greoche, greiche*, and *griais*, meaning a spotted bird, and used in England as "grous" for the red grouse before being applied in North America to grouse in general
Lagopus—from the Greek *lagos*, hare, and *pous*, foot, meaning hare-footed and referring to the similarity between the birds' feathered toes and the densely haired feet of rabbits
    alascensis—of Alaska
    albus—Latin, white
    alexandrae—after Annie M. Alexander, leader of an Alaskan biological expedition during which this form was discovered
    altipetens—from the Latin *altus*, high, and *petens*, seeking
    atkhensis—from Atka, Aleutian Islands
    brevirostris—from the Latin *brevis*, short, and *rostris*, bill
    captus—Latin, captured
    dixoni—after J. S. Dixon, American naturalist
    evermanni—after Professor B. W. Evermann
    gabrielsoni—after Ira Gabrielson, American conservationist
    helveticus—of Switzerland
    hyperboreus—Greek, from beyond the north wind
    islandorum—of Iceland
    japonicus—of Japan
    leucopterus—from the Greek *leukos*, white, and *pteron*, wing
    leucurus—from the Greek *leukos*, white, and *oura*, tail
    maior—Latin, somewhat greater
    millaisi—after J. G. Millais, English naturalist and artist
    muriei—after Olaus Murie (1889–1963), American naturalist

mutus—from the Latin *mutatus*, change or alteration, referring to the variable plumages

nelsoni—after E. W. Nelson, American naturalist and explorer of Alaska

peninsularis—of the (Kenai) peninsula

pyrenaicus—of the Pyrenees

rainierensis—of Mount Ranier

ridgwayi—after Robert Ridgway (1850–1929), American ornithologist

rossicus—of Russia

saturatus—Latin, meaning saturated, in reference to the color

saxatilis—Latin, dwelling among rocks

scoticus—of Scotland

townsendi—after C. H. Townsend (1859–1944), American ornithologist

ungavus—of the Ungava Peninsula

variegatus—Latin, variable

welchi—after George O. Welch, who collected the type specimens

yunaskensis—from Yunaska island

Lyrurus—from the Latin *lyra*, lyre, and the Greek *oura*, tail

Pedioecetes—from the Greek *pedion*, plain, and *oiketes*, inhabitant

Phasianidae—from the Latin *phasianus* (Greek *phasianos*), pheasant, and the familial suffix *-idae*

Ptarmigan—from the Gaelic *tarmichan*, originally applied to the rock ptarmigan. Perhaps also related to the Middle English *termagaunt*, from which termagant derives.

Tetrao—from the Greek *tetraon*, pheasant

britannicus—of Britain

kamtschaticus—of Kamchatka

mlokosiewsiczi—after Louis Mlokosiewicz, Polish natural historian of Georgian Russia, who obtained the first specimens of this species

mongolicus—of Mongolia

parvirostris—from the Latin *parvus*, small, and *rostrum*, bill or beak

taczanowskii—after L. Taczanowski, discoverer of the Caucasian black grouse

tetrix—from the Greek *tetrax*, pheasant

urogallus—from the Greek *oura*, tail, and the Latin *gallus*, cock

ussuriensis—of the Ussuri River

viridanus—from the Latin *viridus*, green

Tetraonine—from the Greek *tetraon*, pheasant, and the subfamilial suffix *-inae*

Tympanuchus—from the Greek *tympanon*, drum, and *echo*, have or hold

attwateri—after Henry P. Attwater, Texas conservationist

campestris—Latin, found in a plain

caurus—Latin, of the northwest

columbianus—of the Columbia River

cupido—from the Latin *Cupid* (referring to the ''Cupid's wings'' on the neck)

jamesi—after Harry C. James, Colorado naturalist

kennicotti—after Robert Kennicott (1835–66), first director of the Chicago Academy of Science

pallidicinctus—from the Latin *pallidus*, pale, and *cinctus*, banded

phasianellus—diminutive of the Latin *phasianus*, pheasant

pinnatus—Latin, plumed

Key to Identification of Grouse and Ptarmigan Species

A.   Tail nearly as long as wing, of 20 strongly graduated and pointed rectrices, tarsus longer than middle toe with claw . . . *Centrocercus urophasianus* (sage grouse).

AA. Tail distinctly shorter than wing, rectrices not distinctly graduated or sharply pointed, tarsus shorter than middle toe with claw.

    B. Tarsus bare of feathers on lower quarter to half, tail more than two-thirds as long as wing . . . *Bonasa*.

        C.   Rectrices 18–20, sides of neck with a conspicuous tuft of broad, soft feathers . . . *B. umbellus* (ruffed grouse).

        CC. Rectrices 16, sides of neck lacking tufts or tufts only rudimentary.

            D.   Two central rectrices brown, rest of tail vermiculated, with a broad subterminal band and a whitish tip . . . *B. bonasia* (hazel grouse).

            DD. Four central rectrices brown, rest of tail barred with black and white and tipped with white . . . *B. severzowi* (black-breasted hazel grouse).

    BB. Tarsus feathered to base of toes or nearly so, tail usually less than two-thirds as long as wing.

        C.   Lateral rectrices longer than middle pair, especially in males . . . *Tetrao* (subgenus *Lyrurus*).

            D.   Black predominating in plumage (males).

                E.   Lateral rectrices curving sharply outward, lower tail coverts white . . . *T. tetrix* (black grouse).

                EE. Lateral rectrices curving gradually outward, lower tail coverts black . . . *T. mlokosiewiczi* (Caucasian black grouse).

DD. Gray or russet tones predominating in plumage (females).

    E.   Length of tail less than two-thirds that of wing, and usually under 130 mm, breast and flanks with broad dark markings . . . *T. tetrix* (black grouse).

    EE. Length of tail about two-thirds that of wing, and usually over 130 mm, breast and flanks with narrow and irregular dark markings . . . *T. mlokosiewiczi* (Caucasian black grouse).

CC. Tail rounded, square tipped, or slightly graduated, never with outer feathers longer than middle feathers.

    D.   Larger (wing at least 260 mm) . . . *Tetrao* (part).

        E.   Tail unbarred, upperparts brown to grayish black (males).

            F.   Bill black, scapulars with white markings . . . *T. parvirostris* (black-billed capercaillie).

            FF. Bill whitish, scapulars without white markings . . . *T. urogallus* (capercaillie).

        EE. Tail barred, upperparts mottled or barred (females).

            F.   Lower foreneck rufous brown, with little or no dark barring . . . *T. urogallus* (capercaillie).

            FF. Lower foreneck and underparts heavily barred with dark brown . . . *T. parvirostris* (black-billed capercaillie).

DD. Smaller (wing less than 255 mm).

    E.   Tail more than half as long as wing, outermost rectrices over four-fifths length of central ones, outer webs of primaries irregularly mottled or uniformly colored.

        F.   Upper tail coverts not extending to tip of tail . . . *Dendragapus*.

            G.   Rectrices 16 (rarely 18), underparts heavily barred.

                H.   Tips of anterior remiges strongly narrowed . . . *D. falcipennis* (sharp-winged grouse).

                HH. Tips of anterior remiges not strongly narrowed . . . *D. canadensis* (spruce grouse).

            GG. Rectrices 18–20 (rarely 16), underparts mostly grayish . . . *D. obscurus* (blue grouse).

        FF. Upper tail coverts extending to tip of tail, normally with 16 rectrices . . . *Lagopus*.

            G.   Lateral rectrices white . . . *L. leucurus* (white-tailed ptarmigan).

            GG. Lateral rectrices dark brown or black.

                H.   Bill (maxilla) distinctly heavier (usually over 9.0 mm high at base; height at base about two-thirds culmen length . . . *L. lagopus* (willow ptarmigan).

          HH. Bill slighter (usually under 8.5 mm high at base) and weaker; height at base under half culmen length . . . *L. mutus* (rock ptarmigan).

EE. Tail less than half as long as wing, outermost rectrices under four-fifths length of central ones, outer webs of primaries regularly patterned with white or buff spots . . . *Tympanuchus*.

    F. Central pair of rectrices considerably longer and different in color from others . . . *T. phasianellus* (sharp-tailed grouse).

    FF. Central pair of rectrices not markedly different from others, neck with tapered, erectile pinnae . . . *T. cupido* (pinnated grouse).

        G. Darker barring of body feathers distinctly bicolored with brown and blackish on sides, flanks, and upperparts . . . *T. c. pallidicinctus* (lesser prairie chicken; often considered a distinct species).

        GG. Darker barring of body feathers unicolored brown or blackish; the total coloration generally darker throughout . . . other races of *T. cupido*.

Hunter Harvest and Population Status Estimates of Grouse and Ptarmigans

*North America (Organized by States and Provinces)**

Alaska: Average harvest from 1952 to 1957 plus 1961 was 93,971 ptarmigans and 59,306 total grouse (blue, spruce, ruffed, and sharp-tailed).

California: Between 1970 and 1979 the average annual kill was about 4,000 sage grouse and 9,000 forest grouse. Of the latter, fewer than 5 percent were ruffed grouse, and the rest were blue grouse. The 1978 sage grouse kill was 3,980.

Colorado: Average 1970–74 harvest was 13,507 sage grouse, 1,054 sharp-tailed grouse, 23,016 blue grouse, and (1970–78) 3,184 white-tailed ptarmigans. The 1978 sage grouse harvest was 3,882.

Connecticut: No statewide estimates on ruffed grouse.

Georgia: The annual ruffed grouse kill was about 2,500 in the 1960s; no current estimates are available.

Idaho: The average 1970–79 grouse harvest was 70,250 sage grouse, 10,800 sharp-tailed grouse, 87,500 ruffed grouse, 83,100 blue grouse, and 5,250 spruce grouse. The 1978 sage grouse kill was 72,000.

Indiana: In 1976, 4,044 ruffed grouse were harvested, and 8,312 were taken in 1977.

Iowa: The current annual ruffed grouse kill is between 5,000 and 10,000 birds.

*Based on information provided by state and provincial agencies.

Kansas:	The average greater prairie chicken harvest from 1962 to 1979 was about 34,000 birds. Lesser prairie chickens have been legally hunted since 1970; the 1979 harvest was about 4,300 birds, while that of the greater prairie chicken was 88,400.
Kentucky:	The annual ruffed grouse harvest between 1964 and 1975 was 35,000 birds.
Maine:	No open season on spruce grouse during the 1970s. The annual ruffed grouse kill between 1970 and 1978 was 259,500 birds.
Maryland:	The 1971–75 annual ruffed grouse harvest was 14,000 birds.
Massachusetts:	A three-year average harvest of ruffed grouse in the 1970s was 41,631 birds.
Michigan:	From 1955 to 1960 the average ruffed grouse kill was 356,000 birds. In recent years the sharp-tailed grouse harvest has been under 500 birds.
Minnesota:	The average ruffed grouse kill (1972–79) is about 568,000 birds. The average sharp-tailed grouse harvest between 1972 and 1979 was 33,000 birds. The first two legal spruce grouse seasons since 1915 were in 1969 and 1970, when the average kill was about 11,500 birds, and more recent seasons (1976–79) have averaged 22,000 birds.
Missouri:	No current season on ruffed or pinnated grouse.
Montana:	Average harvest for 1970 to 1979 was 46,360 sage grouse, 52,306 blue grouse, 25,470 spruce grouse, 51,505 ruffed grouse, and 94,048 sharp-tailed grouse.
Nebraska:	In recent years the average harvest has been 10,000 to 15,000 greater prairie chickens and 30,000 to 45,000 sharp-tailed grouse.
Nevada:	The average harvest from 1970 to 1977 was 19,388 sage grouse and 2,051 blue grouse. Some ruffed grouse are also killed.
New Hampshire:	No data (ruffed grouse only).
New Jersey:	The average annual ruffed grouse kill from 1970 to 1977 was 47,000 birds.
New Mexico:	Between 1958 and 1978 the average harvest was 2,023 blue grouse and 1,125 lesser prairie chickens. The 1978 lesser prairie chicken harvest was 1,248 birds.
New York:	The average harvest of ruffed grouse between 1966 and 1969 was 409,500 birds. More recent surveys do not include the entire state.
North Carolina:	The ruffed grouse kill in 1974 was 38,154 birds, and in 1976 it was 32,196 birds.
North Dakota:	The average sage grouse harvest from 1970 to 1978 was about 200 birds; there was no season in 1979. The average sharp-tailed grouse harvest from 1970 to 1979 was 158,296 birds. About 5,000 ruffed grouse are killed annually.

Ohio:	The 1972 ruffed grouse kill was 130,000 birds, and the 1979 kill was 124,000.
Oklahoma:	The average pinnated grouse harvest from 1959 through 1968 was 7,700. In 1979 the lesser prairie chicken kill was about 134 birds, and that of greater prairie chickens was about 5,105.
Oregon:	The average sage grouse harvest from 1969 to 1975 was 3,090 birds. No seasons were held in 1978 or 1979. The forest grouse (blue, ruffed, and a few spruce) harvest averaged 67,000 during 1969–77.
Pennsylvania:	The annual ruffed grouse harvest (1970–79) was 234,000 birds.
Rhode Island:	The estimated annual ruffed grouse harvest from 1970 to 1979 was about 400 to 1,600 birds.
South Carolina:	The annual ruffed grouse kill is only 100 to 250 birds.
South Dakota:	The average grouse harvest from 1970 to 1978 was 109,700 birds, of which 1 percent consisted of ruffed grouse and 1 percent of sage grouse, about 5–10 percent consisted of greater prairie chickens, and the remainder were sharp-tailed grouse.
Tennessee:	The annual ruffed grouse kill is about 15,000 birds.
Texas:	The average annual lesser prairie chicken harvest from 1965 through 1969 was 275 birds. In 1979 it was 600 birds.
Utah:	The average 1963–78 annual harvest was 14,748 sage grouse, 14,072 ruffed grouse, and 18,623 blue grouse. Limited hunting of sharp-tailed grouse has been allowed since 1974; the 1979 kill was 76 birds.
Vermont:	No data (ruffed grouse only).
Virginia:	The annual ruffed grouse kill from 1973 to 1979 was 108,000 birds.
Washington:	The average annual kill (1974–79) was 788 sage grouse and 365,000 other grouse, of which roughly 224,000 were ruffed grouse, 128,500 were blue grouse, 10,600 were spruce grouse, and 180 were sharp-tailed grouse. The 1978 sage grouse harvest was 380 birds.
West Virginia:	The 1970–71 harvest was 170,000 ruffed grouse, and the 1975–76 kill was 151,000.
Wisconsin:	The average ruffed grouse kill from 1967 to 1976 was 682,000 ruffed grouse. The season on prairie chickens was closed after 1955. The sharp-tailed grouse harvest in the 1970s ranged from 8,160 (1970) to 12,000 (1976).
Wyoming:	The sage grouse harvest from 1977 to 1979 averaged 83,424 birds, with a 1979 high of 94,426. The sharp-tailed grouse harvest during this same period averaged 3,987 birds, the blue grouse kill averaged 19,250 birds, and the ruffed grouse harvest averaged 7,758.

CANADA

Alberta:	The average 1976–78 annual harvest was 59,500 sharp-tailed grouse, 284,000 ruffed grouse, 87,200 spruce grouse, 2,800 blue grouse, and 1,215 sage grouse.
British Columbia:	The average harvest between 1964 and 1968 was 132,030 blue grouse, 133,362 spruce grouse, 361,293 ruffed grouse, and 21,365 sharp-tailed grouse.
Manitoba:	The average harvest between 1971 and 1975 was 25,800 spruce grouse, 9,200 ptarmigans, 82,000 ruffed grouse, and 61,000 sharp-tailed grouse.
New Brunswick:	From 100,000 to more than 300,000 grouse are killed annually. Of these, about 15–20 percent are spruce grouse, and the rest are ruffed grouse.
Newfoundland:	From 25,000 to 50,000 ptarmigans are harvested annually. In recent years ruffed grouse and spruce grouse have been harvested, and the 1976–77 kill of spruce grouse was 12,000 to 15,000 birds.
Nova Scotia:	The annual kill of ruffed grouse ranges from 50,000 to 65,000 birds.
Ontario:	The annual ruffed grouse harvest during the 1970s was 1,208,000 birds. No data available on sharp-tailed or spruce grouse.
Prince Edward Island:	The annual 1971–79 kill of ruffed grouse was 8,740 birds.
Quebec:	No detailed estimates available for any species (ptarmigans, ruffed grouse, spruce grouse), but the annual kill probably exceeds 100,000 birds.
Saskatchewan:	The average harvest between 1970 and 1979 was 8,700 spruce grouse, 52,700 ruffed grouse, and 105,300 sharp-tailed grouse.
Northwest Territories:	About 1,000 sharp-tailed grouse are killed annually; no information is available on other species.
Yukon Territory:	About 300 to 400 sharp-tailed grouse are killed annually; no information is available on other species.

Europe (Organized Alphabetically by Species)*

BLACK GROUSE

Austria:
Declining; alpine population of about 14,000 displaying males reported in late 1960s. In 1977–78 the harvest of males was 1,608.

Czechoslovakia:
Probably several thousand birds present in 1950s and 1960s, apparently fluctuating but decreasing in past few decades.

Denmark:
Protected since 1973, about 100 birds remained by 1978.

East Germany:
No complete counts, but generally declining populations.

Estonia:
No hunting information; surveys from 1953 to 1971 suggest increase since early 1960s, but no total population estimates available.

Finland:
Estimated total population about 900,000 birds during 1964–71. Average harvest declined from 236,000 birds (1930–47) to between 54,500 and 171,500 during 1960s and 1970s.

France:
Protected in Ardennes area; limited hunting (27–55 days) in Alps, but no harvest data available.

Great Britain:
No precise population estimates, but perhaps between 10,000 and 100,000 pairs present, probably closer to 10,000. Hunting traditional on private estates, but no comprehensive harvest data available.

Italy:
Population decreasing, no numerical data available.

Luxembourg:
Now possibly extirpated.

Netherlands:
Protected; estimated population of about 300 males in 1977.

Norway:
Population in 1960 approximately half a million birds, and estimated harvest then about 100,000. At present the population is greatly reduced (possibly 60–80 percent), and 1975 to 1980 harvests averaged 19,000 birds.

Poland:
Total population varied from 29,000 to 40,000 in the late 1970s, and harvests of males during spring shooting ranged from 1,169 to 1,623 during this period.

Rumania:
Marked population decline during past century; now generally scarce, but no numerical data available.

Sweden:
About 300,000 pairs in 1970s; average harvest for 1978 to 1980 seasons was 21,600 birds.

Switzerland:
Current population fluctuating at about half of level of mid-1940s. Average annual harvest during fall hunting season is 1,300 adult males.

Yugoslavia:
Present only in Slovenija Republic, where it is declining. Annual harvests of males (shot during spring) are from 133 to 148 birds.

West Germany:
Scattered populations, generally declining for most of the century.

---

*Population data primarily from Cramp and Simmons (1980) and Glutz (1973); hunting data from various sources.

CAPERCAILLIE

Austria:
Declining; about 10,000 males present in 1966–67; 290 shot in 1977–78. Controlled spring hunting of males.

Bulgaria:
Declining, about 2,600 birds in early 1970s.

Czechoslovakia:
Declining; about 3,400 birds in Slovakia in 1975. Limited spring hunting permitted.

Estonia:
No hunting information; population of about 3,000 birds in 1975.

East Germany:
Nearly extirpated; total male population may be under 200.

Finland:
Decreasing; about 600,000 birds present in late 1960s. Hunting allowed; annual harvests 1959–67 from 47,000 to 104,000 birds; 1969–76, from 14,500 to 33,000 birds.

France:
Limited hunting in the Pyrenees (19 days in 1978); protected or extirpated elsewhere. No harvest data available, but Pyrenees population probably 5,000 birds in mid-1970s, and total French population about 6,000.

Great Britain:
Reintroduced into Scotland and hunted on private estates there since 1842. Total population probably between 1,000 and 10,000 pairs. No comprehensive harvest data.

Norway:
No total population estimates, but 1980 harvest was 13,000 birds. In 1960 the total population may have been 300,000 to 400,000 birds, and the harvest was approximately 40,000 birds, indicating a major recent decline.

Poland:
Total population estimated from 845 to 1,210 in late 1970s, and harvest of males during spring shooting was from 15 to 23 birds annually.

Spain:
The Cantabrian population had a few hundred males in the early 1970s.

Sweden:
Total population possibly 100,000 pairs; average harvest for 1978 to 1980 seasons was 16,000 birds.

Switzerland:
Decreasing, now totally protected. Population probably at least 1,100 males in early 1970s.

West Germany:
No legal hunting since 1973; total population of males may be under 2,000.

Yugoslavia:
Hunting allowed only in Slovenija Republic, where annual spring kill during 1970s was 51 to 105 males.

HAZEL GROUSE

Austria:
About 400 to 1,100 birds shot annually between 1945 and 1954.

Belgium:
Decreasing and protected; possibly about 200 pairs present.

East Germany:
Declining, and nearly extirpated by early 1970s.

Estonia:
No harvest data, but probably about 30,000 birds present in 1975.

Finland:
Recent yearly harvests have varied from 28,200 (1970) to 69,600

(1966). More than 200,000 pairs estimated present in 1950s, average population for 1964–71 about 650,000 birds.

Italy:	Almost extirpated from the western Alps.
Luxembourg:	Decreasing; about 20 pairs present in early 1970s.
Norway:	Annual harvests during 1970s were about 6,000 to 14,000 birds, averaging 10,200 between 1975 and 1980.
Sweden:	About 150,000 pairs present in the 1970s; average harvest for 1978 to 1980 seasons was 18,000 birds.
Switzerland:	About 300 to 1,000 birds were shot annually between 1931 and 1961.
West Germany:	Steady decline in population since 1900, with some areas possibly holding their own.

ROCK PTARMIGAN

Finland:	Population fluctuating around about 4,000 pairs.
France:	Some decline resulting from human activities; 1,000 to 10,000 pairs present.
Great Britain:	No long-term changes in twentieth century; population 1,000 to 10,000 pairs.
Italy:	Marked decrease in population.
Norway:	No information. Annual kill of both ptarmigans averaged 423,000 between 1975 and 1980.
Sweden:	Average harvest for 1978 through 1980 was 11,700 birds.
Switzerland:	Limited data suggest recent population stability with short-term fluctuations.

WILLOW PTARMIGAN

Estonia:	Declining; fewer than 2,000 birds.
Finland:	Marked annual fluctuations, but about 110,000 pairs in 1950s (Merikallio, 1958) and 40,000 shot per year (Rajala 1979).
Great Britain:	Marked decline since 1930s, British and Irish population may total fewer than 500,000 pairs, but Höhn (1980) estimated annual British kill at a minimum of 2.5 million!
Ireland:	Declining since about 1920; see Great Britain for population estimate.
Latvia:	Marked decline since 1900.
Lithuania:	Decreasing; only a few pairs left.
Norway:	Fluctuating populations, peaking every three to four years. Annual kill of both ptarmigans averaged 423,000 between 1975 and 1980.
Sweden:	Fluctuating around about 200,000 pairs. Annual kill from 1978 through 1980 averaged about 19,000 birds.

Literature Cited

Adams, J. L. 1956. A comparison of different methods of progesterone administration to the fowl in affecting egg production and molt. *Poultry Science* 35:323–26.

Ahnlund, H., and Helander, B. 1975. The food of the hazel grouse (*Tetrastes bonasia*) in Sweden. *Viltrevy* 9:221–40.

Aldous, S. E. 1943. Sharp-tailed grouse in the sand dune country of north-central North Dakota. *Journal of Wildlife Management* 7:23–31.

Aldrich, J. W. 1963. Geographic orientation of American Tetraonidae. *Journal of Wildlife Management* 27:529–45.

Aldrich, J. W., and Duvall, A. J. 1955. *Distribution of American gallinaceous game birds*. Circular 34. Washington, D.C.: U.S. Department of Interior, Fish and Wildlife Service.

Aldrich, J. W., and Friedmann, H. 1943. A revision of the ruffed grouse. *Condor* 45:85–103.

Ali, S., and Ripley, S. D. 1969. *Handbook of the birds of India and Pakistan*. Vol. 2. London: Oxford University Press.

Allen, G. A., III. 1968. Keeping and raising blue grouse. *Game Bird Breeders, Aviculturists and Conservationists' Gazette* 16(10–11):6–11.

Allen, H. M. 1977. Abnormal parental behaviour of captive male willow grouse *Lagopus l. lagopus*. *Ibis* 119:199–200.

Allen, H. M.; Boggs, C.; Norris, E.; and Doering, M. 1977. Parental behavior of captive willow grouse *Lagopus l. lagopus*. *Ornis Scandinavica* 8:175–83.

Amadon, D. 1966. The superspecies concept. *Systematic Zoology* 15:245.

American Ornithologists' Union (A.O.U.). 1957. *Check-list of North American birds*. 5th ed. Baltimore: Lord Baltimore Press.

——. 1982. Thirty-fourth supplement to the American Ornithologists' Union Check-list of North American birds. *Auk* 99:1–16CC.

Ammann, G. A. 1944. Determining the age of pinnated and sharp-tailed grouse. *Journal of Wildlife Management* 8:170–71.

———. 1957. *The prairie grouse of Michigan.* Michigan Department of Conservation Technical Bulletin.

———. 1963a. Status and management of sharp-tailed grouse in Michigan. *Journal of Wildlife Management* 27:802–9.

———. 1963b. Status of spruce grouse in Michigan. *Journal of Wildlife Management* 27:591–93.

Anderson, R. K. 1969. Prairie chicken responses to changing booming-ground cover type and height. *Journal of Wildlife Mangement* 33:636–43.

Andreev, A. V. 1977. [Reproductive behavior in black-billed capercaillie in northeast Siberia]. *Ornithologiya* 13:110–16. In Russian.

———. 1979. Reproductive behavior in black-billed capercaillie compared to capercaillie. In Lovel 1979, pp. 135–39.

Andreev, A. V., and Krechmar, A. V. 1976. [Radiotelemetric study of microclimate in snow resting places of *Tetrastes bonasia sibiricus*]. *Zoologicheskii Zhurnal* 55:1113–14. In Russian with English summary.

Angelstam, P. 1979. Black grouse *Lyrurus tetrix* reproductive success and survival rate in peak and crash small-rodent years in central Sweden. In Lovel 1979, pp. 101–11.

Archibald, H. L. 1975. Temporal patterns of spring space use by ruffed grouse. *Journal of Wildlife Management* 39:472–81.

Artmann, J. W. 1970. Spring and summer ecology of the sharptail grouse. Ph.D. dissertation, University of Minnesota.

Aschenbrenner, H. 1982. Keeping and rearing of grouse in enclosures—problems and experiences. In Lovel 1982, pp. 212–17.

Aschenbrenner, H.; Bergmann, H.; and Müller, F. 1978. Gefangenschaftsbrut beim Hazelhuhn (*Bonasia bonasia* L.). *Pirsch* 30:70–75.

Aschenbrenner, H., and Scherzinger, W. 1978. Wenn der Hazelhuhn balzt. *Pirsch—Der Deutsche Jäger,* vol. 19, unpaged.

Aubin, A. E. 1970. Territory and territorial behavior of male ruffed grouse in southwestern Alberta. M.S. thesis, University of Alberta.

———. 1973. Aural communication in ruffed grouse. *Canadian Journal of Zoology* 50:1225–29.

Averin, U. V. 1938. [Caucasian black grouse]. *Transactions of the Caucasian State Game Preserve* 1:57–86. In Russian.

Bailey, A. M., and Niedrach, R. J. 1967. *A pictorial checklist of Colorado birds, with brief notes on the status of each species in neighboring states of Nebraska, Kansas, New Mexico, Utah and Wyoming.* Denver: Denver Museum of Natural History.

Bailey, F. M. 1928. *Birds of New Mexico.* Santa Fe: New Mexico Department of Fish and Game.

Baker, M. F. 1952. Population changes of the greater prairie chicken in Kansas. *Transactions of the Seventeenth North American Wildlife Conference*, pp. 259–366.

———. 1953. *Prairie chickens in Kansas*. University of Kansas Museum of Natural History and State Biological Survey Miscellaneous Publication no. 5.

Ballard, W. B., and Robel, R. J. 1974. Reproductive importance of dominant male greater prairie chickens. *Auk* 91:75–85.

Bannerman, D. A. 1963. *The birds of the British Isles*. Vol. 12. Edinburgh: Oliver and Boyd.

Barth, E. K. 1953. Calculation of egg volume based on loss of weight during incubation. *Auk* 70:151–59.

Batterson, W. M., and Morse, W. B. 1948. *Oregon sage grouse*. Oregon Game Commission Fauna Series no. 1.

Beck, T. D. I. 1977. Sage grouse flock characteristics and habitat selection in winter. *Journal of Wildlife Management* 41:18–26.

Beck, T. D. I., and Braun, C. 1978. Weights of Colorado sage grouse. *Condor* 80:241–43.

Beddard, F. E. 1889. *The structure and classification of birds*. London: Longmans, Green.

Beebe, W. 1926. *Pheasants, their lives and homes*. 2 vols. Garden City, N.Y.: Doubleday, Page.

Beer, J. 1943. Food habits of the blue grouse. *Journal of Wildlife Management* 7:32–44.

Beick, W. 1927. Die Eier von *Tetrastes seversowi* Prezw. *Ornithologische Monatsbericht* 35:176–77.

Bendell, J. F. 1955a. Disease as a control of a population of blue grouse, *Dendragapus obscurus fuliginosus* (Ridgway). *Canadian Journal of Zoology* 33:195–223.

———. 1955b. Age, molt and weight characteristics of blue grouse. *Condor* 57:354–61.

———. 1955c. Age, breeding behavior and migration of sooty grouse, *Dendragapus obscurus fuliginosus* (Ridgway). *Transactions of the Twentieth North American Wildlife Conference*, pp. 367–81.

Bendell, J. F., and Elliott, P. W. 1966. Habitat selection in the blue grouse. *Condor* 68:431–46.

———. 1967. *Behavior and the regulation of numbers in blue grouse*. Canadian Wildlife Service Report Series no. 4.

Bendell, J. F., and Zwickel, F. C. 1979. Problems in the abundance and distribution of blue, spruce and ruffed grouse in North America. In Lovel 1979, pp. 48–63.

Bendire, C. 1892. *Life histories of North American birds*. U.S. National Museum Special Bulletin, vol. 1, no. 1.

Bent, A. C. 1932. *Life histories of North American gallinaceous birds*. U.S. National Museum Bulletin 162.

Berg, L. S. 1950. *Natural regions of the USSR*. New York: Macmillan.

Berger, D. D.; Hamerstrom, F.; and Hamerstrom, F. N., Jr. 1963. The effect of raptors on prairie chickens on booming grounds. *Journal of Wildlife Management* 27:778–91.

Bergerud, A. T. 1970*a*. Vulnerability of willow ptarmigan to hunting. *Journal of Wildlife Management* 34:282–85.

____. 1970*b*. Population dynamics of the willow ptarmigan *Lagopus lagopus alleni* L. in Newfoundland 1955 to 1965. *Oikos* 21:299–325.

____. 1972. Changes in the vulnerability of ptarmigan to hunting in Newfoundland. *Journal of Wildlife Management* 36:104–9.

Bergerud, A. T.; Peters, S. S.; and McGrath, R. 1963. Determining sex and age of willow ptarmigan in Newfoundland. *Journal of Wildlife Management* 27:700–711.

Bergmann, H. H.; Klaus, S.; Müller, F.; and Wiesner, J. 1975. Individualität und Artspezifität in den Gesangsstrophen eineger Population des Haselhuhns (*Bonasa bonasia bonasia* L., Tetraoninae, Phasianidae). *Behaviour* 55:94–114. English summary.

____. 1978. *Das Haselhuhn*. Neue Brehm-Bucherei 77. Wittenberg Lutherstadt: A. Ziemsen Verlag.

Bernard, A. 1982. An analysis of black grouse nesting and brood habitats in the French Alps. In Lovel 1982, pp. 156–72.

Berndt, R., and Meise, W. 1962. *Naturgeschichte der Vogel*. Vol. 2. Stuttgart: Franckh.

Blackford, J. L. 1958. Territoriality and breeding behavior of a population of blue grouse in Montana. *Condor* 60:145–58.

____. 1963. Further observations on the breeding behavior of a blue grouse population in Montana. *Condor* 60:485–513.

Boag, D. A. 1965. Indicators of sex, age, and breeding phenology in blue grouse. *Journal of Wildlife Management* 29:103–8.

____. 1966. Population attributes of blue grouse in southwestern Alberta. *Canadian Journal of Zoology* 44:799–814.

Boag, D. A.; McCourt, K. H.; Herzog, P. W.; and Alway, J. H. 1979. Population regulation in spruce grouse: A working hypothesis. *Canadian Journal of Zoology* 57:2275–84.

Boag, D., and Sumanik, K. M. 1969. Characteristics of drumming sites selected by ruffed grouse in Alberta. *Journal of Wildlife Management* 33:621–28.

Boback, A. W., and Müller-Schwarze, D. *Das Birkhuhn*. Neue Brehm-Bucheri 397. Wittenberg Lutherstadt: A. Ziemsen Verlag.

Bock, W. J., and Ferrand, J., Jr. 1980. The number of species and genera of Recent birds: A contribution to comparative systematics. *American Museum Novitates* 2703:1–29.

Børset, E., and Krafft, A. 1973. Black grouse *Lyrurus tetrix* and capercaillie *Tetrao urogallus* brood habitats in a Norwegian spruce forest. *Oikos* 24:1–7.

Bossert, A. 1980. Winterökologie des Alepnschneehuhns (*Lagopus mutus* Montin) im

Aletschgebiet, Schweize Alpen. *Ornithologische Beobachter* 77:121–61. English summary.

Boyce, M. S., and Tate, J., Jr. 1979. A bibliography of the sage grouse (*Centrocercus urophasianus*). University of Wyoming Agricultural Experiment Station, Science Monograph 38.

Bradbury, W. C. 1915. Notes on the nesting of the white-tailed ptarmigan. *Condor* 17:214–22.

Brander, R. B. 1967. Movements of female ruffed grouse during the mating season. *Wilson Bulletin* 79:28–36.

Braun, C. E. 1969. Population dynamics, habitat, and movements of white-tailed ptarmigan in Colorado. Ph.D. dissertation, Colorado State University.

——. 1970. Distribution and habitat of white-tailed ptarmigan in Colorado and New Mexico. Abstract of paper presented at forty-sixth annual meeting, Southwestern and Rocky Mountain Division, AAAS, April 22–25, 1970, Las Vegas, New Mexico.

——. 1971*a*. Determination of blue grouse sex and age from wing characteristics. Colorado Division of Wildlife, Game Information Leaflet no. 86.

——. 1971*b*. Habitat requirements of Colorado white-tailed ptarmigan. In *Proceedings of the Western Association of State Game and Fish Commissioners*; 9 page reprint.

Braun, C. E.; Hoffman, R. W.; and Rogers, G. E. 1976. *Wintering areas and winter ecology of white-tailed ptarmigan in Colorado.* Colorado Division of Wildlife Special Scientific Report no. 38.

Braun, C. E., and May T. A. 1972. Colorado alpine tundra investigations, 1966–71. *Proceedings of the 1972 Tundra Biome Symposium,* Lake Wilderness Center, University of Washington, pp. 165–68.

Braun, C. E.; Nish, D. H.; and K. M. Giesen. 1978. Release and establishment of white-tailed ptarmigan in Utah. *Southwestern Naturalist* 23:661–68.

Braun, C. E., and Pattie, D. L. 1969. A comparison of alpine habitats and white-tailed ptarmigan occurrence in Colorado and northern Wyoming. Abstract of paper presented at forty-fifth annual meeting, Southwestern and Rocky Mountain Division AAAS, May 8, 1969, Colorado Springs, Colorado.

Braun, C. E., and Rogers, G. E. 1967*a*. Determination of age and sex of the southern white-tailed ptarmigan. Colorado Division of Game, Fish and Parks Department Game Information Leaflet no. 54.

——. 1967*b*. Habitat and seasonal movements of white-tailed ptarmigan in Colorado. Abstract of paper presented at meeting of Colorado-Wyoming Academy of Science, April 28, 1967, Boulder, Colorado.

——. 1971. The white-tailed ptarmigan in Colorado. Colorado Division of Game, Fish and Parks Technical Publication no. 27.

Braun, C. E., and Willers, W. B. 1967. The helminth and protozoan parasites of North American grouse (family: Tetraonidae): A checklist. *Avian Diseases* 11:170–87.

Bremer, P. E. 1967. Sharp-tailed grouse in Minnesota. Minnesota Department of Conservation Informational Leaflet no. 15.

Brodkorb, P. 1964. Catalog of fossil birds. Part 2 (Anseriformes through Galliformes). *Bulletin of the Florida State Museum, Biological Sciences* 8:195–335.

Brooks, A. 1907. A hybrid grouse, Richardson's × sharp-tail. *Auk* 24:167–69.

———. 1926. The display of Richardson's grouse, with some notes on the species and subspecies of the genus *Dendragapus*. *Auk* 43:281–87.

———. 1930. The specialized feathers of the sage hen. *Condor* 32:205–7.

Brown, C. P. 1946. Food of Maine ruffed grouse by seasons and cover types. *Journal of Wildlife Management* 10:17–28.

Brown, D. E., and Smith, R. H. 1980. Winter-spring precipitation and population levels of blue grouse in Arizona. *Wildlife Society Bulletin* 8:136–41.

Browning, M. R. 1979. Distribution, geographic variation and taxonomy of *Lagopus mutus* in Greenland and northern Canada. *Dansk Ornithologisk Forenings Tidsskrift* 73:29–40.

Brüll, H.; Lindner, A.; von Luterotti, L.; and Scherzinger, W. 1977. *Die Waldhühner*. Hamburg and Berlin: Parey.

Buckley, J. L. 1954. Animal population fluctuations in Alaskan history. *Transactions of the Nineteenth North American Wildlife Conference,* pp. 338–54.

Bump, G.; Darrow, R.; Edminster, F.; and Crissey, W. 1947. *The ruffed grouse: Life history, propagation, management.* Albany: New York State Conservation Department.

Bunnell, S. D.; Rensel, J. A.; Kimball, J. F., Jr., and Wolf, M. L. 1977. Determination of age and sex of dusky blue grouse. *Journal of Wildlife Management* 41:662–66.

Campbell, H. 1972. A population study of lesser prairie chickens in New Mexico. *Journal of Wildlife Management* 36:689–99.

Cannon, R. W., and Knopf, F. L. 1980. Distribution and status of the lesser prairie chicken in Oklahoma. In Vohs and Knopf 1980, pp. 71–74.

Carr, R. 1969. Raising ptarmigan and spruce grouse. *Game Bird Breeders, Aviculturalists and Conservationists' Gazette* 17(1):6–9.

Castroviejo, J. 1967. Eine neue Auerhuhnrasse von der Ibersichen Halbinsel. *Journal für Ornithologie* 108:220–21.

———. 1975. El urogallo *Tetrao urogallus* en España. *Monographias Estacion Biologia Doñana,* no. 3, pp. 1–546. English summary.

Chambers, R. E., and Sharp, W. M. 1958. Movement and dispersal within a population of ruffed grouse. *Journal of Wildlife Management* 22:231–39.

Cheng, Tso-hsin, ed. 1963. *China's economic fauna: Birds.* Science Publishing Society, Peiping. Translated 1964 by U.S. Department of Commerce, Washington, D.C.

———. 1978. *Fauna Sinica, series vertebratica: Aves.* vol. 4, *Galliformes.* Peking: Sciences Press, Academia Sinica. In Chinese.

——. 1979*a*. A sketch of the avian fauna of China, with special reference to galliform species. In Lovel 1979, pp. 45–47.

——. 1979*b*. Taxonomic and ecological notes on capercaillies and black grouse in China. In Lovel 1979, pp. 83–86.

Choate, T. S. 1960. Observations on the reproductive activities of the white-tailed ptarmigan (*Lagopus leucurus*) in Glacier Park, Montana. M.S. thesis, Montana State University.

——. 1963. Habitat and population dynamics of white-tailed ptarmigan in Montana. *Journal of Wildlife Management* 27:684–99.

Christisen, D. M. 1967. A vignette of Missouri's native prairie. *Missouri Historical Review* 61:166–86.

——. 1969. National status and management of the greater prairie chicken. *Transactions of the Thirty-fourth North American Wildlife and Natural Resources Conference*, pp. 207–17.

Clark, H. K. 1899. The feather-tracts of North American grouse and quail. *Proceedings of the United States National Museum* 21:641–53.

Coats, J. 1955. Raising lesser prairie chickens in captivity. *Kansas Fish and Game* 13(2):16–20.

Cockrum, E. L. 1952. A check-list and bibliography of hybrid birds in North America north of Mexico. *Wilson Bulletin* 64:140–59.

Collette, R. 1886. On the hybrid between *Lagopus albus* and *Tetrao tetrix*. *Proceedings of the Zoological Society of London,* pp. 224–30.

——. 1898. En ny bastardform blandt Norges Tetraonider, "Fjeldrype-Orre" (*Lagopus mutus* × *Tetrao tetrix*). *Bergens Museums Aarberetning* 1897, no. 7.

Conover, H. B. 1926. Game birds of the Hooper Bay region, Alaska. *Auk* 43:162–80, 303–18.

Copelin, F. F. 1963. *The lesser prairie chicken in Oklahoma.* Oklahoma Wildlife Conservation Department Technical Bulletin no. 6.

Cracraft, J. 1981. Toward a phylogenetic classification of Recent birds. *Auk* 98:681–714.

Cramp, S., and Simmons, K. E. L., eds. 1980. *The birds of the western Palearctic.* Vol. 2. London: Oxford University Press.

Crawford, J. A. 1980. Status, problems and research needs of the lesser prairie chicken. In Vohs and Knopf 1980, pp. 1–7.

Crawford, J. E. 1960. The movements, productivity and management of sage grouse in Clark and Fremont counties, Idaho. M.S. thesis, University of Idaho.

Crichton, V. 1963. Autumn and winter foods of the spruce grouse in central Ontario. *Journal of Wildlife Management* 27:597.

Crunden, C. W. 1963. Age and sex of sage grouse from wings. *Journal of Wildlife Management* 27:846–50.

——, ed. 1959. *The western states sage grouse questionnaire.* Carson City: Nevada Fish and Game Department.

Dalke, P. D.; Pyrah, D. B.; Stanton, D. C.; Crawford, J. E., and Schlatterer, E. F. 1960. Seasonal movements and breeding behavior of sage grouse in Idaho. *Transactions of the Twenty-fifth North American Wildlife Conference*, pp. 396–407.

____. 1963. Ecology, productivity, and management of sage grouse in Idaho. *Journal of Wildlife Management* 27:811–41.

Davis, C. A.; Riley, T. Z.; Smith, R. A.; Suminski, H. R., and Wisdom, M. J. 1979. *Habitat evaluation of lesser prairie chickens in eastern Chaves County, New Mexico*. Las Cruces: New Mexico State Agricultural Experiment Station.

Davis, C. A.; Riley, T. Z.; Smith, R. A.; and Wisdom, M. J. 1980. Spring-summer foods of lesser prairie chickens in New Mexico. In Vohs and Knopf 1980, pp. 75–80.

Davis, D. E., and Domm, L. V. 1942. The sexual behavior of hormonally treated domestic fowl. *Proceedings of the Society for Experimental Biology and Medicine* 48:667–69.

Davis, J. A. 1968. The postjuvenal wing and tail molt of the ruffed grouse (*Bonasa umbellus monticola*) in Ohio. *Ohio Journal of Science* 68:305–12.

____. 1969a. Aging and sexing criteria for Ohio ruffed grouse. *Journal of Wildlife Management* 33:628–36.

____. 1969b. Relative abundance and distribution of ruffed grouse in Ohio, past, present and future. Mimeographed. Ohio Department of Natural Resources, Division of Wildlife In-service Document no. 62.

Davis, J. W.; Anderson, R. C.; Karstad, L.; and Trainer, D. O. 1971. *Infectious and parasitic diseases in wild birds*. Ames: Iowa State University Press.

Davison, V. E. 1940. An 8-year census of lesser prairie chickens. *Journal of Wildlife Management* 4:55–62.

Dawson, W. L. 1923. *The birds of California*. Vol. 3. San Francisco: South Moulton.

Degn, H. J. 1979. The Danish population of black grouse. In Lovel 1979, pp. 27–31.

Delacour, J. 1951. *The pheasants of the world*. London: Country Life.

Dellinger, J. O. 1967. My experiences breeding and raising Mearns quail. *Game Bird Breeders, Aviculturalists and Conservationists' Gazette*, 16(4):9–10.

Dementiev, G. P., and Gladkov, N. A., eds. 1967. *Birds of the Soviet Union*. Vol. 4. Jerusalem: Israel Program for Scientific Translations.

Dixon, J. 1927. Contribution to the life history of the Alaska willow ptarmigan. *Condor* 29:213–23.

Doerr, P. D. 1973. Ruffed grouse ecology in central Alberta: Demography, winter feeding activities, and the impact of fire. Ph.D. dissertation, University of Wisconsin.

Dolbik, M. S. 1968. A study of the ecology of the common Galliformes in Belorussia. pp. 68–74, in *Proceedings of the 4th Baltic Ornithological Congress*, 1960; ''Birds of the Baltic region: Ecology and migrations.'' Translated from Russian by the Israel Program for Scientific Translations, Jerusalem.

Donaldson, J. L., and Bergerud, A. T. 1974. Behavior and habitat selection of an insular population of blue grouse. *Syesis* 7:115–27.

Dorney, R. S. 1959. *Relationships of ruffed grouse to forest cover types in Wisconsin.* Wisconsin Conservation Department Technical Bulletin no. 18.

——. 1963. Sex and age structure of Wisconsin ruffed grouse populations. *Journal of Wildlife Management* 27:599–603.

Dorney, R. S., and Holzer, F. V. 1957. Spring aging methods for ruffed grouse cocks. *Journal of Wildlife Management* 21:268–74.

Dresser, H. E. 1876. Remarks on a hybrid between the black grouse and the hazel grouse. *Proceedings of the Zoological Society of London,* pp. 345–47.

Dwight, J., Jr. 1900. The moult of the North American Tetraonidae (quails, partridge and grouse). *Auk* 17:34–51, 143–66.

Dzieciolowski, R., and Matuszewski, G. 1982. Habitat preferences of capercaillie in lowland forests of Poland. In Lovel 1982, pp. 139–47.

Edminster, F. C. 1947. *The ruffed grouse: Its life story, ecology and management.* New York: Macmillan.

——. 1954. *American game birds of field and forest.* New York: Charles Scribner's Sons.

Elliot, D. G. 1864–65. *A monograph of the Tetraoninae, or family of the grouse.* New York.

Ellison, L. 1966. Seasonal foods and chemical analysis of winter diet of Alaskan spruce grouse. *Journal of Wildlife Management* 30:729–35.

——. 1968*a*. Sexing and aging Alaskan spruce grouse by plumage. *Journal of Wildlife Management* 32:12–16.

——. 1968*b*. Movements and behavior of Alaskan spruce grouse during the breeding season. Mimeographed. Transactions of the meeting of the California-Nevada Section of the Wildlife Society.

——. 1972. Role of winter food in regulating numbers of Alaskan spruce grouse. Ph.D. dissertation, University of California, Berkeley.

——. 1973. Seasonal social organization and movements of spruce grouse. *Condor* 75:375–85.

——. 1974. Population characteristics of Alaskan spruce grouse. *Journal of Wildlife Management* 38:383–95.

——. 1975. Density of Alaskan spruce grouse before and after fire. *Journal of Wildlife Management* 39:468–71.

——. 1979. Black grouse population characteristics on a hunted and three nonhunted areas in the French Alps. In Lovel 1979, pp. 64–73.

Ellison, L. N., and Weeden, R. B. 1979. Seasonal and local weights of Alaskan spruce grouse. *Journal of Wildlife Management* 43:176–83.

Eng, R. L. 1955. A method for obtaining sage grouse age and sex ratios from wings. *Journal of Wildlife Management* 19:267–72.

——. 1959. A study of the ecology of male ruffed grouse (*Bonasa umbellus* L.) on the Cloquet Forest Research Center, Minnesota. Ph.D. dissertation, University of Minnesota.

____. 1963. Observations on the breeding biology of male sage grouse. *Journal of Wildlife Management* 27:841–46.

____. 1971. Two hybrid sage grouse × sharp-tailed grouse from Montana. *Condor* 73:491–93.

Eng, R. L., and Schladweiler, P. 1972. Sage grouse winter movements and habitat use in central Montana. *Journal of Wildlife Management* 36:141–46.

Evans, K. E., and Gilbert, D. L. 1969. A method for evaluating greater prairie chicken habitat in Colorado. *Journal of Wildlife Management* 33:643–49.

Evans, R. M. 1961. Courtship and mating behavior of sharp-tailed grouse *Pedioecetes phasianellus jamesi*. M.S. thesis, University of Alberta.

____. 1969. Territorial stability in sharp-tailed grouse. *Wilson Bulletin* 81:75–78.

Eyre, S. R. 1968. *Vegetation and soils: A world picture.* New York: Aldine.

Farner, D. S. 1955. Birdbanding in the study of population dynamics. In *Recent studies in avian biology*. Urbana: University of Illinois Press.

Fay, L. D. 1963. Recent success in raising ruffed grouse in captivity. *Journal of Wildlife Management* 27:642–47.

Fenna, L., and Boag, D. A. 1974. Adaptive significance of the caeca in Japanese quail and spruce grouse. *Canadian Journal of Zoology* 52:1577–84.

Fjeldså, J. 1977. *Guide to the young of European precocial birds.* Copenhagen: Skarv.

Flint, V. 1978. [Birds]. In *Red data book for the USSR*, USSR Ministry of Agriculture, pp. 90–149. Moscow: Lesnaya Promyshlennost. In Russian.

Fowle, C. D. 1960. A study of the blue grouse (*Dendragapus obscurus* [Say]) on Vancouver Island, British Columbia. *Canadian Journal of Zoology* 38:701–13.

Gabrielson, I. N., and Jewett, S. G. 1940. *Birds of Oregon.* Corvallis: Oregon State University.

Gajdar, A. A. 1974. [Ringing *Tetrastes bonasia* L. and its results]. *Byulleten Moskovskogo Obshchestve Ispytatelei Prirody Otdelenie Biologischeskii* 78:120–24. In Russian, English summary.

Gajdar, A. A., and Zhitkov, B. M. 1974. [Method of age determination of *Tetrastes bonasia*]. *Ekologiya* 1974(3):102–3. In Russian, translated in *Soviet Journal of Ecology* 5(3):290–91.

Game bird geography. 1961. *Utah Fish and Game* 17(11):4–7.

Gaunt, A. S., and Wells, S. K. 1973. Models of syringeal mechanism. *American Zoologist* 13:1227–47.

Gavrin, V. F. 1969. (The ecology of hazel grouse in the Bielowicza forest). *Gosud. Zapov. Ochotn. Choz. Belovezskaja Pusca Minsk* 3:146–72. (In Russian).

Geist, V. 1977. A comparison of social adaptations in relation to ecology in gallinaceous bird and ungulate societies. *Annual Review of Ecology and Systematics* 8:193–207.

Giesen, K. M., and Braun, C. E. 1979a. Nesting behavior of female white-tailed ptarmigan in Colorado. *Condor* 81:215–17.

____. 1979*b*. A technique for age determination of juvenile white-tailed ptarmigan. *Journal of Wildlife Management* 43:508–11.

____. 1979*c*. Renesting of white-tailed ptarmigan in Colorado. *Condor* 81:217–18.

Giesen, K. M.; Braun, C. E.; and May, T. A. 1980. Reproduction and nest-site selection by white-tailed ptarmigan in Colorado. *Wilson Bulletin* 92:188–99.

Gilfillan, M. C., and Bezdek, H. 1944. Winter foods of the ruffed grouse, in Ohio. *Journal of Wildlife Management* 8:208–10.

Gill, R. B. 1966. A literature review on the sage grouse. Colorado Department of Game, Fish and Parks and Colorado Cooperative Wildlife Research Unit, Report no. 6.

Gindre, R. 1979. Status of capercaillie *Tetrao urogallus* and black grouse *Lyrurus tetrix* in France. In Lovel 1979, pp. 35–44.

Girard, G. L. 1937. Life history, habits and food of the sage grouse. *University of Wyoming Publications* 3:1–56.

Gladfelter, H. L., and McBurney, R. S. 1971. Mating activity of ruffed grouse. *Auk* 88:176–77.

Glutz, U. N. von Blotzheim, ed. 1973. *Handbuch der Vögel Mitteleuropas*. Vol. 5. *Galliformes und Gruiformes*. Frankfurt: Akademische Verlagsgesellschaft.

Godfrey, G. A. 1975. Home range characteristics of ruffed grouse broods in Minnesota. *Journal of Wildlife Management* 39:287–98.

Godfrey, W. E. 1966. *The birds of Canada*. Ottawa: Queen's Printer.

Gower, W. C. 1939. The use of the bursa of Fabricius as an indication of age in game birds. *Transactions of the Fourth North American Wildlife Conference,* pp. 426–30.

Grange, W. W. 1948. *Wisconsin grouse problems*. Wisconsin Conservation Department Publication no. 338.

Gray, A. P. 1958. *Bird hybrids*. Farnham Royal: Commonwealth Agricultural Bureau.

Greenberg, D. B. 1949. *Raising gamebirds in captivity*. Princeton: Van Nostrand.

Greenewalt, C. H. 1968. *Bird song: Acoustics and physiology*. Washington, D.C.: Smithsonian Institution Press.

Grieg, J. A. 1889. *Lagopus urogallus-albus:* Ein neuer Moorschneehuhnbastard. *Bergens Museums Aarberetning* 1889(5):1–13.

Griner, L. A. 1939. A study of the sage grouse, *Centrocercus urophasianus*, with special reference to life history, habitat requirements and numbers and distribution. M.S. thesis, Utah State Agricultural College.

Grinnell, J.; Bryant, H. C.; and Storer, T. I. 1918. *The game birds of California*. Berkeley: University of California Press.

Grinnell, J., and Miller, A. H. 1944. *The distribution of the birds of California*. Cooper Ornithological Club, Pacific Coast Avifauna no. 27.

Grinnell, J., and Storer, T. I. 1924. *Animal life in the Yosemite*. Berkeley: University of California Press.

Gromme, O. J. 1963. *Birds of Wisconsin*. Madison: University of Wisconsin Press.

Gross, A. O. 1928. The heath hen. *Memoirs of the Boston Society of Natural History* 6:491–588.

Gross, W. B. 1964. Voice production by the chicken. *Poultry Science* 43:1005–8.

——. 1968. Voice production by the turkey. *Poultry Science* 47:1101-5.

Grzimek, B., ed. 1972. Grouse. In *Grzimek's animal life encyclopedia*, 7:451–72. New York: Van Nostrand Reinhold.

Guiguet, C. J. 1955. *The birds of British Columbia: Upland game birds*. British Columbia Provincial Museum Handbook no. 10.

Gullion, G. W. 1967*a*. Selection and use of drumming sites by male ruffed grouse. *Auk* 84:87–112.

——. 1967*b*. *The ruffed grouse in northern Minnesota*. University of Minnesota Forest Wildlife Relations Project, Cloquet Forest Research Center, Cloquet, Minnesota.

——. 1969. Aspen-ruffed grouse relationships. Abstract of paper presented at Thirty-first Midwest Wildlife Conference, December 8, 1969, St. Paul, Minnesota.

——. 1976. Reevaluation of "activity clustering" by male grouse. *Auk* 93:192–93.

——. 1981. Non-drumming in a ruffed grouse population. *Wilson Bulletin* 93:372–82.

Gullion, G. W., and Christensen, G. C. 1967. A review of the distribution of gallinaceous game birds in Nevada. *Condor* 58:128–38.

Gullion, G. W., and Gullion, A. M. 1961. Weight variations of captive Gambel quail in the breeding season. *Condor* 63:95–97.

Gullion, G. W., and Marshall, W. H. 1968. Survival of ruffed grouse in a boreal forest. *Living Bird* 7:117–67.

Haas, G. H. 1974. Habitat selection, reproduction, and movements in female spruce grouse. Ph.D. dissertation, University of Minnesota.

Hagen, Y. 1935. [Ringing of willow ptarmigan at Rauland and Tinn]. *Meddelelser Fra Statens Viltenderoksler*. 1:1–46. In Norwegian.

Hagen, A. 1980. A change in leadership in the forest. *Wildlife*: 22(12):18–20.

Hainard, R., and Meylan, O. 1935. Notes sur le grand tetra. *Alauda* 7:282–327.

Haker, M., and Myrberget, S. 1969. Browsing by *Lyrurus tetrix* on *Pinus mugo* var. *arborea*. *Sterna* 8:243–47. In Norwegian, English summary.

Hale, J. B., and Dorney, R. S. 1963. Seasonal movements of ruffed grouse in Wisconsin. *Journal of Wildlife Management* 27:648–56.

Hale, J. B.; Wendt, R. F.; and Halazon, G. C. 1954. *Sex and age criteria for Wisconsin ruffed grouse*. Wisconsin Conservation Department Technical Wildlife Bulletin no. 9.

Hamerstrom, F.; Berger, D. D.; and Hamerstrom, F. N., Jr. 1965. The effect of mammals on prairie chickens on booming grounds. *Journal of Wildlife Management* 29:536–42.

Hamerstrom, F. N., Jr. 1939. A study of Wisconsin prairie chicken and sharp-tailed grouse. *Wilson Bulletin* 51:105–20.

——. 1963. Sharptail brood habitat in Wisconsin's northern pine barrens. *Journal of Wildlife Management* 27:793–802.

Hamerstrom, F. N., Jr., and Hamerstrom, F. 1949. Daily and seasonal movements of Wisconsin prairie chickens. *Auk* 66:313–37.

——. 1951. Mobility of the sharptailed grouse in relation to its ecology and distribution. *American Midland Naturalist* 46:174–226.

——. 1960. Comparability of some social displays of grouse. *Proceedings of the Twelfth International Ornithological Congress, 1958*, pp. 274–93.

——. 1961. Status and problems of North American grouse. *Wilson Bulletin* 73:284–94.

——. 1973. *The prairie chicken in Wisconsin: Highlights of a twenty-two-year study of counts, behavior, movements, turnover and habitat.* Wisconsin Department of Natural Resources Technical Bulletin no. 64.

Hamerstrom, F. N., Jr.; Mattson, O. E.; and Hamerstrom, F. 1957. *A guide to prairie chicken management.* Wisconsin Conservation Department Technical Wildlife Bulletin no. 15.

Harju, H. J. 1971. Spruce grouse copulation. *Condor* 73:380–81.

——. 1974. An analysis of some aspects of the ecology of dusky grouse. Ph.D. dissertation, University of Wyoming.

Harper, F. 1953. Birds of the Nueltin Lake expedition, Keewatin, 1947. *American Midland Naturalist* 49:1–116.

——. 1958. *Birds of the Ungava peninsula.* University of Kansas Museum of Natural History Miscellaneous Publication no. 17.

Harris, C. L.; Gross, W. B.; and Robeson, A. 1968. Vocal acoustics of the chicken. *Poultry Science* 47:107–12.

Hart, C. M.; Lee, O. S.; and Low, J. B. 1952. *The sharp-tailed grouse in Utah, its life history, status and management.* Utah Department of Fish and Game Publication no. 3.

Hartman, F. A. 1955. Heart weight in birds. *Condor* 57:221–38.

Hass, G. H. 1974. Habitat selection, reproduction and movements in female spruce grouse. Ph.D. dissertation, University of Minnesota, Minneapolis.

Heinroth, O., and Heinroth, K. 1928–31. *Die Vogel Mitteleuropas.* 4 vols. Berlin: H. Bermühler.

Helminen, M. 1963. Composition of the Finnish populations of capercaillie and black grouse in the autumns of 1952–1961, as revealed by a study of wings. *Papers on Game Research* 23:123–39.

Henderson, F. R. 1964. Grouse and grass, twin crops. *South Dakota Conservation Digest* 31(1):16–19.

Henderson, F. R.; Brooks, F. W.; Wood, R. E.; and Dahlgren, R. B. 1967. Sexing of prairie grouse by crown feather patterns. *Journal of Wildlife Management* 37:764–69.

Herman, M. F. 1980. Spruce grouse habitat requirements in western Montana. Ph.D. dissertation, University of Montana.

Herzog, P. W. 1978. Food selection by female spruce grouse during incubation. *Journal of Wildlife Management* 42:632–36.

Herzog, P. W., and Boag, D. A. 1978. Dispersal and mobility in a local population of spruce grouse. *Journal of Wildlife Management* 42:853–66.

Herzog, P. W., and Keppie, D. E. 1980. Migration in a local population of spruce grouse. *Condor* 82:366–72.

Hewson, R. 1973. The moults of captive Scottish ptarmigan (*Lagopus mutus*). *Journal of Zoology* 171:177–87.

Hickey, J. J. 1955. Some American population research on gallinaceous birds. In *Recent studies in avian biology*. Urbana: University of Illinois Press.

Higby, L. W. 1969. A summary of the Longs Creek sagebrush control project. *Proceedings of the Sixth Biennial Western States Sage Grouse Workshop*, pp. 164–68 (Wyoming Game and Fish Commission).

Hillman, C. N., and Jackson, W. W. 1973. *The sharp-tailed grouse in South Dakota*. South Dakota Department of Game, Fish and Parks Technical Bulletin no. 3.

Hjorth, I. 1967. Fortplantningsbeteende inom Hönfagelfamiljen Tetraonidae. [Reproductive behavior in male grouse]. *Vår Fågelvärld* 26:193–243.

——. 1970. Reproductive behaviour in Tetraonidae, with special reference to males. *Viltrevy* 7:183–596.

——. 1976. The divalent origin and adaptive radiation of grouse songs. *Ornis Scandinavica* 7:147–57.

——. 1982. Attributes of capercaillie display grounds and the influence of forestry: A progress report. In Lovel 1982, pp. 26–35.

Hobmaier, A. 1932. The life history and control of the cropworm, *Capillaria capillaria*, in quail. *California Fish and Game* 18:290–96.

Hoffman, D. M. 1963. The lesser prairie chicken in Colorado. *Journal of Wildlife Management* 27:726–32.

Hoffmann, R. S. 1956. Observations on a sooty grouse population at Sage Hen Creek, California. *Condor* 58:321–36.

——. 1961. The quality of the winter food of blue grouse. *Journal of Wildlife Management* 25:209–10.

Hoffmann, R. W., and Braun, C. E. 1975. Migration of a wintering population of white-tailed ptarmigan in Colorado. *Journal of Wildlife Management* 39:485–90.

——. 1977. Characteristics of a wintering population of white-tailed ptarmigan in Colorado. *Wilson Bulletin* 89:107–15.

Hoffmann, R. 1927. *Birds of the Pacific states*. Boston: Houghton Mifflin.

Hofstad, M. S., ed. 1972. *Diseases of poultry*, 6th ed. Ames: Iowa State University Press.

Högland, N. H. 1955. Body temperature, activity and reproduction in the capercaillie. *Viltrevy* 1:1–87. In Swedish, with English summary.

——. 1956. On sex-distinguishing characters in capercaillie chicks. *Viltrevy* 1:150–57. In Swedish, with English summary.

——. 1980. Studies on the winter ecology of the willow grouse (*Lagopus lagopus lagopus* L.). *Viltrevy* 11:249–70.

Höhn, E. O. 1953. Display and mating behaviour of the black grouse *Lyrurus tetrix* (L.). *British Journal of Animal Behaviour* 1:48–58.

——. 1957. Observations on display and other forms of behavior of certain arctic birds. *Auk* 74:203–14.

——. 1977. The "snowshoe effect" of the feathering on ptarmigan feet. *Condor* 79:380–82.

——. 1980. *Die Schneehühner*. 2d. ed. Neue Brehm-Bucherei no. 408. Wittenberg Lutherstadt: A. Ziemsen Verlag.

Höhn, E. O., and Braun, C. E. 1977. Hormones and seasonal plumage color in ptarmigans. *Proceedings of the Physiological Society* 1977:34–35.

Holman, J. A. 1964. Osteology of gallinaceous birds. *Quarterly Journal of the Florida Academy of Science* 27:230–52.

Hopkinson, E. 1926. *Records of birds bred in captivity*. London: Witherby.

Horkel, J. D. 1979. Cover and space requirements of Attwater's prairie chicken (*Tympanuchus cupido attwateri*) in Refugio County, Texas. Ph.D. dissertation, Texas A & M University.

Høst, P. 1942. Effect of light on the molts and sequences of plumage in the willow ptarmigan. *Auk* 58:388–403.

Howard, H. 1966. Two fossil birds from the Lower Miocene of South Dakota. *Los Angeles County Museum Contributions in Science*, no. 107, pp. 1–8.

Howes, J. R. 1968. Raising game birds: Principles of good management. *Game Bird Breeders, Aviculturalists and Conservationists' Gazette* 16(2):8–11.

Hudson, G. E. 1955. An apparent hybrid between the ring-necked pheasant and the blue grouse. *Condor* 57:304.

Hudson, G. E.; Lanzillotti, P. J.; and Edward, G. D. 1959. Muscles of the pelvic limb in galliform birds. *American Midland Naturalist* 61:1–67.

——. 1964. Muscles of the pectoral limb in galliform birds. *American Midland Naturalist* 71:1–113.

Hudson, G. E.; Parker, R. A.; Berge, J. V.; and Lanzillotti, P. J. 1966. A numerical analysis of the modifications of the appendicular muscles in various genera of gallinaceous birds. *American Midland Naturalist* 76:1–73.

Huff, D. D. 1973. A preliminary study of ruffed grouse-aspen nutrient relationships. Ph.D. dissertation, University of Minnesota.

Humphrey, P. S., and Parkes, K. C. 1959. An approach to the study of molts and plumages. *Auk* 76:1–31.

Irving, L. 1960. Birds of Anaktuvuk Pass, Kobuk and Old Crow: A study in arctic adaptation. *United States National Museum Bulletin* 217:1–409.

Irving, L.; West, C. G.; Peyton, L. J.; and Paneak, S. 1967. Migration of willow ptarmigan in arctic Alaska. *Arctic* 20:77–85.

Jackson, A. S., and DeArment, R. 1963. The lesser prairie chicken in the Texas Panhandle. *Journal of Wildlife Management* 27:733–37.

Janossy, D. 1976. Plio-Pleistocene bird remains from the Carpathian basin. I. Galliformes. 1. Tetraonidae. *Aquila* 82:13–36.

Janson, R. 1953. Prairie chickens in South Dakota. *Conservation Digest* 20(2):11, 15–16.

Jehl, J. R. 1969. Fossil grouse of the genus *Dendragapus*. *Transactions of the San Diego Society of Natural History* 15:165–74.

Jenkins, D.; Watson, A.; and Miller, G. R. 1963. Population studies on red grouse, *Lagopus lagopus scoticus* (Lath.) in north-east Scotland. *Journal of Animal Ecology* 32:317–76.

——. 1967. Population fluctuations in the red grouse *Lagopus lagopus scoticus*. *Journal of Animal Ecology* 36:97–122.

Jewett, S. G.; Taylor, W. P.; Shaw, W. T., and Aldrich, J. W. 1953. *Birds of Washington State*. Seattle: University of Washington Press.

Johansen, H. 1956. Revision und Entstehund der arktischen Vögelfauna. *Acta Arctica* 8:1–98.

——. 1961. Die Vögelfauna Westsibiriens. Part 3. Non-Passeres. *Journal für Ornithologie* 102:237–69.

Johnsgard, P. A. 1968. *Animal behavior*. Dubuque: Wm. Brown.

——. 1973. *Grouse and quails of North America*. Lincoln: University of Nebraska Press.

——. 1982. Etho-ecological aspects of hybridization in the Tetraonidae. *World Pheasant Association Journal* 7:42–57.

Johnsgard, P. A., and Wood, R. W. 1968. Distributional changes and interactions between prairie chickens and sharp-tailed grouse in the Midwest. *Wilson Bulletin* 80:173–88.

Johnson, M. D. 1964. Feathers from the prairie: A short history of upland game birds. North Dakota Game and Fish Department Project Report W-67-R-5.

Johnson, R. E., and Lockner, J. R. 1968. Heart size and altitude in ptarmigan. *Condor* 70:185.

Johnston, D. W. 1963. Heart weights of some Alaska birds. *Wilson Bulletin* 75:435–46.

Johnston, G. D. 1967. Black game and capercaillie in relation to forestry in Britain. *Forestry*, supplement 1967:68–77.

Johnston, G. W. 1969. Ecology, dispersion and arena behaviour of black grouse *Lyrurus tetrix* (L.) in Glen Dye, N.E. Scotland. Ph.D. dissertation, Aberdeen University.

Johnston, R. F. 1964. The breeding birds of Kansas. *University of Kansas, Publications of the Museum of Natural History* 12:575–655.

Jollie, M. 1955. A hybrid between the spruce grouse and the blue grouse. *Condor* 57:213–15.

Jones, A. M. 1981. A field study of capercaillie *Tetrao urogallus* (abstract). *Ibis* 123:579.

——. 1982. Capercaillie in Scotland: Toward a conservation strategy. In Lovel 1982, pp. 60–74.

Jones, R. 1963. Identification and analysis of lesser and greater prairie chicken habitat. *Journal of Wildlife Management* 27:757–58.

——. 1964a. The specific distinctness of the greater and lesser prairie chicken. *Auk* 81:65–73.

——. 1964b. Habitat used by lesser prairie chickens for feeding related to seasonal behavior of plants in Beaver County, Oklahoma. *Southwestern Naturalist* 9:111–17.

——. 1966. Spring, summer and fall foods of the Columbian sharp-tailed grouse in eastern Washington. *Condor* 68:536–40.

Jones, R. 1969. Hormonal control of incubation patch development in the California quail, *Lophortyx californicus*. *General and Comparative Endocrinology* 13:1–14.

Jonkel, C. J., and Greer, K. R. 1963. Fall food habits of spruce grouse in northwest Montana. *Journal of Wildlife Management* 27:593–96.

Judakov, A. G. 1972. [A contribution to the biology of *Falcipennis falcipennis* in the Amur district]. *Zoologicheskii Zhurnal* 51:620–23. In Russian, English summary.

Judd, S. 1905. The grouse and wild turkeys of the United States, and their economic value. *United States Biological Survey Bulletin* 24:1–55.

Juhn, M., and Harris, P. C. 1955. Local effects on the feather papilla of thyroxine and of progesterone. *Proceedings of the Society for Experimental Biology and Medicine* 90:202–4.

——. 1968. Molt of capon feathering with prolactin. *Proceedings of the Society for Experimental Biology and Medicine* 90:699–72.

Junco Rivera, E. 1975. [The capercaillie of the Cordillera Cantabrica]. *Vida Silvestre* 16:216–25. In Spanish.

June, J. 1967. A comprehensive report on the Wyoming grouse family. *Wyoming Wildlife* 10(1):19–24.

Kaase, J. 1959. En undersokelse over naerigan hos orrfulgen i Norge. *Meddelser Fra Statens Viltunderokelser* 2/4:1–112.

Kealy, R. D. 1970. Storage and incubation of game bird eggs. *Modern Game Breeding* 6(4):19–21.

Keith, L. B. 1963. *Wildlife's ten-year cycle*. Madison: University of Wisconsin Press.

Keller, H.; Pauli, H. R.; and Glutz von Blotzheim, U. 1979. [Feeding ecology in winter of the black grouse in the Bernese Alps, Switzerland]. *Ornithologische Beobachter* 76:9–32. In German, English summary.

Keller, R. J.; Shepherd, H. R.; and Randall, R. H. 1941. *Survey of 1941: North Park, Jackson County, Moffat County, including comparative data of previous seasons*. Colorado Game and Fish Commission Sage Grouse Survey 3.

Keppie, D. M. 1975a. Clutch size of the spruce grouse, *Canachites canadensis franklinii*, in southwest Alaska. *Condor* 77:91–92.

——. 1975b. Dispersal, overwinter mortality and population size of spruce grouse (*Canachites canadensis franklini*). Ph.D. dissertation, University of Alberta, Edmonton.

392

____. 1979. Dispersal, overwinter mortality and recruitment of spruce grouse. *Journal of Wildlife Management* 43:117–27.

Keppie, D. M., and Herzog, P. W. 1978. Nest site characteristics and nest success of spruce grouse. *Journal of Wildlife Management* 42:628–32.

Kermott, L. H., and Oring, L. W. 1975. Acoustical communication of male sharp-tailed grouse (*Pedioecetes phasianellus*) on a North Dakota dancing ground. *Animal Behaviour* 23:375–86.

Kessel, B., and Schaller, G. B. 1960. Birds of the upper Sheenjek valley, northeastern Alaska. *Biological Papers, University of Alaska* 4:1–59.

Kessler, W. B. 1979. The Attwater's greater prairie chicken: Endangered grouse of the Texas coastal prairie. *Proceedings of the Welder Wildlife Foundation Symposium* 1:189–98.

Kihlen, G. 1914. Om bastarder mellan Moripa och Dalripa. *Fauna och Flora* 9:112–17. (In Swedish.)

Kirikov, S. V., and Shubinkova, I., eds. 1968. [*Resources of gallinaceous birds in the USSR*]. Proceedings of a Congress, 2–4 April, 1968, Moscow Society of Naturalists and Institute of Geography of the Academy of Sciences of the USSR, Moscow. In Russian.

Kirpichev, S. P. 1958. [On hybrids of *Tetrao urogallus* and *Tetrao parvirostris*]. *Uchenye Zapiski Moskovskogo Gosudarstvennyi Universitet* 197:217–21. In Russian.

____. 1972. [The molt of the capercaillie]. *Ornitologiya* 10:303–19. In Russian.

Klebenow, D. A. 1969. Sage grouse nesting and brood habitat in Idaho. *Journal of Wildlife Management* 33:649–62.

Klebenow, D. A., and Gray, G. M. 1968. Food habits of juvenile sage grouse. *Journal of Wildlife Management* 21:80–83.

Klonglan, E. D., and Hlavka, G. 1969. Iowa's first ruffed grouse hunting season in 45 years. *Proceedings Iowa Academy of Science* 76:226–30.

Kobayashi, H. 1958. On the induction of molt in birds by 17 α oxyprogesterone-17 capronate. *Endocrinology* 63:420–30.

Kobriger, G. D. 1965. Status, movements, habitats, and foods of prairie grouse on a sandhills refuge. *Journal of Wildlife Management* 29:788–800.

____. 1980. Habitat use and brooding sharp-tailed grouse in southwestern North Dakota. *North Dakota Outdoors* 43(1):2–5.

Koivisto, I. 1965. Behavior of the black grouse, *Lyrurus tetrix* (L.) during the spring display. *Finnish Game Research* 26:1–60.

Koivisto, I., and Pirkola, M. 1961. (Behavior and numbers of capercaillie and black grouse on display grounds.) *Suomen Riista* 14:53–64. (In Finnish, English summary).

Kolb, H. H. 1971. Development and aggressive behaviour of red grouse (*L. lagopus scoticus* Lath.) in captivity. Ph.D. dissertation, Aberdeen University.

Korschgen, L. T. 1962. Food habitats of greater prairie chickens in Missouri. *American Midland Naturalist* 68:307–18.

Koskimies, J. 1957. Flocking behaviour in capercaillie *Tetrao urogallus* (L.) and blackgame *Lyrurus tetrix* (L.). *Papers in Game Research* (Helsinki), no. 18, pp. 1–32.

——. 1966. Foods and nutrition of ruffed grouse in Missouri. *Journal of Wildlife Management* 30:86–100.

Krahn, P. J. 1980. Sharp-tailed–spruce grouse mixture. *Blue Jay* 38:119.

Kratzig, H. 1939. Untersuchungen zur Biologie und Ethologie des Hazelhuhns (*Tetrastes bonasia ruprestris* Brehm) wahrend der Jungentwicklung. *Berichte der Vereinigung Schlesischer Ornithologen* 24:1–25.

Kruijt, J. P.; de Vos, G. T.; and Bossema, I. 1972. The arena system of black grouse. *Proceedings of the Fifteenth International Ornithological Congress,* The Hague, 1970, pp. 399–423.

Kruijt, J. P., and Hogan, J. A. 1964. Organization of the lek in black grouse. *Archives Neerlandaises de Zoologie* 16(1):156–57.

——. 1967. Social behavior on the lek in the black grouse, *Lyrurus tetrix tetrix* (L.). *Ardea* 55:203–40.

Kubisiak, J. F.; Moulton, J. C.; and McCaffery, K. R. 1980. Ruffed grouse density and habitat relationships in Wisconsin. Wisconsin Department of Natural Resources Technical Bulletin 118.

Lack, D. 1939. The display of the blackcock. *British Birds* 32:290–303.

——. 1966. *Population studies of birds.* London: Oxford University Press.

——. 1968. *Ecological adaptations for breeding in birds.* London: Methuen.

Lance, A. N. 1970. Movements of blue grouse on the summer range. *Condor* 72:437–44.

——. 1978. Territories and the food plants of individual red grouse. 2. Territory size compared with an index of nutrient supply in heather. *Journal of Animal Ecology* 47:307–13.

Larrison, E. J., and Sonnenberg, K. G. 1968. *Washington birds: Their location and identification.* Seattle: Seattle Audubon Society.

Larson, B. B.; Wegge, P.; and Storass, T. 1982. Spacing behaviour of capercaillie cocks during spring and summer as determined by radio telemetry. In Lovel 1982, pp. 124–30.

Lauckhart, J. B. 1957. Animal cycles and food. *Journal of Wildlife Management* 21:230–33.

Lawrence, J. S., and Silvy, N. J. 1980. Status of the Attwater's prairie chicken—an update. In Vohs and Knopf 1980, pp. 23–33.

Leach, H. R., and Browning, B. M. 1958. A note on the food of sage grouse in the Madeline Plains area of California. *California Fish and Game* 44:73–76.

Leach, H. R., and Hensley, A. L. 1954. The sage grouse in California, with special reference to food habits. *California Fish and Game* 40:385–94.

Lee, L. 1950. Kill analysis for the lesser prairie chicken in New Mexico, 1949. *Journal of Wildlife Management* 14:475–77.

Lehmann, V. W. 1941. Attwater's prairie chicken: Its life history and management. United States Department of the Interior, Fish and Wildlife Service, North American Fauna no. 57.

___. 1968. The Attwater prairie chicken, current status and restoration opportunities. *Transactions of the Thirty-third North American Wildlife Conference*, pp. 398–407.

Lehmann, V. W., and Mauermann, R. G. 1963. Status of Attwater's prairie chicken. *Journal of Wildlife Management* 27:713–25.

Lemburg, W. W. 1962. Rearing sharp-tailed grouse. *Game Bird Breeders, Pheasant Fanciers and Aviculturalists' Gazette* 13(9):10–11.

Leopold, A. 1949. *A sand county almanac.* New York: Oxford University Press.

___. 1953. Intestinal morphology of gallinaceous birds in relation to food habits. *Journal of Wildlife Management* 17:197–203.

Lewin, V. 1963. Reproduction and development of young in a population of California quail. *Condor* 65:249–78.

Lewis, J. B.; McGowan, J. D.; and Baskett, T. S. 1968. Evaluating ruffed grouse reintroduction in Missouri. *Journal of Wildlife Management* 32:17–28.

Lewis, R. A. 1981. Characteristics of persistent and transient territorial sites of male blue grouse. *Journal of Wildlife Management* 45:1048–51.

Ligon, J. S. 1961. *New Mexico birds and where to find them.* Albuquerque: University of New Mexico Press.

Linden, H. 1981. Estimation of juvenile mortality in the capercaillie *Tetrao urogallus*, and the black grouse, *Tetrao tetrix*, from indirect evidence. *Finnish Game Research* 39:35–51.

Little, T. W. 1978. Populations, distributions and habitat selection by drumming male ruffed grouse in central Minnesota. Ph.D. dissertation, University of Minnesota.

Lofts, B., and Murton, R. K. 1968. Photoperiodic and physiological adaptations regulating avian breeding cycles and their ecological significance. *Journal of Zoology* 155:327–94.

Lovel, T. W. I., ed. 1979. *Woodland grouse, 1978.* Proceedings of a symposium held December 4–8, 1978, Inverness, Scotland. Lamarsh: World Pheasant Association.

___. 1982. *Grouse.* Proceedings of the Second International Symposium on Grouse, March 16–20, 1981, Dalhousie Castle, Edinburgh. Lamarch: World Pheasant Association.

Lumsden, H. G. 1961a. Displays of the spruce grouse. *Canadian Field-Naturalist* 75:152–60.

___. 1961b. The display of the capercaillie. *British Birds* 54:257–72.

___. 1965. *Displays of the sharptail grouse.* Ontario Department of Lands and Forests Technical Series Research Report no. 66.

——. 1968. *The displays of the sage grouse.* Ontario Department of Lands and Forests Research Report (Wildlife) no. 83.

——. 1969. A hybrid grouse, *Lagopus × Canachites*, from northern Ontario. *Canadian Field-Naturalist* 83:23–30.

——. 1970. The shoulder-spot display of grouse. *Living Bird* 9:65–74.

Lumsden, H. G., and Weeden, R. B. 1963. Notes on the harvest of spruce grouse. *Journal of Wildlife Management* 27:587–91.

McCabe, R. A., and Hawkins, A. S. 1946. The Hungarian partridge in Wisconsin. *American Midland Naturalist* 36:1–76.

McColm, M. 1970. Ruffed grouse. *Nevada Outdoors* 4(3):27.

McCourt, K. H. 1969. Dispersion and dispersal of female and juvenile Franklin's grouse in southwestern Alberta. M.S. thesis, University of Alberta.

McCourt, K. H.; Boag, D. A.; and Keppie, D. M. 1973. Female spruce grouse activities during laying and incubation. *Auk* 90:619–23.

McCourt, K. H., and Keppie, D. M. 1975. Age determination of juvenile spruce grouse. *Journal of Wildlife Management* 39:790–94.

MacDonald, S. D. 1968. The courtship and territorial behavior of Franklin's race of the spruce grouse. *Living Bird* 7:4–25.

——. 1970. The breeding behavior of the rock ptarmigan. *Living Bird* 9:195–238.

McEwen, L. C.; Knapp, D. B.; and Hilliard, E. A. 1969. Propagation of prairie grouse in captivity. *Journal of Wildlife Management* 33:276–83.

McGowan, J. E. 1973. Fall and winter foods of ruffed grouse in interior Alaska. *Auk* 90:636–40.

McNicholl, M. K. 1978. Behaviour and social organization in a population of blue grouse on Vancouver Island. Ph.D. dissertation, University of Alberta.

Maher, W. J. 1959. Habitat distribution of birds breeding along the upper Kaolak River, northern Alaska. *Condor* 61:351–68.

Malafeev, Y. M. 1970. [Characteristics of the nesting of the capercaillie in 1969 in the southern taiga of the Sverdlovsk region]. *Ekologiya* 1(3):76–8. In Russian.

Marcstrom, V. 1979. A review of the tetraonid situation in Sweden. In Lovel 1979, pp. 13–16.

Marcstrom, V., and Hoglund, N. H. 1981. Factors affecting reproduction of willow grouse (*Lagopus lagopus*) in two highland areas of Sweden. *Viltrevy* 11:285–314.

Marsden, H. M., and Baskett, T. S. 1958. Annual mortality in a banded bobwhite population. *Journal of Wildlife Management* 22:414–19.

Marshall, W. H. 1946. Cover preferences, seasonal movements, and food habits of Richardson's grouse and ruffed grouse in southern Idaho. *Wilson Bulletin* 58:42–52.

——. 1965. Ruffed grouse behavior. *Bio-Science* 15:92–94.

Marshall, W. H., and Jensen, M. S. 1937. Winter and spring studies of the sharp-tailed grouse in Utah. *Journal of Wildlife Management* 1:87–99.

Martin, A. C.; Zim, H. S.; and Nelson, A. L. 1951. *American wildlife and plants.* New York: McGraw-Hill.

Martin, N. S. 1970. Sagebrush control related to habitat and sage grouse occurrence. *Journal of Wildlife Management* 34:313–20.

Martin, S. A., and Knopf, F. L. 1980. Distribution and numbers of greater prairie chickens in Oklahoma. In Vohs and Knopf 1980, pp. 68–70.

Martinka, R. R. 1972. Structural characteristics of blue grouse territories in southwestern Montana. *Journal of Wildlife Management* 36:498–510.

Masson, W. V., and Mace, R. U. 1962. *Upland game birds*. Oregon State Game Commission Wildlife Bulletin no. 5.

Maxon, S. J. 1977. Activity patterns of female ruffed grouse during the breeding season. *Wilson Bulletin* 89:439–55.

——. 1978. Spring home range and habitat use by female ruffed grouse. *Journal of Wildlife Management* 42:61–71.

May, T. A. 1970. Seasonal foods of white-tailed ptarmigan in Colorado. M.S. thesis, Colorado State University.

May, T. A., and Braun, C. E. 1969. Observations on winter foods of Colorado white-tailed ptarmigan. Abstract of paper presented at meeting of the Southwestern and Rocky Mountain Division, AAAS, May 8, 1969, Colorado Springs, Colorado.

——. 1972. Seasonal foods of adult white-tailed ptarmigan in Colorado. *Journal of Wildlife Management* 36:1180–86.

Mayr, E. 1942. *Systematics and the origin of species*. New York: Columbia University Press.

Mayr, E., and Amadon, D. 1951. A classification of Recent birds. *American Museum Novitates* 1496:1–42.

Mayr, E., and Short, L. L., Jr. 1970. *Species taxa of North American birds: A contribution to comparative systematics*. Nuttall Ornithological Club Publication no. 9.

Mercer, E., and McGrath, R. 1963. *A study of a high ptarmigan population on Burnette Island, Newfoundland, in 1962*. Saint Johns: Newfoundland Wildlife Division, Department of Mines and Resources.

Merikallio, E. 1958. Finnish birds, their distribution and numbers. *Fauna Fennica* 5:1–181.

Miller, G. C., and Graul, W. D. 1980. Status of sharp-tailed grouse in North America. In Vohs and Knopf 1980, pp. 8–17.

Moffit, J. 1938. The downy young of *Dendragapus*. *Auk* 55:589–95.

Mohler, L. L. 1944. Distribution of upland game birds in Nebraska. *Nebraska Bird Review* 12:1–6.

——. 1963. Winter surveys of Nebraska prairie chickens and management implications. *Journal of Wildlife Management* 27:737–38.

Monson, G., and Phillips, A. R. 1964. The species of birds in Arizona. In *The vertebrates of Arizona*. Tucson: University of Arizona Press.

Moody, A. F. 1932. *Waterfowl and game-birds in captivity, with notes on habits and management*. London: Witherby.

Mosby, H. S., ed. 1963. *Wildlife investigational techniques,* 2d ed. Blacksburg, Virginia: Wildlife Society.

Moss, R. 1968. Food selection and nutrition in ptarmigan (*Lagopus mutus*). *Symposia of the Zoological Society of London* 21:207–16.

——. 1969. Rearing red grouse and ptarmigan in captivity. *Avicultural Magazine* 75:256–61.

——. 1980. Why are capercaillies so big? *British Birds* 73:440–47.

Moss, R., and Lockie, T. 1979. Infrasonic components in the song of the capercaillie *Tetrao urogallus. Ibis* 121:95–97.

Moss, R., and Watson, A. 1980. Inherent changes in the aggressive behavior of a fluctuating red grouse *Lagopus lagopus scoticus* Lath. population *Ardea.* 68:113–19.

Moss, R.; Watson, A.; and Parr, R. 1975. Maternal nutrition and breeding success in red grouse (*Lagopus lagopus scoticus*). *Journal of Animal Ecology* 44:233–44.

Moss, R.; Weir, D.; and Jones, A. 1979. Capercaillie management in Scotland. In Lovel 1979, pp. 140–45.

Moulton, J. C., and Vanderschaegen, P. V. 1974. *Bibliography of the ruffed grouse.* Madison: Wisconsin Department of Natural Resources.

Moyles, D. L. J., and Boag, D. A. 1981. Where, when and how male sharp-tailed grouse establish territories on arenas. *Canadian Journal of Zoology* 59:1576–81.

Müller, F. J. 1974. [Territorial behaviour and distribution structure of a capercaillie population]. Ph.D. dissertation, Phillips University of Marburg, Germany. In German.

——. 1979. A 15-year study of a capercaillie lek in the western Rhon Mountains (W. Germany). In Lovel 1979, pp. 120–30.

Mussehl, T. W. 1960. Blue grouse production, movements, and populations in the Bridger Mountains, Montana. *Journal of Wildlife Management* 24:60–68.

——. 1963. Blue grouse brood cover selection and land-use implications. *Journal of Wildlife Management* 27:547–55.

Mussehl, T. W., and Leik, T. H. 1963. Sexing wings of adult blue grouse. *Journal of Wildlife Management* 27:102–6.

Mussehl, T. W., and Schladweiler, P. 1969. *Forest grouse and experimental spruce budworm insecticide studies.* Montana Fish and Game Department Technical Bulletin no. 4.

Myers, J. A. 1917. Studies on the syrinx of *Gallus domesticus. Journal of Morphology* 29:165–214.

Myrberget, S. 1972. Fluctuations in a north Norwegian population of willow grouse. *Proceedings of the Fifteenth International Ornithological Congress,* The Hague, pp. 107–20.

——. 1974. Variations in the production of willow grouse *Lagopus lagopus* (L.) in Norway, 1963–1972. *Ornis Scandinavica* 5:163–72.

——. 1975. Age determination of willow grouse chicks. *Norwegian Journal of Zoology* 23:165–71.

___. 1976. Lirypas reirhabat. *Meddelelser fra Norsk Viltforskning.* Series 3(1):1–29. English summary.

___. 1979. Winter food of willow grouse in two Norwegian areas. *Meddelelser fra Norsk Viltforskning,* ser. 3(7):1–32.

Nakata, Y. 1973. [My experiments in breeding the white-tailed ptarmigan and spruce grouse in captivity]. *Bulletin of the Ornamental Pheasant and Waterfowl Society of Japan* 6(2):8–13. In Japanese.

National Survey of Fishing and Hunting. 1965. United States Bureau of Sport Fisheries and Wildlife, Resource Publication 27.

Nelson, A. L., and Martin, A. C. 1953. Gamebird weights. *Journal of Wildlife Management* 17:36–42.

Nelson, O. C. 1955. A field study of the sage grouse in southeastern Oregon, with special reference to reproduction and survival. M.S. thesis, Oregon State University.

Noska, M. 1895. Das kaukasische Birkhuhn, *Tetrao mloskosiewisei* Tac. *Ornithologische Jahrbuch* 6:209–43.

Ohmart, R. B. 1967. Comparative molt and pterylography in the quail genera *Callipella* and *Lophortyx. Condor* 69:535–48.

Olson, S. L., and Feduccia, A. 1980. *Presbyornis* and the origin of the Anseriformes (Aves: Charadriomorphae). *Smithsonian Contributions to Zoology* 323:1–24.

Oring, L. W. 1982. Avian mating systems. In *Avian biology,* 6:1–92. New York: Academic Press.

Ouellet, H. 1974. An intergeneric grouse hybrid (*Bonasa* × *Canachites*). *Canadian Field-Naturalist* 88:183–86.

Palmer, R. S. 1949. Maine birds. *Harvard Museum of Comparative Zoology Bulletin* 102:1–656.

Palmer, W. L. 1954. Unusual ruffed grouse density in Benzie County, Michigan. *Journal of Wildlife Management* 18:542–43.

___. 1963. Ruffed grouse drumming sites in northern Michigan. *Journal of Wildlife Management* 27:656–63.

Parker, H. 1981. Renesting biology of Norwegian willow ptarmigan. *Journal of Wildlife Management* 45:858–64.

Parmelee, D. F.; Stephens, H. A.; and Schmidt, R. H. 1967. The birds of southeastern Victoria Island and adjacent small islands. *National Museum of Canada Bulletin* 222:1–229.

Parr, R. 1975. Aging red grouse chicks by primary molt and development. *Journal of Wildlife Management* 39:188–90.

Patterson, R. L. 1949. Sage grouse along the Oregon trail. *Wyoming Wild Life* 13(8):1–16.

___. 1952. *The sage grouse in Wyoming.* Denver: Sage Books.

Pauli, H. R. 1974. [On the winter ecology of the black grouse in the Swiss Alps]. *Ornithologische Beobachter* 71:247–78. In German, with English and French summaries.

——. 1979. Zur Bedeutung von Nährstoffgehalt und Verdaulichkeit der wichtigsten Nahrungspflanzen des Birkhuhns *Tetrao tetrix* in den Schweizer Alpen. *Ornithologische Beobachter* 75:57–84. English abstract in Lovel 1979, p. 99.

Pendergast, B. A., and Boag, D. A. 1970. Seasonal changes in diet of spruce grouse in central Alberta. *Journal of Wildlife Management* 34:605–11.

——. 1971*a*. Maintenance and breeding of spruce grouse in captivity. *Journal of Wildlife Management* 35:177–79.

——. 1971*b*. Nutritional aspects of the diet of spruce grouse in central Alberta. *Condor* 73:437–43.

——. 1973. Seasonal changes in the internal anatomy of spruce grouse in Alberta. *Auk* 90:307–17.

Pepper, G. W. 1972. *The ecology of sharp-tailed grouse during spring and summer in the aspen parklands of Alberta.* Saskatchewan Department of Natural Resources Wildlife Report no. 1.

Peterle, T. J. 1951. Intergeneric galliform hybrids; A review. *Wilson Bulletin* 63:219–24.

Peters, J. L. 1934. *Check-list of the birds of the world.* Vol. 2. Cambridge: Harvard University Press.

Peters, S. S. 1958. Food habits of the Newfoundland willow ptarmigan. *Journal of Wildlife Management* 22:384–94.

Petersen, B. E. 1980. Breeding and nesting ecology of female sage grouse in North Park, Colorado. M.S. thesis, Colorado State University.

Peterson, J. G. 1970. The food habits and summer distribution of juvenile sage grouse in central Montana. *Journal of Wildlife Management* 34:147–55.

Petrides, G. A. 1942. Age determination in American gallinaceous game birds. *Transactions of the Seventh North American Wildlife Conference*, pp. 308–28.

——. 1949. Viewpoint on the analysis of open season sex and age ratios. *Transactions of the Fourteenth North American Wildlife Conference*, pp. 391–410.

Phillips, A. R.; Marshall, J.; and Monson, G. 1964. *Birds of Arizona.* Tucson: University of Arizona Press.

Phillips, R. L. 1967. Fall and winter food habits of ruffed grouse in northern Utah. *Journal of Wildlife Management* 31:827–29.

Pleske, T. 1887. Beschreibung einiger Vogelbastarde. *Memoirs of the Academy of Science, St. Petersburg*, 35(5):1–8.

Porath, W. R., and Vohs, P. A. 1972. Population ecology of ruffed grouse in northeastern Iowa. *Journal of Wildlife Management* 36:793–802.

Potapov, R. L. 1969. [The courtship display of *Falcipennis falcipennis*]. *Zoologicheskii Zhurnal* 48:864–70. In Russian.

Potapov, R. L., and Pavlova, E. 1977. [Specific features of the mating display by the Caucasian black grouse *Lyrurus mlokosiewisei*]. *Ornitologiya* 13:117–26. In Russian.

Principal game birds and mammals of Texas. 1945. Austin: Texas Game, Fish and Oyster Commission.

Pukinsky, Y. B., and Roo, S. S. 1966. [Behavior of the capercaillie during the mating season]. *Vestnik Leningradskogo Universiteta*, Biological Ser., no. 4, pp. 22–28. In Russian.

Pullianinen, E. 1970*a*. Color variation and sex identification in the rock ptarmigan (*Lagopus mutus*) in Finland. *Annales Academiae Scientiarum Fennicae*, ser. A, no. 4, pp. 1–6.

——. 1970*b*. Composition and selection of winter food by capercaillie (*Tetrao urogallus*) in northeastern Finnish Lapland. *Suomen Riista* 22:67–73.

——. 1971. Behaviour of a nesting capercaillie (*Tetrao urogallus*) in northeastern Lapland. *Annales Zoologici Fennici* 8:456–62.

——. 1980. Heart/body weight ratio in the willow grouse *Lagopus lagopus* and rock ptarmigan *Lagopus mutus* in Finnish Lapland. *Ornis Fennica* 57:88–90.

Pynnönen, A. 1954. Beiträge zur Kenntnis der Lebenweise des Haselhuhns *Tetrastes bonasia* (L.). *Papers on Game Research* 12:1–90.

Pyrah, D. B. 1954. A preliminary study toward sage grouse management in Clark and Fremont counties based on seasonal movement. M.S. thesis, University of Idaho.

——. 1963. *Sage grouse investigations*. Idaho Fish and Game Department, Wildlife Restoration Division Job Completion Report, project W 125-R-2.

——. 1964. Sage chickens in captivity. *Game Bird Breeders, Pheasant Fanciers and Aviculturists' Gazette* 13(9):10–11.

Quick, H. F. 1947. Winter food of white-tailed ptarmigan in Colorado. *Condor* 49:233–35.

Raitt, R. J., Jr. 1961. Plumage development and molts of California quail. *Condor* 63:294–303.

Rajala, P. 1974. The structure and reproduction of Finnish populations of capercaillie *Tetrao urogallus,* and black grouse, *Lyrurus tetrix,* on the basis of late summer data from 1963–66. *Finnish Game Research,* no. 35.

——. 1979. Status of tetraonid populations in Finland. In Lovel, 1979, pp. 32–34.

Ramharter, B. G. 1976. Habitat selection and movements by sharp-tailed grouse (*Pedioecetes phasianellus*) hens during the nesting and brood rearing periods in a fire maintained brush prairie. Ph.D. dissertation, University of Minnesota.

Rasmussen, D. I., and Griner, L. A. 1938. Life history and management studies of the sage grouse in Utah, with special reference to nesting and feeding habits. *Transactions of the Third North American Wildlife Conference*, pp. 852–64.

Rawley, E. V., and Bailey, W. J. 1964. *Utah upland game birds*. Utah State Department of Fish and Game Publication 63-12.

Redfield, J. A. 1973. Variations in weight of blue grouse (*Dendragapus obscurus*). *Condor* 75:312–21.

——. 1978. Growth of juvenile blue grouse *Dendragapus obscurus*. *Ibis* 120:55–61.

Redfield, J. A., and Zwickel, F. C. 1976. Determining the age of young blue grouse: A correction for bias. *Journal of Wildlife Management* 40:349–51.

Redmond, G. W., Keppie, D. M., and Herzog, P. W. 1982. Vegetative structure, concealment and success at nests of two races of spruce grouse. *Canadian Journal of Zoology* 60:670–75.

Ricklefs, R. E. 1969. *An analysis of nesting mortality in birds*. Smithsonian Contributions to Zoology no. 9.

Ridgway, R., and Freidmann, H. 1946. The birds of North and Middle America. Part 10. *Galliformes. Smithsonian Institution Bulletin* 50:1–484.

Rippin, A. P. 1970. Social organization and recruitment on the arena of sharp-tailed grouse. M.S. thesis, University of Alberta.

Rippin, A. P., and Boag, D. A. 1974a. Recruitment to populations of male sharp-tailed grouse. *Journal of Wildlife Management* 38:616–21.

——. 1974b. Spatial organization among male sharp-tailed grouse on arenas. *Canadian Journal of Zoology* 52:591–97.

Robel, R. J. 1965. Quantitative indices to activity and territoriality of booming *Tympanuchus cupido pinnatus* in Kansas. *Transactions of the Kansas Academy of Science* 67:702–12.

——. 1966. Booming, territory size and mating success of the greater prairie chicken (*Tympanuchus cupido pinnatus*). *Animal Behaviour* 14:328–31.

——. 1967. Significance of booming grounds of greater prairie chickens. *Proceedings of the American Philosophical Society* 111:109–14.

——. 1969. Nesting activities and brood movements of black grouse in Scotland. *Ibis* 111:395–99.

——. 1970. Possible role of behavior in regulating greater prairie chicken populations. *Journal of Wildlife Management* 34:306–12.

Robel, R. J.; Briggs, J. N.; Cebula, J. J.; Silvey, N. J.; Viers, C. E.; and Watt, P. G. 1970. Greater prairie chicken ranges, movements, and habitat usage in Kansas. *Journal of Wildlife Management* 34:286–306.

Robel, R. J.; Henderson, F. R.; and Jackson, W. 1972. Some sharp-tailed grouse population statistics from South Dakota. *Journal of Wildlife Management* 36:87–98.

Roberts, H. S. 1963. Aspects of the life history and food habitats of rock and willow ptarmigan. M.S. thesis, University of Alaska.

Roberts, T. S. 1932. *The birds of Minnesota*. Vol. 1. Minneapolis: University of Minnesota Press.

Robertson, K. 1980. Changes occurring in Nebraska's prairie grouse range. In Vohs and Knopf 1980, pp. 52–54.

Robinson, W. L. 1969. Habitat selection by spruce grouse in northern Michigan. *Journal of Wildlife Management* 33:113–20.

——. 1980. *Fool hen: The spruce grouse on the Yellow Dog Plains*. Madison: University of Wisconsin Press.

Robinson, W. L., and Maxwell, D. E. 1968. Ecological study of the spruce grouse on the Yellow Dog Plains. *Jack-Pine Warbler* 46:75–83.

Rogers, G. E. 1964. *Sage grouse investigations in Colorado*. Colorado Game, Fish and Parks Department, Game Research Division Technical Publication no. 16.

____. 1968. *The blue grouse in Colorado*. Colorado Game, Fish and Parks Department, Game Research Division Technical Publication no. 21.

____. 1969. *The sharp-tailed grouse in Colorado*. Colorado Game, Fish and Parks Department, Game Research Division Technical Publication no. 23.

Rogers, G. E., and Braun, C. E. 1967. Ptarmigan. *Colorado Outdoors* 16(4):22–28.

____. 1968. Ptarmigan management in Colorado. Mimeographed. Proceedings of the Forty-eighth Annual Conference of the Western Association of Game and Fish Commissioners, July 9, 1968, Reno, Nevada.

Romanoff, A. L.; Bump, G.; and Holm, E. 1938. *Artificial incubation of some upland game bird eggs*. New York Conservation Department Bulletin no. 2.

Roussel, Y. E., and Ouellet, R. 1975. A new criterion for sexing Quebec ruffed grouse. *Journal of Wildlife Management* 39:443–45.

Rusch, D. H., and Keith, L. B. 1971. Ruffed grouse–vegetation relationships in central Alberta. *Journal of Wildlife Management* 35:417–29.

Salo, L. J. 1971. Autumn and winter diet of the hazel grouse (*Tetrastes bonasia* L.) in northeastern Finnish Lapland. *Annales Zoologici Fennici* 8:543–46.

____. 1978. Characteristics of ruffed grouse drumming sites in western Washington and their relevance to management. *Annales Zoologici Fennici* 15:261–78.

Salomonsen, F. 1939. Moults and sequence of plumages in the rock ptarmigan (*Lagopus mutus* [Montin]). *Dansk Ornithologisk Forenings Tidsskrift* 103:1–491.

Samuel, D. E. 1974. Ruffed grouse vocalizations at the drumming log. *Wilson Bulletin* 86:131–35.

Sands, J. L. 1968. Status of the lesser prairie chicken. *Audubon Field Notes* 22:454–56.

Savory, C. J. 1977. The food of red grouse chicks *Lagopus l. scoticus*. *Ibis* 119:1–9.

Schaanning, H. T. L. 1920. Myt fun av Fjellrype-Orre. *Lagopus mutus* × *Lyrurus tetrix*. *Stavanger Museums Aarsh.*, vol. 29/30, no. 3.

____. 1920–23. Orre-typer. *Norsk orn. Tidsskrift,* ser. 1, nos. 1/4, pp. 239–69.

Scherzinger, W. 1976. *Rauhfuss-hühner*. Nationalpark Bayerischer Wald, Arbeiten no. 2.

____. 1981. Stimminventar und Forpflanzungsverhalten des Haselhuhnes *Bonasia bonasia*. *Ornithologische Beobachter* 78:57–86.

Schiller, R. J. 1973. Reproductive ecology of female sharp-tailed grouse (*Pedioecetes phasianellus*) and its relationship to early plant success in northwestern Minnesota. Ph.D. dissertation, University of Minnesota.

Schlatterer, E. F. 1960. Productivity and movements of a population of sage grouse in southeastern Idaho. M.S. thesis, University of Idaho.

Schlotthauer, P. H. 1967. All about quail and grouse. *Game Bird Breeders, Conservationists and Aviculturists' Gazette* 16(3):9–11.

Schmidt, R. K., Jr. 1969. Behavior of white-tailed ptarmigan in Colorado. M.S. thesis, Colorado State University.

Schneegas, E. R. 1967. Sage grouse and sagebrush control. *Transactions of the Thirty-second North American Wildlife Conference,* pp. 270–74.

Schönwetter, M. 1929. Vogeleier aus Kansu. *Journal für Ornithologie* 77:35–40.

Schwartz, C. W. 1945. The ecology of the prairie chicken in Missouri. *University of Missouri Studies* 20:1–99.

Scott, D. A. 1976. The Caucasian black grouse (*Lyrurus mlokosiewiczi*) in Iran. *World Pheasant Association Journal* 1975–76, pp. 66–68.

Scott, J. W. 1942. Mating behavior of the sage grouse. *Auk* 59:472–98.

——. 1950. A study of the phylogenetic or comparative behavior of three species of grouse. *Annals of the New York Academy of Science* 51:477–98.

Scott, W. E. 1943. The Canada spruce grouse in Wisconsin. *Passenger Pigeon* 5:61–72.

——. 1947. The Canada spruce grouse (*Canachites canadensis canace*). *Wisconsin Conservation Bulletin* 12(3):27–30.

Seiskari, P. 1962. On the winter biology of the capercaillie, *Tetrao urogallus,* and the black grouse, *Lyrurus tetrix,* in Finland. *Papers on Game Research* 22:1–119.

Selous, E. 1909–10. An observational diary on the nuptial habits of the blackcock (*Tetrao tetrix*) in Scandinavia and England. *Zoologist* 13:401–13; 14:23–29, 51–56, 176–82, 248–65.

Semenov-Tian-Schanski, O. I. 1959. Ecologiya Tetervinykh Ptitis. [Ecology of Tetraonids]. Transactions of Laplandsk State National Park, Moscow. Translated from Russian in 1979 by the Al Ahram Center for Scientific Translations, for the United States Department of Agriculture and the National Science Foundation (TT 76-59023).

Seth-Smith, D. 1929. Grouse (Tetraonidae). *Avicultural Magazine*, ser. 4, 7:96–98.

Severson, K. E. 1978. A bibliography of blue grouse 1889–1977. General Technical Report R.M.-59, Rocky Mountain Forest and Range Experiment Station, United States Forest Service, Fort Collins.

Sexton, D. A. 1979. Off-lek copulation in sharp-tailed grouse. *Wilson Bulletin* 91:150–51.

Shaffner, C. S. 1955. Progesterone induced molt. *Poultry Science* 34:840–42.

Sharp, W. M. 1963. The effects of habitat manipulation and forest succession on ruffed grouse. *Journal of Wildlife Management* 27:665–71.

Sharpe, R. S. 1968. The evolutionary relationships and comparative behavior of prairie chickens. Ph.D. dissertation, University of Nebraska.

Shoemaker, H. H. 1961. Rearing of young prairie chickens in captivity. *Illinois Wildlife* 16(4):1–4.

——. 1964. Report on studies of captive prairie chickens. *Illinois Wildlife* 19(3):6–8.

Short, L. L., Jr. 1967. A review of the genera of grouse (Aves, Tetraoninae). *American Museum Novitates* 2289:1–39.

Shrader, T. A. 1944. Ruffed and spruce grouse. *Conservation Volunteer* 7(40):36–40.

Sibley, C. G. 1957. The evolutionary and taxonomic significance of sexual dimorphism and hybridization in birds. *Condor* 59:166–91.

——. 1960. The electrophoretic patterns of avian egg-white proteins as taxonomic characters. *Ibis* 102:215–84.

404

Sibley, C. G., and Ahlquist, J. E. 1972. A comparative study of the egg white proteins of non-passerine birds. *Peabody Museum of Natural History Bulletin* 39:1–276.

Siivonen, L. 1952. On the reflection of short-term fluctuations in numbers in the reproduction of tetraonids. *Papers on Game Research* 9:1–43.

———. 1957. The problem of the short-term fluctuations in numbers of tetraonids in Europe. *Papers on Game Research* 19:1–44.

Simpson, G. 1935. Breeding blue grouse in captivity. *Transactions of the American Game Conference* 21:218–19.

Sisson, L. H. 1970. Distribution and selection of sharptailed grouse dancing grounds in the Nebraska sand hills. *Proceedings of the Eighth Conference of the Prairie Grouse Technical Council,* 1969.

———. 1976. *The sharp-tailed grouse in Nebraska: A research study.* Lincoln: Nebraska Game and Parks Commission.

Smith, N. D., and Buss, I. O. 1963. Age determination and plumage observations of blue grouse. *Journal of Wildlife Management* 27:566–78.

Snyder, L. L. 1935. *A study of the sharp-tailed grouse.* University of Toronto Studies, Biological Series no. 40.

———. 1957. *Arctic birds of Canada.* Toronto: University of Toronto Press.

Sparling, D. W. 1979. Reproductive isolating mechanisms and communication in greater prairie chickens (*Tympanuchus cupido*) and sharp-tailed grouse (*Pedioecetes phasianellus*). Ph.D. dissertation, University of North Dakota.

———. 1980. Hybridization and taxonomic status of greater prairie chickens and sharp-tailed grouse. *Prairie Naturalist* 12:92–101.

———. 1983. Quantitative analysis of prairie grouse vocalizations. *Condor* 85:30–42.

Stafford, S. K., and Dimminck, R. W. 1979. Autumn and winter foods of ruffed grouse in the southern Appalachians. *Journal of Wildlife Management* 43:121–27.

Stanton, D. C. 1958. A study of breeding and reproduction in a sage grouse population in southeastern Idaho. M.S. thesis, University of Idaho.

Stempel, M. E., and Rogers, S., Jr. 1961. History of prairie chickens in Iowa. *Proceedings Iowa Academy of Science* 68:314–22.

Stenlund, M. H., and Magnus, L. T. 1951. The spruce hen comes back. *Conservation Volunteer* 14(84):20–24.

Stenman, O., and Helminen, M. 1974. [Aging method for hazel grouse (*Tetrastes bonasia*) based on wings]. *Suomen Riista* 25:90–96. In Finnish, with English summary.

Stepanyan, L. S. 1962. [The systematic relationships between the sharp-winged grouse and the spruce grouse]. *Ornitologiya* 5:368–71. In Russian.

Steward, P. A. 1967. Hooting of Sitka blue grouse in relation to weather, season, and time of day. *Journal of Wildlife Management* 31:28–34.

Stirling, I. 1968. Aggressive behavior and the dispersion of female blue grouse. *Canadian Journal of Zoology* 46:405–8.

Stirling, I., and Bendell, J. F. 1966. Census of blue grouse with recorded calls of a female. *Journal of Wildlife Management* 30:184–87.

———. 1970. The reproductive behavior of blue grouse. *Syesis* 3:161–71.

Stoddard, H. L. 1931. *The bobwhite quail: Its habits, preservation, and increase.* New York: Charles Scribner's Sons.

Stokkan, K. A. 1979. Testosterone and daylength-dependent development of comb size and breeding plumage of male willow ptarmigan (*Lagopus lagopus lagopus*). *Auk* 96:106–15.

Stoll, R. J., Jr.; McClain, M. W.; Boston, R. L.; and Honchul, G. P. 1979. Ruffed grouse drumming site characteristics in Ohio. *Journal of Wildlife Management* 43:324–33.

Stoneberg, R. P. 1967. A preliminary study of the breeding biology of the spruce grouse in northwestern Montana. M.S. thesis, University of Montana.

Stonehouse, B. 1966. Egg volume from linear dimensions. *Emu* 65:227–28.

Stresemann, E. 1966. Die Mauser der Vogel. *Journal für Ornithologie* 107 (Sonderheft):1–448.

Stresemann, E.; Meise, W.; and Schönwetter, M. 1938. Aves Beichianae. *Journal für Ornithologie* 88:171–221.

Stupka, A. 1963. *Notes on the birds of the Great Smoky Mountains National Park.* Knoxville: University of Tennessee Press.

Sturkie, P. D. 1965. *Avian physiology,* 2d ed. Ithaca: Cornell University Press.

Sumner, L., and Dixon, J. S. 1953. *Birds and mammals of the Sierra Nevada.* Berkeley: University of California Press.

Sutherland, C. A., and McChesney, D. S. 1965. Sound production in two species of geese. *Living Bird* 4:99–106.

Sutter, E. 1972. Duration and pattern of post-juvenile wing molt in black grouse and pheasant. *Proceedings of the Eleventh International Ornithological Congress,* The Hague, 1970, p. 692 (abstract).

Sutton, G. M. 1967. *Oklahoma birds.* Norman: University of Oklahoma Press.

———. 1968. The natal plumage of the lesser prairie chicken. *Auk* 85:679.

Sutton, G. M., and Parmelee, D. F. 1956. The rock ptarmigan in southern Baffin Island. *Wilson Bulletin* 68:52–62.

Svedarksy, W. D. 1979. Spring and summer ecology of female greater prairie chickens in northwestern Minnesota. Ph.D. dissertation, University of Minnesota, Minneapolis.

Symington, D. F., and Harper, T. A. 1957. *Sharp-tailed grouse in Saskatchewan.* Saskatchewan Department of Natural Resources Conservation Bulletin no. 4.

Tachenko, V. I. 1966. [The ecology of game birds in the alpine region of northwestern Caucasus]. *Trudy Teberdinskavo Gusudarstvennave Zapovednika* 6:5–144. In Russian.

Taylor, M. 1980. Lesser prairie chicken use of man-made leks. *Southwestern Naturalist* 24:706–7.

Taylor, M. A., and Guthery, F. S. 1980. *Status, ecology and management of the lesser prairie chicken*. United States Forest Service, General Technical Report Rm-77, Rocky Mountain Forest and Range Experiment Station, United States Department of Agriculture, Fort Collins.

Thurman, J. R. 1966. Ruffed grouse ecology in southeastern Monroe County, Indiana. M.S. thesis, Purdue University.

Todd, W. E. C. 1940. Eastern races of ruffed grouse. *Auk* 57:390–97.

——. 1963. *Birds of the Labrador peninsula and adjacent areas*. Toronto: University of Toronto Press.

Trippensee, R. E. 1948. *Wildlife management: Upland game and general principles*. New York: McGraw-Hill.

Trobec, R. J., and Oring, L. W. 1972. Effects of testosterone proprionate implantation on lek behavior of sharp-tailed grouse. *American Midland Naturalist* 87:531–36.

Trueblood, R. W. 1954. The effect of grass reseeding in sagebrush lands on sage grouse populations. M.S. thesis, Utah State Agricultural College.

Tuck, L. M. 1968. Recent Newfoundland bird records. *Auk* 85:304–11.

Tufts, R. W. 1961. *The birds of Nova Scotia*. Halifax: Nova Scotia Museum.

——. 1975. Intergeneric grouse hybrids (*Bonasa* × *Canachites*). *Canadian Field-Naturalist* 89:72.

Upland game birds of Idaho. 1951. Idaho Fish and Game Commission, Boise.

van Rossem, A. J. 1925. Flight feathers as age indicators in *Dendragapus*. *Ibis,* ser. 12, 1:417–22.

Vaurie, C. 1965. *The birds of the Palearctic fauna: Non Passeriformes*. London: H. F. and G. Witherby.

Vohs, P. A., Jr., and Knopf, F. L., eds. 1980. Proceedings of the Prairie Grouse Symposium, September 17–18, 1980, Oklahoma State University, Stillwater.

Vorobiev, K. A. 1954. (*Birds of the Ussuri area*.) Far Eastern Branch, Academy of Sciences of the USSR, Moscow. In Russian.

Voous, K. H. 1960. *Atlas of European birds*. Amsterdam: Thomas Nelson.

Voronin, R. N. 1971. [Methods of determining the sex of the willow ptarmigan in winter plumage]. *Ekologiya* 2:93–94. In Russian.

Vos, G. J., ed. 1979. Adaptiveness of arena behaviour in black grouse (*Tetrao tetrix*) and other grouse species. *Behaviour* 68:277–314.

——. 1983. Social behaviour of black grouse. An observational and experimental field study. *Ardea* 71:1–103.

Waddell, B., and Hanslick, B. 1978. The vanishing sandsage prairie. *Kansas Fish and Game* 35:17–23.

Wallestad, R. 1975. *Life history and habitat requirements of sage grouse in central Montana*. Helena: Montana Department of Fish and Game.

Wallestad, R., and Pyrah, D. 1974. Movement and nesting of sage grouse hens in central Montana. *Journal of Wildlife Management* 38:630–33.

Warner, N. L., and Szenberg, A. 1964. The immunological function of the bursa of Fabricius in the chicken. *Annual Review of Microbiology* 18:253–68.

Watson, A. 1965. A population study of ptarmigan (*Lagopus mutus*) in Scotland. *Journal of Animal Ecology* 34:135–72.

———. 1972. The behaviour of the ptarmigan. *British Birds* 65:6–26, 93–117.

———. 1973. Moults of wild Scottish ptarmigan, *Lagopus mutus,* in relation to sex, climate and status. *Journal of Zoology* 171:207–23.

Watson, A., and Jenkins, D. 1964. Notes on the behaviour of the red grouse. *British Birds* 57:137–70.

Watson, A., and Moss, R. 1972. A current model of population dynamics in red grouse. *Proceedings of the Fifteenth International Ornithological Congress,* The Hague, pp. 134–49.

———. 1979. Population cycles in the Tetraonidae. *Ornis Fennica* 56:87–109.

Watson, A.; Parr, R.; and Lumsden, H. G. 1969. Differences in the downy young of red and willow grouse and ptarmigan. *British Birds* 62:150–53.

Weeden, R. B. 1959. Ptarmigan research project, final report. Mimeographed. Arctic Institute of North America.

———. 1961. Outer primaries as indicators of age among rock ptarmigan. *Journal of Wildlife Management* 25:337–39.

———. 1963. Management of ptarmigan in North America. *Journal of Wildlife Management* 27:673–83.

———. 1964. Spatial separation of sexes in rock and willow ptarmigan in winter. *Auk* 81:534–41.

———. 1965*a*. Breeding density, reproductive success, and mortality of rock ptarmigan at Eagle Creek, Alaska, from 1960–1964. *Transactions of the Thirtieth North American Wildlife Conference,* pp. 336–48.

———. 1965*b*. Grouse and ptarmigan in Alaska: Their ecology and management. Mimeographed. Alaska Department of Fish and Game.

———. 1967. Seasonal and geographic variation in the foods of adult white-tailed ptarmigan. *Condor* 69:303–9.

———. 1979. Relative heart size in Alaskan tetraonids. *Auk* 96:306–18.

Weeden, R. B., and Theberge, J. B. 1972. The dynamics of a fluctuating population of rock ptarmigan in Alaska. *Proceedings of the Fifteenth International Ornithological Congress,* The Hague, pp. 90–106.

Weeden, R. B., and Watson, A. 1967. Determining the age of rock ptarmigan in Alaska and Scotland. *Journal of Wildlife Management* 31:825–26.

Wegge, P. 1979. Status of capercaillie and black grouse in Norway. In Lovel 1979, pp. 17–26.

West, G. C., and Meng, M. S. 1966. Nutrition of willow ptarmigan in northern Alaska. *Auk* 83:603–15.

Westemeier, R. L. 1980. Greater prairie chicken status and management, 1968–1979. In Vohs and Knopf 1980, pp. 18–28.

Westerskov, K. 1956. Age determination and dating nesting events in the willow ptarmigan. *Journal of Wildlife Management* 20:274–79.

Wetmore, A. 1960. A classification for the birds of the world. *Smithsonian Miscellaneous Collections* 139(11):1–37.

Wiesner, J.; Bergmann, H. H.; Klaus, S.; and Müller, F. 1977. [Population density and habitat structure of the hazel hen in the woodlands of Bialowieza]. *Journal für Ornithologie* 118:1–20. In German.

Wiley, R. H. 1973*a*. The strut display of the male sage grouse: A "fixed" action pattern. *Behaviour* 47:129–52.

——. 1973*b*. Territoriality and non-random mating in sage grouse, *Centrocercus urophasianus*. *Animal Behaviour Monographs* 6:85–169.

——. 1974. Evolution of social organization and life-history patterns among grouse. *Quarterly Review of Biology* 49:201–27.

——. 1978. The lek mating system of the sage grouse. *Scientific American* 238(5):114–25.

Williams, J. B.; Best, D.; and Warford, C. 1980. Foraging ecology of ptarmigan at Meade River, Alaska. *Wilson Bulletin* 92:341–51.

Williams, R. 1979. Oddities of the grouse clan. *Wyoming Wildlife* 43(10):4–5.

Wing, L. 1946. Drumming flight in the blue grouse and courtship characters in the Tetraonidae. *Condor* 48:154–57.

Wing, L.; Beer, J.; and Tidyman, W. 1944. The brood habits and growth of "blue grouse." *Auk* 61:426–40.

Witherby, H. F.; Jourdain, F. C. R.; Ticehurst, N. F.; and Tucker, B. W. 1944. *The handbook of British birds*. Vol. 5. Revised ed. London: H. F. and G. Witherby.

Wittenberger, J. F. 1978. The evolution of mating systems in grouse. *Condor* 80:126–37.

Yamashina, Y., and Yamada, S. 1934. [On a collection of some birds from Saghalein]. *Tori* 8:304–25. In Japanese.

——. 1935. [The habits of *Falcipennis falcipennis* and an experience of the species in captivity]. *Tori* 9:13–18. In Japanese.

Yeatter, R. E. 1943. The prairie chicken in Illinois. *Illinois Natural History Bulletin* 22:377–416.

——. 1963. Population responses of prairie chickens to land-use changes in Illinois. *Journal of Wildlife Management* 27:739–57.

Zbinden, N. 1979. Zur Okologie des Hazelhuhns *Bonasia bonasia* in den Buchenwaldern des Chasseral, Faltenjura. *Ornithologische Beobachter* 76:169–214. In German, English summary.

——. 1980. [On the digestibility and metabolizable energy of tetraonid winter food and on the energy requirements for maintenance of black grouse in captivity, with remarks on digestibility trials]. *Vogelwelt* 101:1–18. In German, English summary.

Zettel, J. 1974. [Feeding ecology investigations of the black grouse *Tetrao tetrix* in the Swiss Alps]. *Ornithologische Beobachter* 71:186–246. In German, English summary.

Zwickel, F. C. 1966*a*. Sex and age ratios and weights of capercaillie from the 1965–66 shooting season in Scotland. *Scottish Birds* 4:209–13.

———. 1966*b*. Early mortality and the numbers of blue grouse. Ph.D. dissertation, University of British Columbia.

———. 1967. Early behavior in young blue grouse. *Murrelet* 48:2–7.

———. 1973. Dispersion of female blue grouse during the brood season. *Condor* 75:114–19.

———. 1975. Nesting parameters of blue grouse and their relevance to populations. *Condor* 77:423–30.

Zwickel, F. C., and Bendell, J. F. 1967. Early mortality and the regulation of numbers in blue grouse. *Canadian Journal of Zoology* 45:817–51.

———. 1972. Blue grouse, habitat, and populations. *Proceedings of the Fifteenth International Ornithological Congress,* The Hague, pp. 150–69.

Zwickel, F. C.; Brigham, J. H.; and Buss, I. O. 1966. Autumn weights of blue grouse in north-central Washington, 1954 to 1963. *Condor* 68:488–96.

———. 1975. Autumn structure of blue grouse populations in north-central Washington. *Journal of Wildlife Management* 39:461–67.

Zwickel, F. C.; Buss, I. O.; and Brigham, J. H. 1968. Autumn movements of blue grouse and their relevance to populations and management. *Journal of Wildlife Management* 32:456–68.

Zwickel, F. C., and Carveth, R. G. 1978. Desertion of nests by blue grouse. *Condor* 80:109–11.

Zwickel, F. C., and Dake, J. A. 1977. Primary molt of blue grouse (*Dendragapus obscurus*) and its relation to reproductive activity and migration. *Canadian Journal of Zoology* 55:1782–87.

Zwickel, F. C., and Lance, A. N. 1965. Renesting in blue grouse. *Journal of Wildlife Management* 29:202–4.

———. 1966. Determining the age of young blue grouse. *Journal of Wildlife Management* 30:712–17.

Zwickel, F. C., and Martinsen, C. F. 1967. Determining the age and sex of Franklin spruce grouse by tails alone. *Journal of Wildlife Management* 31:760–63.

INDEX

This index is limited to the English vernacular and Latin names of grouse species and subspecies. Complete indexing is limited to entries for the English vernacular names of species as used in this book. The principal account of each species is indicated by italics. The appendixes are not indexed.